Health Problems

*Philosophical Puzzles about
the Nature of Health*

ELIZABETH BARNES

OXFORD
UNIVERSITY PRESS

OXFORD
UNIVERSITY PRESS

Great Clarendon Street, Oxford, OX2 6DP,
United Kingdom

Oxford University Press is a department of the University of Oxford.
It furthers the University's objective of excellence in research, scholarship,
and education by publishing worldwide. Oxford is a registered trade mark of
Oxford University Press in the UK and in certain other countries

Published in the United States of America by Oxford University Press
198 Madison Avenue, New York, NY 10016, United States of America

British Library Cataloguing in Publication Data
Data available

Library of Congress Control Number: 2023934088

ISBN 978–0–19–288347–6

DOI: 10.1093/oso/9780192883476.001.0001

Printed and bound in the UK by
TJ Books Limited

For Ross Cameron

In Sickness and in Health

Contents

Foreword

This is not the book I wanted to write. It has turned out an altogether stranger, fuzzier, and more ambivalent beast than I had hoped it would be. A friend once told me that philosophy, when it goes well, is like jazz—you begin to see how the pieces fit, and then you let them flow smoothly together into the work they want to be. Writing this book was not like jazz. It was like tapdancing with a porcupine.

I began to think carefully about health while finishing my first book, *The Minority Body*. I felt there was so much about health that was relevant to what I was saying in that book, but relevant only in ways that made everything more complicated and difficult. So I locked health away in the corner of my mind where I hide the philosophy monsters, to be let out another day.

The late, great Anita Silvers, the last time I saw her, talked with me about her own frustrations with the way health is discussed in philosophy. "Your next book is on health, of course," she told me, in that special way she had—where it was unclear whether she was making a descriptive statement or a command. Yes, Anita—as you foretold or possibly decreed. This book is on health.

But the book I wanted to write solves problems. It gives clear, crisp arguments that resolve apparent tensions into an elegant precision. Where there is confusion, it brings illumination.

This little porcupine doesn't do that. The deeper I dove into the philosophical complexities of health, the more I realized that it would take a better philosopher than me to analyze them into clarity. I was stuck marveling at the complexities— the sense that so much of what we care about when we care about health is real, interdependent, and yet often pulls us in different directions in ways we can't reconcile.

What I slowly became convinced of, in this process, was that as philosophers we're perhaps too quick with our attempts to resolve inconsistencies. We want to replace unclarity with precision; we want to provide resolution where ideas conflict. And there's value to all of that. But what I was struck with, the more I thought about health, was my own sense of wonder.

I often come to ideas with a sense of wonder. And in forcing weird ideas into neat solutions or elegant theories, it can be easy for the wonder to slowly ebb away. What I found myself wanting, in writing this, was to find my way to a book that would preserve that sense of wonder. I don't know what health is. I don't think that anyone does, or that anyone can. But I'm fascinated by that difficulty.

So I let my little porcupine dance. I hope she has her own rhythm.

And then, the closest thing a philosophy book can have to a plot twist. In the final stages of completing this manuscript, I was diagnosed with young-onset Parkinson's disease. I have lived with a medically complex disability all my life, so to some degree the subject of health was always personal. But I write this with the knowledge that by the time it is in print, my Parkinson's will no longer be something I'm keeping secret. And it has been hard not to edit the book in this light. But I've resisted this urge—I have added an Afterword, but everything else remains the same. There are bits that now seem personal, even semi-autobiographical, in a way that I never intended them to, and in a way they weren't when I wrote them. I feel odd about that. But in a way, this very experience speaks to so much of what interests me about health, and what is unique about health.

Health is something we care about. Health is something that matters to our lives in distinctive ways. But for each and every one of us, our health will fail us. We don't know when, but we know it will happen, sooner or later. And the present might be later than we think.

Acknowledgments

Philosophy takes a village—in this case, a mostly Zoom-based village, as I worked on the most substantial portions of this book during the height of the COVID-19 pandemic. (There is little that better sums up the life of a philosopher than sitting around wondering "but what is health *really*?" in the midst of international lockdown measures due to a global pandemic.) I'm grateful to audiences, Q&As, random conversations—if you spoke with me about any of the ideas in this book, I am grateful to you. But I would be remiss not to give a few people some extra-special, extra-mile mentions.

My heartfelt thanks and gratitude to Sean Aas, Liam Kofi Bright, Tom Dougherty, Maegan Fairchild, Michael Fara, Shane Glackin, Sukaina Hirji, Sophie Horowitz, Katharine Jenkins, Matt Lindauer, Michaela McSweeney, Trenton Merricks, Sarah Moss, Daniel Nolan, Assya Passinsky, Jason Raibley, Jenny Saul, Ted Sider, Eric Swanson, and Daniel Wodak. Each of them contributed to this manuscript in substantial ways and deserve a lot of credit for any bits of it that are good. (The bits that are bad are all on me.)

Immense thanks and gratitude to editor-extraordinaire Peter Momtchiloff, for his endless supply of good sense and support. You make this whole process so much better, Peter. Additional immense thanks to RA-extraordinaire, Elyse Oakley. Master of the index, queen of the bibliography, empress of being incredibly helpful—Elyse, you're the best.

I owe a very special thank you to Mike Rea. He read every part of this book and provided me with detailed, incredibly helpful feedback on everything he read. And he also encouraged and supported me when I really needed it. I don't have the right words to express my gratitude, but a great thing about Mike is his fondness for plain explanations that cut to the chase. And so, Mike: thanks for everything, friend.

This book also wouldn't exist in the form it does—and my career wouldn't exist in the form it does—without the influence of three remarkable women who I can't thank personally. Katharine Hawley was my dissertation supervisor, my hero, the kind of philosopher I want to be. Anita Silvers told me to write this, and without her work on disability I wouldn't have the space in the profession that I occupy, and that allows me to write like this. Delia Graff Fara provided the philosophical tools I needed to draw my own view together, and she also provided an extraordinary kind of role modeling that I'm only beginning to fully appreciate. I would not have been able to write this book without these women. And the end result is poorer for not having been able to discuss it with them.

But the book also owes much of its current form to a woman I *can* thank personally—Sally Haslanger. Sally provided extraordinarily helpful feedback to me as I tried to formulate some of the central ideas in this book. But her thinking—as will be evident to anyone who knows her work—also pervades and impacts almost every aspect of how I do philosophy. Thank you, Sally, for what you are and what you do.

And then there are my forever besties—Jason Turner and Robbie Williams. They read the whole damn thing, even though it's far outside their areas of interest. And even though it's far outside their areas of interest, they still had the most incisive, smart things to say. They're really annoying like that. They're also the kind of people I'm grateful every day to have in my life.

And finally—Ross Cameron. Without his constant conversation and commentary, this book wouldn't have made it past a blinking cursor. Without his love, I wouldn't have made it—simpliciter. I don't know if he's been enough to make the book good. But he's been enough to make my life wonderful.

Introduction

In early January, friends and strangers alike wish each other a "happy and healthy" New Year. Your friends raise a glass to you on your birthday and say, warmly, "here's to your health!". Excited new parents proclaim their baby "healthy, and that's all that matters." At an especially low point in someone's life, a friend attempts to reassure them that "at least you still have your health."

Health is something we care about. Indeed, it's one of the few things that we assume almost *everyone* cares about, at least to some extent. Whatever else divides us, we've all experienced the unpleasantness of a cold, and we're all going to die. Health is central to our moral and practical reasoning. And impacts on health—our own and our community's—have the potential to shape our lives in dramatic ways. In the simplest case, of course, an extreme reduction in health can end your life. Likewise, the person who didn't care about health at all—who gave no thought to their health in any way—would in all likelihood wind up dead quite quickly. But illness, as well as what we do to avoid illness, can also have a dramatic effect on how we live.

For something so central to our lives, though, we're fairly confused about what exactly health *is*—at least when you scratch the surface. At the time of this writing, much of the Western world is emerging from, and dealing with the fallout of, lockdown measures put in place to slow the COVID-19 pandemic. These measures were enacted both to protect health and to prevent the devastating consequences—economic, social, political—of widespread loss of health. But as debate emerged over which measures were warranted, and for how long, it became clear that health, and loss of health, can often be hard to quantify.

There is perhaps no clearer or more obvious threat to health than infectious disease. You get a virus. You get sick. You might die. And so to preserve health, we try to prevent the spread of the disease. And yet some of the measures put in place to prevent the spread of disease have negative consequences—short and long term—for our health. Stuck at home, we don't exercise as much. We don't see friends, and struggles with depression and isolation worsen, especially for those living alone. We don't make routine medical appointments, or schedule preventive testing. Jobs and income are lost, widening socioeconomic divides that are major predictors of poor health outcomes. Schools go online, raising fears of a widening socioeconomic divide in educational attainment—when education is one of the best predictors of health status across a person's lifetime.[1]

---

[1] See especially Cutler and Lleras-Muney. (2006).

Health Problems: Philosophical Puzzles about the Nature of Health. Elizabeth Barnes, Oxford University Press.
© Elizabeth Barnes 2023. DOI: 10.1093/oso/9780192883476.003.0001

We have a relatively good understanding of what it takes to slow the spread of infectious disease. The question of what it takes to promote health is far murkier. When does the toll of isolation and loneliness shift from compromising your wellbeing to genuinely harming your health? What's the relationship between mental health and physical health? How, if at all, can we distinguish between the social determinants of health (socioeconomic status, education) and health itself? How do we quantify the toll that acute conditions take on health compared to the toll that long-term chronic conditions take on health?

We need to evaluate questions like these to establish good policy—to establish which measures are justified, in what context, and for how long. But in order to do this, we need to have some workable sense of what health is. We need to be able to quantify it, to compare it, to weigh up different aspects of it. And if the morass of public debate surrounding various lockdown measures is any indication, we don't have much idea of how to do that—even though we all care about health, and we're all familiar with loss of health.

To be very clear, I am not a policy expert, and the goal of this book is not to provide practical, concrete recommendations for policy. The task, rather, is to explore the foundational philosophical issues that I think underlie the web of confusion we find when it comes to quantifying and measuring health. I am not attempting to provide a cure for that confusion, but rather a diagnosis.

In a nutshell, the problem is this: health is weird. You can think of this book as an extended argument for the claim that health is philosophically distinctive, and indeed philosophically weird. As such, it's striking that philosophical discussions of health are so often relegated to the margins of "applied" philosophy. We all care about health—our own and others. But we don't know how to define it, how to quantify it, or how to relate it to other familiar things we care about (wellbeing, happiness, etc.). Philosophy, in short, has health problems. And if I can convince you of nothing else in the course of reading this book, I hope to convince you that those problems are fascinating.

1. The Distinctive Biological Importance of Health

Part of what makes health so interesting is that it has distinctive, sometimes competing axes along which it matters to us. To begin with, health has obvious biological significance. The normal function of the immune system is to attack pathogens within the body. If you have multiple sclerosis (MS), your immune system—for reasons we don't yet understand—attacks your central nervous system. The resulting process of demyelination is a clear case of physiological dysfunction. Your body's central nervous system—and thus its overall functioning—is objectively impaired.

Processes like these are of clear relevance to understanding the biology of the human species. Part of what we talk about when we talk about what is healthy for a human organism is how our bodies maintain their functional status, their homeostasis, and their viability over time. Part of what we talk about when we talk about loss of health are reductions in objective functional status, threats or compromises to homeostasis, and reduced longevity.

There is a biological reality to health—and to loss of health—that is independent of how we think and speak about it. Even when some cases of MS were classified as a form of hysteria, even when neurological disorders in general were blamed on demonic possession, the underlying demyelination process was still there, and it still reduced the health of the person involved. And even if society completely changed our attitudes to MS—if we somehow decided that the process should be sought after or celebrated—it would still compromise the function of the central nervous system, still reduce life expectancy, and so on. A degenerative demyelinating disease reduces the health of the person who has it simply because of the type of process it is—regardless of how we conceptualize it and regardless of our attitudes to it.

2. The Distinctive Normative Importance of Health

While the biology of the human organism is a major part of what matters to us when we talk about health, it's far from the only thing that matters. And that's why we can't simply defer to scientists (or philosophers of science) to tell us what health *really* is. Health also has distinctive normative significance. If you have MS, you lose function in parts of your central nervous system. But you also lose something harder to quantify—something related to value, to goodness, or perhaps to wellbeing. Many changes in biological status are morally neutral. As you age, your skin loses elasticity and slows its rate of regeneration. That's why a baby's skin is soft as their own proverbial bottom, and yours is not, no matter how many gallons of moisturizer you slather on it. Your hair follicles will also produce less melanin over time, creating that lovely salt-and-pepper effect that you may or may not cover up with hair dye. Changes like these might be socially devalued, they might be minor annoyances, they might even turn you into an absolute silver fox. But absent social stigma or social meaning, they are not themselves morally significant. Loss of health, however, *is* morally significant. It's not just a change in functional status; it's a way of being harmed.

As debates raged over which lockdown measures were appropriate and for how long, one thing always stayed relatively fixed in the background: people care about health. We may have different ways of prioritizing health relative to other things we care about, we may have different attitudes to risk, we may have different beliefs about what we owe to others, and we may disagree about the practical implications of policy, but there is near-universal consensus that health is important to us. And surprisingly few things share this level of endorsement. Beautiful music can transport and move us, but some people just aren't into music. Friendships can enrich our lives, but some people are natural loners. Sex can be fun, but asexual people often prefer to go without it. You'd be hard pressed, though, to find someone who simply doesn't care about their health *at all*. Indeed, the active desire to be unhealthy—and to reduce one's own health—is often classified as a form of mental illness (e.g., Munchausen's syndrome). People can choose to disregard their health for the sake of something else they value, and people can want to safeguard their own health yet act akratically. But it's often taken as something like a Moorean fixed point that people care about their own health, at least to some degree.

Health, in short, is something we almost always value. It's not the only thing we value, and you can live a good life in reduced health or a terrible life in glorious health. But health is distinctively valuable—bound up as it is in how we feel, what we can do, our physical and psychological integrity, how we're viewed by others, and our place in society.

3. The Distinctive Political Importance of Health

Health is also something that matters in distinctive ways to our societies, as well as to us as individuals. There is a distinctive political axis of importance to our discussions of health: part of what matters to us when we talk about health are political issues of *public health*. Major political problems within our societies—socioeconomic disadvantage, lack of access to education, lack of access to fair housing, pollution and environmental degradation—often have health disparities among their bleakest consequences. It is not a coincidence that black and brown Americans died from COVID-19 at significantly higher rates (when adjusted for age) than white Americans, for example. Likewise, many public policy interventions—lowering carbon emissions, raising the minimum wage, providing free school lunches—have positive health outcomes among their key goals. People lose health partly in response to social problems, and this is generally understood as one of the worst aspects of those social problems. (The issue in Flint, MI isn't that the water tasted bad; it's that kids got sick.) And similarly, improving health and reducing health inequalities is generally understood to be one of the best results that can come from public interventions.

Importantly, though, the political significance of health doesn't derive merely from its normative significance, or from the role it plays in the wellbeing of individuals. Health also has distinctively public, collective value, much of which stems as much from its biological significance (its role in the functioning of the human organism) as from its normative significance. A pandemic like COVID-19 didn't threaten countries merely by threatening to harm their citizens (by making them ill, and possibly killing them). It also threatened massive political destabilization. If people are acutely ill, they can't go to work. If large numbers of people can't go to work, supply chains and basic services are interrupted. Prices increase. Jobs are cut. Resources become scarcer. You're a few short steps away from widespread chaos, no toilet paper, and a black market for hand sanitizer.

Health is inevitably political. You do not maintain your health on your own. You need basic goods and services, you need housing, you need healthcare. And the more we learn about health, the more we understand that you also need things such as education, socioeconomic security, community and supportive relationships, and freedom from constant social stressors.

Likewise, your own behavior affects the health of others—whether they like it or not. Measures such as mask-wearing and vaccination aren't important, during an outbreak of infectious disease, merely to protect yourself. If you get vaccinated, you increase the risk that you spread disease to someone at higher risk of serious illness than you are—and so that person is affected by your choices. If you don't vaccinate your kids as a "personal choice," the kids they go to school with are at increased risk of disease outbreak. And so on.

Part of what matters to us when we talk about health is this public dimension, this *political* dimension. Health is not just something we view as a personal good; it's also something we view as a public good, and as something that's shaped by our public life.

4. The Distinctive Phenomenological Importance of Health

There is also a curiously distinctive phenomenological aspect to what we care about when we talk about health. This is, of course, related to the normative significance of health (people value feeling good and disvalue feeling bad), as well as the biological significance of health (specific physiological processes make you feel specific kinds of ways). But it isn't fully captured by either. If the idea of catching COVID-19 weighed on your mind, you probably thought about existential issues such as the possibility of death, what would happen to your family if you were hospitalized and couldn't see them, how you'd manage being out of work, and whether you'd recover your strength and functioning. But you probably also had, at the forefront of your mind, *what it would feel like* to be ill.

There are specific and immediately salient experiences we associate with loss of health—nausea, fatigue, lightheadedness, pain. Not all loss of health causes such sensations (an illness might be symptomless) and not all such sensations are caused by loss of health (you might feel a twinge of nausea when you're embarrassed). But we associate specific ways of feeling with reduced health, and those distinctive experiences are part of what's most salient to us when we talk about health.

Moreover, they're salient to us in ways that go beyond either the biological or normative importance of health. If you get punched in the face, it's going to hurt. A normally functioning nervous system will (absent a ton of adrenaline or other defeating factors) induce pain after a solid punch to the face. But how much it hurts, and how much the pain affects your functioning, isn't something for which we can give a fully reductive physiological explanation. Maybe you're a boxer, so the pain is expected, and voluntary, and something you're very used to—you get punched in the face a lot. Maybe you're into that kind of thing and getting punched in the face, in controlled and consensual circumstances, is your idea of good fun. Factors like these will affect how you feel, and how you function. Likewise, while we can all grant that causing someone pain is in general a way of harming them, there isn't a neat correlation between amount of pain and reduction of wellbeing. The boxer might take not-so-secret pleasure in the pain of a good punch—it shows her toughness, it reminds her that she can take whatever's coming her way, it makes her feel alive. Sensations like pain—how intense they are, how bad they are, how distressing they are, how they feel to us—are invariably subjective.

And yet they're also a major part of what we care about when we talk about health. When told we have a disease, we want to know what it will *feel like*. Will it hurt? Will our bodies feel strange and alien? Will our limbs feel heavy and wooden? And while the answers might be different for everyone, nearly everyone is invested in those answers.

5. Ameliorative Skepticism

Spoiler alert: the main thesis of this book is that there is no coherent, unified account of health that can do all of this work for us. That is, there is nothing that can adequately explain all the ways in which health matters to us, at least in part because the ways that health matters to us often pull against each other. And there is no one thing in the world—no physiological state of the human organism, no natural resemblance class, no relationship between an organism and its environment—that we're tracking when we talk about health. Our concept of health is confused and lacks internal consistency. We talk about health in many different, often incompatible ways, depending on the context and our goals.

But the upshot shouldn't be error theory. Error theory is best suited to cases in which we have wholesale reference failure. "Witches" attempted to pick out women in league with the devil, and it emerged upon investigation that there are no such women. We shouldn't then use "witch" to refer to single women who live with multiple cats, and say that this is in fact the thing we were tracking with our use of "witch" all along, just with some false beliefs attached. Rather, we should grant that there are no witches—nothing in the world plays the role that our beliefs and language required in order for there to be witches.

In contrast, the things we care about when we talk about health are very real, and obviously so. "Philosopher discovers there is no such thing as health" is a headline that borders on parody. We all experience fluctuations in health. We all care about health. We have all seen the devastation that can occur when health is lost.

We're tracking real parts of the world when we talk about health. More specifically, we're talking about parts of the world that are interwoven in distinctive ways, which is why we need to talk about health as a unified whole, rather than just focusing on the underlying factors that determine it. Moreover, health has an ingrained moral and political importance. People are not going to stop speaking about health, or caring about health, because philosophers point out tensions in our basic understanding of it.

The goal of this book is thus to argue for something I am calling *ameliorative skepticism*. Following the influential work of Sally Haslanger (2012a), there has been growing philosophical interest in the idea of ameliorative projects. There are many different ways of understanding the scope of ameliorative projects, but the basic idea is that some of our ways of speaking, concepts, and theories need to be altered to better suit our legitimate goals (especially political goals). There is, for example, probably no one thing that is the reference magnet of all our different ways of speaking and thinking about gender. But Haslanger argues that we can move from that observation to the question "what do we want gender to be?". That is, what is the best way of understanding gender *given our moral and political goals*?

What I hope to motivate over the course of this book, however, is the idea that sometimes the best route forward is the skeptical path. I am going to argue that there is no specific, coherent thing that health is; no specific theory or concept or definition of health that can do everything we require. But I also don't think that there's a conceptual fix for this—I don't think there's something better we should mean by "health," or some more effective way we should understand what it means to be healthy. Rather, I think that health is an irretrievable mess, and yet it is also real, and something we need to theorize. We're tracking real aspects of the world—many of which are entangled and interdependent on each other in complex ways, such that we need a theory of how they fit together—when we talk about health. But the things we care about are confusing, imprecise, and often in tension with each other.

Our best way forward, I'll argue, is to accept this messiness, rather than try to clean it up or replace it with something neater. An ameliorative skepticism tries to explain *why* our understanding of health is a mess, and how it might still be useful and helpful to talk about health in spite of this inherent messiness.

6. In Praise of Messiness

As philosophers, we value clarity and consistency. We favor theories that are elegant, precise, and parsimonious. But part of taking social reality and the nature of the social world seriously involves allowing that some of the ways the world is are that way because they have been shaped by the way we think and speak about them. And the way that we think and speak is often inconsistent, imprecise, or otherwise confused. It's thus a mistake, I suggest, to expect that our theories of what social categories and social kinds are will yield simple and elegant results.

We can aim for precision, consistency, clarity, and parsimony in *how* we theorize about social categories such as health. But that is different than expecting that we should wind up with a theory that says that *what we are theorizing about* is precise, consistent, clear, or simple. Many things in the social world are a mess. They're vague, they're strange, they're complicated. And that's because we made them that way. To take seriously the idea that our thought and talk shapes social reality means taking seriously the idea that social reality might inherit some of the vices of how we think and speak.

To fully understand a socially embedded category such as health, we need to be able to theorize the ways in which health is *messy*. We can give clear and precise accounts of some categories—what it is for a number to be prime, what it is for an element to have an atomic weight. We can give clear and elegant explanations even of categories that admit of borderline cases (what it is to be a mammal, what it is to be conscious). We have a clear understanding of what these categories are, and what it takes to be a member, even if it's not always clear whether those conditions are met.

Health, I'll argue, doesn't work like that. And part of a successful theory of health needs to explain the way in which health is a mess. Appreciating this messiness, rather than trying to eliminate it, will give us a better understanding of health—what it is and why it matters to us.

In pursuing this tactic of ameliorative skepticism, I hope to offer a model for messy theories of social reality. This book is about health specifically, but you can also read it as a kind of illustrative example case of a more general strategy for how we might approach the philosophy of a messy, fragmented, amorphous social world. That is, you can think of it as an attempt (no doubt imperfect) to model clean theorizing about messy reality.

7. A Reader's Guide: Outline and Overview

This book can be divided into roughly three main sections. Section 1 (Chapter 1) gives a general overview of the current philosophical literature on health, Section 2 (Chapters 2–4) motivates some specific puzzles about the nature of health, and Section 3 (Chapters 5–6) uses those puzzles to motivate my own view.

We'll begin (Chapter 1) by looking at extant theories of health and examining their shortcomings. This section lays the necessary groundwork for the argument I'm making—if I want to make the case that we need to take a new approach to talking about health, I first need to argue that current approaches are lacking. (Be forewarned: it's a strange and complicated literature to wade through, and if generating counter-examples isn't where you philosophical joy lies, it's okay to skim over this section.)

After looking at the places where extant views get into trouble, we'll use those difficulties as a springboard toward a more general discussion (Chapters 2, 3, and 4) of some of the major puzzles for any adequate theory of health. The puzzles I'll focus on are the relationship between health and wellbeing, the relationship between objective and subjective dimensions of health, and the relationship between health and disability.

And finally (Chapters 5 and 6), I'll outline my own favored approach to theorizing health, given the complexities involved. These two chapters, taken together, outline the approach I'm calling *ameliorative skepticism*. And while they're focused, in this context, on health specifically, I'm hopeful that the model of dealing with philosophical messiness—especially in the social world—can be of broader interest to a range of debates.

Here, then, is an overview in slightly more detail.

Chapter 1: Theories of Health

In this chapter, I survey the philosophical landscape of theories of health—accounts of what it means to be healthy, to have a disease, and so on. I aim to show that each major family of theories has significant problems, and that these problems can't be solved simply by combining or slightly modifying those theories.

Chapter 2: Health and Wellbeing

One of the most perplexing aspects of health is its relationship to wellbeing. Contra some accounts, I argue that health and wellbeing are not the same thing, nor is good health a necessary condition of wellbeing. And yet the two are intimately related, much more so than many other things we value and that contribute to our wellbeing. I then examine the common claim that health is valuable because and insofar as it contributes to our wellbeing, and argue that this claim is also too simplistic. A plausible account of the relationship between health and wellbeing needs to be able to say that a person can have high levels of wellbeing but be in poor health (or be in good health but have poor wellbeing). It also, more strongly, needs to be able to say that someone who has lost health has lost something valuable, even if they still have a high quality of life. To argue this point, I make an analogy to grief and the relationship between grief and overall wellbeing.

Chapter 3: Health, Subjectivity, and Capability

Another perplexing aspect of health is its relationship to subjectivity and objectivity. "Objective" and "subjective" can mean many different things in different contexts, and in this chapter I'll first try to specify what I'll mean by them in this context. Then I'll argue that they have a distinctive relationship to how we think about health. Loss of health can objectively impair our body's functioning, and some have argued that the distinctive value of health comes from its role in preserving functional capability. But whether and to what extent a person's functional capacity is limited by a health condition is partly determined by their subjective experience of (and reaction to) that health condition. And, more strongly, some aspects of our health—including those that are objectively important to our overall health and that can be important limitations on our ability to function, such as pain and fatigue—are inherently subjective. The net result is that subjective and objective aspects of health are often interdependent, and ignoring either (or the connections between them) leads to misjudgments about health. I argue that in order to fully understand the relationship between subjective and objective components of health, we need to develop an understanding not just of the social *determinants* of health but also of the social *constituents* of health.

Chapter 4: Health and Disability

In the third "it's complicated" chapter, I examine the relationship between health and disability. Not all loss of health leads to disability, and disability can't be understood or explained simply in terms of loss of health. And yet there is an intimate connection between being disabled and having reduced health. An increasingly common view within the disability rights community is that disability is a "mere difference"—it is something that makes you different but not something that (by itself) makes you worse off. A "mere-difference" view of health, however, looks both implausible and politically disastrous. This chapter attempts to find a way through this apparent tension. I appeal to the puzzle of the statue and the clay, arguing that the same physiological condition can appear to have a different character depending on whether it is viewed as a socially embedded phenomenon or a biomedical pathology. I then argue that David Lewis' proposed solution to the statue/clay puzzle can be usefully adapted to the case of social kinds such as disability, arguably with fewer drawbacks than Lewis' own application of it to material constitution.

As an Appendix to this chapter, I survey the extant empirical literature on the relationship between disability and quality of life/subjective wellbeing. The basic takeaway message of this survey is that asking "what is the relationship between disability and quality of life?" is in general not a very helpful question, since a wide range of things can be meant by "disability," which seem to have substantially different impacts on subjective wellbeing, at least according to current empirical evidence.

Chapter 5: Ameliorative Skepticism and the Nature of Health

In the final two chapters, I develop my own approach to theorizing health—based on the idea of *ameliorative skepticism*. I defend a characterization of health that falls between views such as error theory, fictionalism, and eliminativism, on the one hand, and metaphysical realism, on the other. Our understanding of health, I argue, is inherently unstable. And this is both because of how we conceptualize health and because of the underlying nature of the things we're tracking when we talk about health. More strongly, health is, I suggest, a case in which tensions and confusions in our concepts might plausibly create tensions and confusions in social reality itself. If we take seriously the idea that social reality is in a significant way shaped by our collective norms, beliefs, and practices, and we also take seriously the idea that sometimes those norms, beliefs, and practices can be incomplete, confused, or inconsistent, then we should expect some aspects of social reality to inherit those confusions. Health, I argue, is one of those confusing places. We should reject error theory or eliminativism because the things we're tracking when we talk about health are real, and entangled in such a way that we wouldn't be better off replacing talk of health with talk of some more specific determinants or realizers. And yet, because of the nature of what we're trying to capture, there is no stable, consistent account of health we can offer that will adequately answer the question "what is health?" or give us everything we need from a theory of health.

Chapter 6: Ameliorative Skepticism, Shifting Standards, and the Measure of Health

There is no one standard by which we can evaluate how healthy someone is, or whether one person is healthier than another. But this doesn't mean we should abandon the idea of health or replace it with something more precise. We can best understand discussions of health, instead, as serving particular pragmatic interests in particular contexts. Whether someone is healthy—and how healthy they are— depends on why we care about health in that context, and what work we need our talk of health to do. To explain this idea, I appeal to Delia Graff Fara's approach to vagueness. A salient feature of Fara's model is that imprecision can sometimes be helpful to communication. And, likewise, in making things more precise, we can sometimes lose as much information as we gain, such that we're not always making things better or clearer.

Using the insights of Fara's model, I argue that often what matters most for particular evaluations or discussions of health is the salient role we need health to play in that context. On the view I'm defending, there are no deep or context-independent facts about what health is. It's also possible for two people with different aims and goals to disagree with each other about health, and yet for neither to be making a mistake. And yet, I argue, this is compatible with health being something that's real, valuable, and important, and compatible with there being facts about health.

The overall picture this book seeks to defend is one in which health is a mess—but an important, interesting, and useful mess. And the best way forward, I suggest, is to appreciate the ineradicable messiness rather than to precisify. Philosophical skepticism is often viewed as an entirely negative enterprise—it says why every answer is wrong, why every view has a counterexample, but provides no better alternatives. What I hope to model in this book is a way in which a kind of skepticism can be therapeutic rather than antagonistic. It's helpful to be clear about why things are so confusing, and instructive to see why debates can be so intractable. Step one, as they say, is admitting you have a problem. And so I see this book as a loving exploration of philosophy's health problems.

1

Theories of Health

1. Welcome to the Jungle

The philosopher W. V. Quine famously argued that good theories should resemble desert landscapes—bare, austere, uncomplicated. The philosophical literature on health is a jungle. And it's a jungle where it can be very hard to see the forest for the trees (and the vines, and the quicksand, and the giant mosquitos). Nor is there a single debate here. Philosophers come at the question "what is health?" from various different literatures, including philosophy of science, bioethics, philosophy of wellbeing, and political philosophy.

So many traditions have a stake in this conversation because health is an issue that matters so broadly. It matters to public policy, to scientific practice, to ethical understanding, and to personal flourishing. And perhaps more distinctively, health is a place where the theory of a thing matters substantially to the practice of that thing. Philosophers might not be able to agree on the necessary and sufficient conditions for something being a table, and yet we can all sit down and eat our dinner just fine—theoretical questions notwithstanding. With health, though, we cannot separate theory and practice quite so easily. Your doctor can, of course, prescribe antibiotics for your infection or monitor your blood pressure without giving much thought to the question "what is health?." But health is, invariably, something we need to quantify and measure. We need to assess whether a given health intervention is effective. We need to compare the health of different groups and different populations. And so on. Which invariably leads down the path of finding a workable answer to the question "what is health?."[1]

In what follows, I attempt a big-picture overview of the extant philosophical literature on health. I will, in doing this, by no means be able to address all approaches or discuss all views, nor is an exhaustive literature survey the goal of this chapter. What I am attempting, instead, is to motivate the idea that it is hard—surprisingly hard—to give anything like an adequate, informative answer to the question "what is health?." Most extant approaches have obvious virtues but also serious—and on my view irremediable—vices. To make this case, I'm going to examine the broad landscape of views: the ways in which philosophers have

---

[1] That's not to say that a successful measure of health must exactly mimic our best philosophical theory of health. But theories of health inform which measures of health we find most plausible, and different measures of health provide strikingly different results.

Health Problems: Philosophical Puzzles about the Nature of Health. Elizabeth Barnes, Oxford University Press.
© Elizabeth Barnes 2023. DOI: 10.1093/oso/9780192883476.003.0002

tended to approach the question "what is health?" and the kinds of answers they have tended to give. My aim, in doing this, is skeptical. I want to show that the current approach to the question "what is health?"—in which we attempt to define or give necessary and sufficient conditions for health—is doomed to failure.

In order to motivate this skeptical project, the discussion in this chapter explores pretty far into the weeds of some theories of health. This is, after all, a jungle. If you're like me and you like getting into the weeds, come along for the ride. But if theory-chopping details don't hold your interest, it's fine to take the moral of this section to be 'it's complicated' and go straight to Chapter 2.

Much of the current dialectic surrounding philosophical theories of health is, in many ways, a long series of responses to the attempt to say that health (or, more specifically, pathology/disease) is something purely naturalistic or biological—something that we can explain without reference to our values, our feelings, our norms, or our social conventions. A series of views—mostly based around the idea of *normal species functioning*—attempt to explain health in these terms. So we'll first look at such views, the key problems they run into, and some of the main options they offer for addressing these problems. We'll look, that is, at the best case to be made for naturalistic theories of health.

Finding such theories wanting, however, many in this debate have argued that we have to add something normative or evaluative in order to explain health. As we'll see, though, "add something normative" doesn't give us a recipe for turning water into wine. There are many different places we could look for the normative secret sauce that will give us a satisfactory theory of health, and each of them, I'll argue, faces problems as significant as those encountered by naturalistic approaches. The result is a dialectic littered with views that seem to be getting at *something* interesting and *something* significant about health, but failing—in serious and substantial ways—to actually provide a workable theory of health.

Here's the gameplan. After a methodological overview (section 1), I'll begin with a deep dive into species function-based accounts (section 2). The appeal of these accounts is that they explain health in biological terms—making it an appropriate source of study and discovery for the natural sciences. But as we'll see, serious worries arise about their explanatory adequacy. From there, I examine attempts to fill this explanatory gap, looking first at views that attempt to explain what health is by appealing to wellbeing (section 3), then at views that appeal to a distinctive type of phenomenological experience (section 4), and then at views that appeal to social norms and conventions (section 5). Finally, I consider so-called "hybrid" views, which attempt to combine various aspects of other views to offer a more explanatorily rich account (section 6).

If my analysis is correct, then the reason that each of these families of views gets into so much trouble isn't that we need to tweak one of them until it works. Rather, it's that they're engaged in an explanatory project that, by its nature, isn't going to work.

1.1 Targets of Analysis

It's important to note from the outset that there are many different debates, and different philosophical conversations, that can be seen as attempting to answer the question "what is health?." There is not a single body of literature here or a single shared set of terms or methodological assumptions.

Given how divergent they can be in approaches and assumptions, it might be tempting to suggest that the various philosophical attempts to explain what health is are simply talking past each other.[2] They don't mean the same thing by "health," and they aren't engaged in the same project. But I think this would be a mistake. Philosophical theories of health start from a wide range of presuppositions, differ on terminology, and often lack a shared literature. They're relatively united, though, in the basic phenomena that they're trying to explain.

To begin with, they have the same basic constraints of extensional adequacy and agree on most paradigm cases. Someone with stage 4 liver cancer is not healthy—indeed, they are very unhealthy. Someone with moderate coronary artery disease has reduced health because of that condition, but their health is not as compromised as the person with liver cancer. While there will be debates about some famous test cases that present puzzles for specific models, philosophical theories of health tend to converge in judgments about such paradigm cases. More strongly, they take themselves to be *constrained* by such paradigm cases. The naturalist and the phenomenologist alike, for example—although they differ dramatically in methodology, and even in what they take the primary target of their analysis to be—agree that amyotrophic lateral sclerosis (ALS) compromises health, and that any theory that delivered the result that someone with ALS was fully healthy would therefore be an unsuccessful theory.

More substantially, I think that we can see commonality in theories of health by looking at what *work* these theories are trying to do. Philosophical theories of health diverge in their methodological approaches and in the concepts they employ. But the distinctive roles we need health to play still constrain the success of any such theory. As discussed in the Introduction, there are distinctive ways that health matters to us and distinctive work we need an answer to the question "what is health?" to do for us. We can thus see any theory of health as being constrained by its ability to ground explanations of:

---

[2] For an in-depth argument that such theorists are, in fact, talking past each other, see especially Plutynski (2018). I'm very sympathetic to Plutynski's view, since we agree—as I'll argue in defending my own view in Chapter 6—that what matters most are our aims and purposes in talking about health, and these can shift dramatically depending on the context. I don't think these differences in aims and purposes, however, are sufficient to make it the case that theorists are in fact trying to describe different things when giving a theory of health—I just think this variation is a feature of the inherently slippery thing all parties are trying to explain.

(1) The distinctive biological importance of health
(2) The distinctive normative importance of health
(3) The distinctive political importance of health
(4) The distinctive phenomenological importance of health.

Any successful theory of health needs to be able to explain the biological significance of health—why our objective functional status as organisms is directly
related to our health; why compromise in health can often lead to objective loss of
functional capacity (up to and including death); why conditions that affect our
health can be studied by epidemiologists, geneticists, virologists, and so on.
A theory that renders health entirely subjective, or entirely orthogonal to the
investigation of the natural sciences, would simply change the subject. And in
doing so it would fail to engage with much of what is obviously relevant and
obviously matters to us about our health. But an answer to the question "what is
health?" also needs to be able to ground the normative significance of health—why
health is something that is valuable to us, why reduction is health is often a
distinctive form of harm, why health has a close relationship to wellbeing, and so
on. A theory of health that says that health is something that is biologically
significant but not something that has any particular connection to what we
care about, to what harms or benefits us, or to our wellbeing has again simply
changed the subject.

Likewise, any successful theory of health needs to provide grounds for the
distinctive political and the distinctive phenomenological importance of health.
Part of what we care about when we talk about health is the public role that health
plays—both in terms of the social impacts that loss of health can have and in terms
of the distinctive political implications of health inequalities. A theory of health
that renders health entirely individual or personal—with no connection to public
impact—would be a non-starter. Similarly, part of what we care about when we
talk about health is our own subjective experience—how we feel, how we experience our minds and bodies, how much pain or distress we undergo. And we want
some account of the connection between these experiences and our health.

To be clear, theories of health don't need to directly engage these questions.
A purely naturalistic theory of health, for example, won't talk about our values or
our political goals, and that's not an obvious mark against it. But it's a constraint
on the success of any account of what health is that it can serve as an adequate
grounding for the distinctive roles that health occupies.

It's insofar as they're addressing that same basic issue that they share common
constraints. An epidemiologist, a hospitalist, and a public health policy wonk will
have very different approaches and perspectives in thinking about health, but
when they come together on a panel to discuss what COVID-19 mitigation
measures will be best for overall health outcomes, we don't think their differing
approaches and methodologies mean they're talking past each other or engaged in

different conversations. Rather, they're all interested in the same core issues, but coming at them from slightly different directions. Similarly, I'm suggesting that what unifies—and constrains the success of—the various philosophical theories of health is the subject matter. Insofar as they're all trying to explain what health is, they're all trying to address the same basic issue, though from different perspectives and with different presuppositions about what's needed to tackle the issue. Likewise, insofar as they're all trying to explain what health is, they're constrained by the extent to which they can account for the distinctive biological, normative, political, and phenomenological importance of health.

1.2 Terminology and Methodology

Before proceeding further, let's work through a few more preliminary clarifications. First and perhaps most importantly, as already noted, this discussion is by no means an exhaustive literature survey. I'm not intending to offer a completionist discussion of every attempt to explain the nature of health. Rather, I'm attempting to give an overview of the major *styles* of view available, and to discuss (what I take to be) the most influential theories. My aim is to show, in broad strokes, how the various styles of view get into trouble with at least some aspects of the work we need a theory of health to do.

In doing this, I will try, whenever possible, to offer objections that can be seen as objections within the framework of the view in question. It's not a particularly compelling objection to a naturalistic view that such a view doesn't appeal to 'lived experienced' in characterizing health; nor is it a particularly interesting objection to a phenomenological view that it doesn't present a theory of illness that would be useful to epidemiologists. That's not the project either approach is engaged in— even if you think that ultimately it's the project they *should* be engaged in. So where possible, I will try to offer objections to the views under discussion that can be seen as objections by the lights of people who hold and defend those views.

This means that, throughout the subsequent discussion, there will be some slight shifts in terminology as well as in focus. I'm going to highlight a few of the major ones here—though readers should note that this is mostly to explain how I'll be using these terms as I discuss these views, and I don't take all that much to hang on any of these distinctions.

For some theories, the primary project is to explain what health is, and then to characterize ways in which we can have lost or reduced health. For others, it's to explain what disease is, and then to understand health as the relative absence of disease. I'll call the former projects *positive* theories of health and the latter projects *negative* theories of health. These two approaches can lead to interestingly different conceptions of health. Positive theories of health try to give some sort of specific criteria for or analysis of being healthy, over and above the absence of pathology. For negative theories, you're healthy just in case there's nothing wrong with you. The only way you can be healthy to greater or lesser degrees, on these views, is how substantially or seriously you are affected by pathology.

It's also worth clarifying what we're trying to characterize when we talk about what detracts from health. Some authors use the term "disease," but disease is a relatively specific concept, at least as it's employed in biomedical contexts. Lots of things can detract from health that are not, strictly speaking, diseases—if you are poisoned, for example, you will be in poor health, but you won't thereby have a disease, at least in the usual understanding of "disease."[3] Some opt for "illness" instead—you don't have a disease if you're poisoned but you're most definitely ill. Again, though, "illness" is probably too restrictive. Many things that affect health—injuries, most especially—aren't well classified as illnesses. And so for what follows I'm going to use the expansive term "pathology"—which can be used to mean, roughly, any deviation from healthy condition, encompassing diseases, injuries, malformations, and similar.[4] Thus, as I understand it, the first major divide in theories of health is between theories that are trying to give an account of pathology and theories that are trying to give an account of health, where health is understood as something more substantial than a relative absence of pathology.[5]

The second main divide in theories of health is between those that seek to give a purely non-evaluative or descriptive account of health and those that maintain that we need at least some normative or evaluative concepts to explain what health is. Let's call the former *non-evaluative* theories of health and the latter *evaluative* theories of health. I am not foolhardy enough to attempt to define what it takes for a theory to count as evaluative (or non-evaluative). In the literature on health, non-evaluative theories are often described as those that appeal only to things that are within the remit of the biomedical sciences. But I think it's a mistake to assume that the practice of science—especially social and biomedical science—is value-free. For the purposes here, I'll simply assume that normatively laden concepts are things like: good, better, right, beneficial, harmful, and so on. That is, they're concepts that are explicitly evaluative rather than descriptive. As I'm using the terms, a theory of health succeeds in being non-evaluative if it doesn't need to appeal—explicitly or implicitly—to normative concepts to explain what health is.

---

[3] See Kingma (2010) and Hausman (2011) for discussion.

[4] Unsurprisingly, the term "disability" is often included alongside illness, disease, and pathology when theorizing health, but as I'll explain in Chapter 4, I think the relationship between health and disability is far more complicated. So I'm going to omit discussion of disability specifically for the remainder of this chapter, and focus on it in more detail later. I'll be discussing plenty of physical conditions that give rise to disability, but not disability per se.

[5] One further clarification to make is that I am using the term 'pathology' in a morally neutral way. Something might be pathological in the biomedical sense without being bad or harmful to the overall wellbeing of the person who has it, for example. We'll discuss (Chapters 2 and 4) the relationship between health and wellbeing in far more detail later.

Likewise, I'll assume that an account is evaluative if it explains what health is at least partly in normative or evaluative terms.[6]

A final important distinction in theories of health is between subjective and objective theories. According to subjective theories, what makes it the case that you are healthy (or that some condition is pathological) is primarily your own reaction to that condition or the interaction between that condition and your specific goals, desires, and preferences. According to objective theories of health, there is a fact of the matter about whether a condition is pathological (or about whether you are healthy) that goes beyond your own reactions, goals, or desires. This is, importantly, not the same thing as saying that there is a *biological* fact of the matter.[7] You might define health as an aspect of wellbeing, for example, and be an objectivist about wellbeing. You would then think that there are objective facts about whether someone is healthy, but they aren't purely biological facts. Likewise, holding an objective view of health isn't the same thing as denying that subjective reactions play a role in determining whether someone is healthy. To take an obvious case, a person's subjective mental state plays a role in whether they are depressed, and whether a person is depressed matters to how we assess their mental health. But you can allow this without holding the view that whether someone has depression, or whether they have a mental health condition, is determined entirely by their own subjective reactions to that condition. And finally, saying that health is subjective is not the same thing as saying it is contextually variable. Social constructionists, for example, think that which things count as diseases is determined by social norms and practices. And so, which things count as diseases will vary from place to place and time to time, as norms and practices surrounding health and illness shift. But it doesn't follow that which things are diseases is subjective. In each case, there's a fact of the matter about whether something is a disease that's determined by something—in this case, a broad network of social practices—over and above private, first-person reactions. It's just that the objective facts about which things are diseases can and do vary.

In summary, I've identified three main axes on which theories of health can be differentiated: positive and negative theories, evaluative and non-evaluative

---

[6] Again, clarifications are in order. A theory can be evaluative—can insist that we need some kind of language about harm or badness, for example, to define pathology—without maintaining that pathology automatically reduces a person's overall wellbeing or that there is a neat correlation between amount of pathology and loss of wellbeing (see Chapters 2 and 4). I have, for example, in previous work argued for a *mere-difference* view of physical disabilities, according to which physical disability does not by itself make a person worse off (see Barnes (2016)). This is a claim about overall wellbeing, one that I think is consistent with at least some normative accounts of pathology. Explaining why is complicated, though—again, see Chapter 4.

[7] See Broadbent (2019) for elaboration on this point and a discussion of the common confusion between objective and non-evaluative theories of health.

theories, and objective and subjective theories.[8] These distinctions often cut across each other—whether you're a subjectivist doesn't determine whether your approach to health is positive or negative, for example. So the classification of theories of health gets complicated fairly quickly.

These axes of difference do nothing to assuage the worry that philosophers debating the nature of health are simply talking past each other and trying to analyze different things. Again, though, I'm going to assume that everyone agrees on the basic target of analysis. (They also seem to be meaningfully disagreeing with each other, rather than simply talking past each other.) In giving a theory of health, we are attempting to give a philosophical theory of whatever it is we care about when we talk about public health initiatives, whatever we are concerned with when we go to the doctor because we don't feel well, whatever it is that we think is distinctively at stake in public health crises like the Flint, MI water crisis.

And this is why I think it's especially important to hold paradigm judgments fixed. Although there can be disagreement on marginal cases, most everyone agrees on the paradigm examples of things that detract from health—cancer, infection, autoimmunity, and so on. These basic points of agreement constrain debate. If your theory suggests that cancer is not unhealthy so long as you have a positive mental attitude about it, we have good reason to reject your theory; likewise, if your model suggests that once a deadly virus spreads widely enough to be common it is no longer unhealthy, your model is getting something wrong. In what follows, I'm going to assume that these basic parameters—and basic points of agreement—are sufficient to make debates about health genuine, and not merely verbal, disputes. All parties, I think, are trying to give an account of roughly the same thing; they just have very different approaches to it.

2. Function-Based Theories

In the philosophical literature on health, by far the most influential and widely discussed family of theories are those that appeal to the idea of *function*—either

---

[8] A further important distinction is the divide between individual and population health. Some things that are unhealthy for a population have no effect on the health of individuals, and some things that might be unhealthy for an individual are perfectly normal in healthy populations. I'm here primarily concerned with individual health, and the theories I'm examining are theories of individual health (which dominate the philosophical landscape on health). Authors often assume that individual health is the more foundational notion and that we can understand the health of populations in terms of the health of their individual members and the conditions likely to promote the health of their members. Some argue that this model is overly simplistic—see especially Valles (2018)—and that we should understand individual health via population health, and not the other way around. I take no stand on the relationship between individual and population health.

normal function[9] or *proper* function.[10] The basic gist of such views is that the healthy organism is the organism that is functioning in a species-typical or species-appropriate way. The normal functioning immune system is the immune system that attacks pathogens, for example. Pathology can then be understood as a particular way of functioning abnormally. Perhaps your immune system doesn't attack anything, making you highly vulnerable to infection. Or perhaps it attacks parts of your body that is not pathogens, resulting in an autoimmune disorder. Either way, your immune system is not functioning in a species-typical way—it's not doing its job—and as a result your overall health is compromised.

The primary aim of most function-based accounts is to give a broadly naturalistic, non-evaluative[11] account of health.[12] Whether your immunocompromised state is pathological doesn't depend on how we feel about it or our cultural attitudes to it, and so on—it's a matter of its role in your biology. In searching for a naturalistic theory of health, function-based theorists are often interested in giving an account of health that is relevant—and perhaps of use—to epidemiologists, the biomedical sciences, and so on.[13] And typically, they do this via giving an account of pathology, construing health as the relative absence of pathology (that is, they are "negative" theories in my sense). Such accounts are also typically objective—whether an individual has compromised health, on such views, is a matter of objective fact determined by their relationship to biological norms for their species, and would remain constant no matter what norms we attached to it or how the individual felt about it.

This is, as we'll see, an ambitious project, and perhaps something of a tall order. But it's worth taking the time to examine function-based theories in detail, since most other major theories of health can be understood at least in part via how they depart from the naturalistic ambitions of this family of views.

The fundamental challenge for any normal function theory of health, of course, is that not all "abnormal" functioning is pathological (nor is it obvious that all pathological functioning is abnormal). One way to function abnormally is to function especially well. Other abnormalities are mere statistical variations—unusual but not pathological. Most people can't "wiggle" their ears, and so those

---

[9] See especially Boorse (1977), (1997), (2011), and (2014).

[10] See especially Millikan (1989a) and (1989b) and Neander (1991a), (1991b), and (1995).

[11] For some defenders of function-based views—especially Neander and Millikan—the idea of function is inherently normative, but in a way that they argue is compatible with naturalism (contra many of the other approaches to normative accounts of health). We might quibble whether "non-evaluative" is a good descriptor for such views, but the main point here is that they eschew thickly evaluative terminology like "bad," "unlucky," "wrong," and so on.

[12] There is also a family of views—often called "hybrid" views—that take a naturalistic notion of normal function as a necessary but non-sufficient condition for pathology, arguing that a further evaluative condition is needed. We'll discuss such views in section 6.

[13] Another major tradition that appeals to the idea of function are Aristotelian theories of health—but they mean something more strongly normative in their conception of "normal function." We'll discuss these views in section 3.2.

who can are functioning abnormally in some sense. But they don't have ear-wiggling disease; they just have a delightful third-grade party trick. We need to have some way of singling out which abnormalities are *dysfunctions*. And if function-based accounts are to succeed in their non-evaluative ambitions, we need to be able to do so without simply saying that the dysfunctions are the ones that are bad, the ones that have a negative impact on wellbeing, or similar.

Although there's a wide variety of individual approaches to this puzzle, there are two main strands in the literature: explaining dysfunction via statistical typicality and explaining dysfunction via natural selection. I'm going to argue that a key way of understanding the difference between these two views is by looking at the difference in how they respond to what I will call the *counterfactual problem*. Most naturalistic accounts of pathology appeal to something like the idea of fitness. The things that are pathological (i.e., dysfunctions rather than abnormalities) are the things that impair our fitness—with fitness broadly understood as our capacity to survive and reproduce. The counterfactual problem, in a nutshell, is this: there are plenty of things that are in fact pathological, but do not, in actual, present circumstances, impair survival or reproduction. So function-based views must tell us what makes it true that these pathologies *would have*—in some naturalistic sense that is relevant to our current circumstances—impaired our fitness, even though they don't seem to in our actual circumstances.

2.1 Biostatistical Theory

By far the most widely known and influential version of the normal function-based approach to health is Christopher Boorse's *biostatistical theory*. Although the details have evolved over time, the basic picture has remained relatively constant since Boorse first proposed it. And while other defenders of normal function accounts depart from Boorse in various ways, this has largely been a matter of quibbling with details. The central idea is that health and pathology can be understood in terms of what's statistically typical for the species. When a dog can't digest chocolate, for example, we may think it's a shame for them and perhaps annoying for us if they're constantly trying to eat it anyway, but it's not a disease—dogs just can't eat chocolate, and that's normal for the species. But if a human can't digest chocolate—if they become ill the way a dog does when eating chocolate—we think something has gone wrong, precisely because that's *not* typical for the species.

For Boorse, normal functioning is defined in terms of statistical typicality for a specific comparison class. An organism functions normally if it functions in a way that is statistically typical for age- and sex-matched comparison classes of its particular species. The comparison class that determines statistical typicality has to be age- and sex-matched for the statistical typicality to get off the ground as a basis for normal function. Absence of menstrual cycles, for example, is abnormal in a twenty-five-year-old female (assuming she isn't pregnant) but normal in a twenty-five-year-old male and normal in a sixty-five-year-old female.[14]

A function, for Boorse, is a goal-oriented process. Boorse construes the function of the entire organism—that is, its basic biological goals or purposes—as survival and reproduction. And so, an organism is functioning normally if it is able to accomplish its basic biological goals of survival and reproduction at a level that is statistically typical for age- and sex-matched members of its species. Boorse then zeroes in on the specific functions of parts or processes within the organism. The functioning of a part or process, for Boorse, is the ability of that part or process to accomplish its basic biological goals or purposes. So, for example, the function of the kidneys is to filter waste, the function of the cardiovascular system is to circulate blood, and so on.

The normal function of an organism can be understood, according to Boorse, in terms of the normal functioning of its parts and processes. A part/process is functioning normally if it is making a species-typical contribution (for age- and sex-matched comparison classes) to the organism's ability to survive and reproduce. Building up from there, an organism is functioning normally if all its parts function normally. Likewise, an organism is dysfunctioning if one or more of its parts/processes is making a lower-than-typical contribution to its ability to survive and reproduce.[15]

Boorse then argues that we can construe pathology as *negative* departure from normal function—departure from species-typical function that hinders the organism's ability to survive and reproduce.[16] For a part or process to be pathological, it must interfere with an organism's ability to survive and reproduce, in the sense that its contribution to the organism's survival and reproduction falls *below* the species-typical norm.

---

[14] Already, though, we will begin to see problems emerge. How specific should the comparison classes be? It seems like we need to be more fine-grained than "adult" when it comes to something like age in order to explain why, e.g., the absence of a menstrual cycle or lowered bone density that would be pathological in a younger female are perfectly normal in an older female. But once we restrict to a group of older adults, certain pathological conditions—arthritis, coronary artery disease, diabetes—are incredibly common, affecting as much as 25–30 per cent of adults over sixty-five within the US. So it becomes harder to say that such pathologies are statistically abnormal for that age-matched comparison class. We'll return to this issue shortly.

[15] This claim will clearly need a bit of modification, however, since you can function abnormally by *missing* a part. We'll return to the issue of what is meant by a part or process making a "lower" contribution to survival and reproduction

[16] There's some unclarity in the various presentations of Boorse's account over whether we should read the claim about goals as a disjunction ("survival or reproduction") or a conjunction ("survival and reproduction"). The conjunctive reading has been objected to, since ostensibly some things that enhance your chances at reproducing might limit your chances at surviving. In light of this, Boorse (2014) explicitly revises to the disjunctive reading. But the disjunctive reading is arguably too expansive. For example, there are many things that might detract from an organism's ability to reproduce that aren't pathological (same-sex attraction, preference for non-penetrative sex, and so on).

Boorse emphasizes that the goals of survival and reproduction are to be understood in evolutionary terms—they are the "goals" of an organism trying to pass on its DNA. So the physical organism that is your body can be said to have the goal of reproduction, according to Boorse, even if you spend a lot of effort trying to avoid reproduction. And states that interfere with this "goal" count as pathological in Boorse's sense even if they are desired and acquired by medical intervention (e.g., tubal ligation or vasectomy). What matters for an evaluation of health or pathology is an organism's ability to meet these goals, which might come apart from whether the goals are actually attained. So, for example, you don't count as diseased if you simply remain celibate and never reproduce; in such a scenario, you could still have the *ability* to reproduce, and ability to reproduce— rather than actual reproduction—is what matters to a normal function analysis.

Let's take a brief summary. For Boorse, health is the absence of pathology. You're healthy just in case there's nothing much wrong with you. Pathology is a negative departure from normal function. And negative departures from normal function are departures from species-, age-, and sex-typical function that reduce the ability to survive and/or reproduce. The function of your immune system, for example, is to target pathogens; that's the role it plays in your overall ability to survive and reproduce. Your immune system is functioning normally if it per- forms this function at species-typical levels—which doesn't mean you won't ever get sick, or that human immune systems are ideal. It just means that it's perform- ing its function in a way that is not pathological. Some people have especially strong immune systems. Good for them. They're statistically atypical, but nothing about that difference is pathological because it doesn't interfere with their overall ability to survive and reproduce (if anything, it enhances it). Some people, though, have compromised immune systems, or immune systems that target things other than pathogens. *This* is pathology—the function of their immune system is falling below what's typical for the species in a way that interferes with their ability to survive and reproduce.

Does this account give a workable, extensionally adequate definition of pathology? There are several major objections in the literature that argue that it does not. One serious problem, for example, is the presence of common pathology—especially given that Boorse's analysis must incorporate statistical normalcy for sex- and age- matched reference classes.[17] This creates problems when we consider health conditions that are incredibly common in aging populations. Almost half of adults age sixty-five or older, for example, have been diagnosed with osteoarthritis, as have a substantial percentage of adults over forty. Given that statistical atypicality for age- and sex-matched comparison classes is a necessary condition for pathol- ogy, it doesn't look like Boorse's model can consider arthritis in older adults

---

[17] See, e.g., Cooper (2002), Kingma (2010), and Schwartz (2007) for discussion.

pathology, even though it's a significant health problem.[18] A closely related problem is the *line-drawing* problem;[19] it looks like it's difficult for a model like Boorse's to specify the difference between mere abnormality and pathology, especially in cases where pathology represents one end of a spectrum of variation.

But for the purposes here, I'm going to focus on the counterfactual problem—a problem I think affects all versions of the normal function approach to health. The problem is rooted in the obvious fact that many common pathologies don't *actually* reduce lifespan or limit reproduction for the individuals that have them. That is, there are plenty of things we consider health problems such that an individual can have those things and yet live to old age and have children without difficulty. And so in order to treat these conditions as pathology, Boorse must be able to say that they reduce the *ability* to survive and reproduce even though the individual's *actual* survival and reproduction is unimpaired. We thus need some explanation of the relevant sense in which such conditions would have or might have or could have impaired survival and reproduction, even when in actual circumstances they don't. This, in a nutshell, is the counterfactual problem.

If you break your ankle, that's clearly an injury—and clearly biomedical pathology. And yet, at least if we're comparing you to the comparison class of non-elderly adults, breaking an ankle doesn't make you any less likely to live a normal lifespan or reproduce successfully. Many diseases associated with advanced age—including even some subtypes of neurodegenerative disorders like Parkinson's disease[20]—have normal life expectancies because of their late onset, and likewise don't impair reproduction given how late in life they occur. Alopecia is an autoimmune disorder—so, again, a clear case of pathology—that causes you to lose your hair (sometimes over your entire body), but it does not affect life expectancy and doesn't impair reproduction.[21] Deafness is one of Boorse's paradigm examples of pathology, but in modern Western contexts

---

[18] What's more, it doesn't look like a viable option for Boorse to simply say that common health problems like arthritis are not really pathology, since he adamantly maintains that the "target concept" of his analysis is the understanding of pathology employed in medical practice, and medical practice certainly understands arthritis and other common ailments as pathology.

Boorse (2014) discusses the possibility of making age-comparison classes less fine-grained, and so having "adult" simply be the comparison class—but grants that this removes the possibility of distinguishing normal aging (including menopause, graying hair, etc.) from age-related pathology such as arthritis.

[19] See especially Schwartz (2007) and Plutynski (2018).   [20] See Backstrom et al. (2018).

[21] It might be argued that alopecia does impair reproductive capacity—even if it doesn't impair the function of the reproductive organs—because having it will make a person appear less "choiceworthy" as a mate and therefor render them less likely to be able to reproduce. If this is the case, however, then many things that are clearly not biomedical pathology—but that might be statistically atypical—will count as things that impair our ability to reproduce. More strongly, it seems like lots of contingent, evaluative social norms will play a key role in determining which physical features make a person seem more or less physical attractive in a context. As Quill Kukla (2019) argues, reproductive capacity and (in)fertility are inherently social.

being deaf does not reduce life expectancy and obviously deaf people can have children just fine.[22] And so on.

Thus for the biostatistical theory to have any hope as an extensionally adequate theory of health, we need to find a reading of "ability" for which *ability* to survive and reproduce can be diminished even when *actual* survival and reproduction are not impaired. That is, we need to find an answer to the counterfactual problem. Let's examine some potential ones.

2.1.1 Functional efficiency

Consider three common conditions—hypothyroidism, psoriasis, and benign prostatic hyperplasia. Hypothyroidism is a condition in which the body fails to produce adequate amounts of thyroid hormones, and it is an encouraging example of medical progress. Although it can be debilitating if left untreated, response to hormone replacement tends to be excellent and life expectancy for those taking supplemental thyroid is normal. Psoriasis is an inflammatory skin condition, which—perhaps surprisingly—is statistically correlated with reduced life expectancy. But it's generally believed that psoriasis itself doesn't lead to an early death. Rather, psoriasis is a risk factor—it is *correlated* with more systematic inflammatory conditions that themselves promote early mortality. And finally, benign prostatic hyperplasia (BPH) is the condition—very common in males over age fifty—in which the prostate enlarges, often causing urinary dysfunction. BPH can cause pain, lifestyle disruption, and infection, but it doesn't interfere with life expectancy or reproduction.

Each of these conditions are examples of pathology. But each of them poses difficulties for something like a biostatistical theory. In modern Western contexts, hypothyroidism doesn't statistically reduce life expectancy or reproduction— simply because hypothyroidism is easily treated. Psoriasis *does* statistically correlate with reduced life expectancy, but it seems odd to say that it interferes with an organism's ability to survive (it's simply correlated with other conditions that do). And BPH is not correlated—even if untreated—with early death or lowered fertility. How, then, are we to interpret the claim that such conditions impair an organism's *ability* to survive and reproduce?

---

[22] There is correlation between deafness and lower life expectancy in some populations, but these correlations appear to be mitigated by two factors: the cause of deafness and the socioeconomic impact of deafness. So, e.g., infectious disease is a major cause of deafness internationally (especially in the global south), but in these cases it is the infectious diseases, and not the deafness, that are causing shorter life expectancy. Similarly, deafness is often associated with poverty and poverty is frequently associated with lower life expectancy, but again it seems that it is the health impact of poverty, and not deafness, that causes the reduction in life expectancy. A major UK study comparing deaf and hearing populations found that deaf people often have poorer health outcomes, but for health factors such as blood pressure and diabetes that are medically unrelated to deafness. They did not find evidence of variation in life expectancy between deaf and non-deaf populations in the UK. See Emond et al. (2015).

For Boorse, pathology is dysfunction of a part or process. An organism can be doing reasonably well overall but have a dysfunctional part or process. Dysfunction of an individual part or process, for Boorse, needn't be defined simply as overall reduction in reproductive capacity or life expectancy. Rather, part-dysfunction is explained in terms of the *contribution* of the individual part or process to the ultimate goals of survival and reproduction. As Boorse says, the "normal function of a part or process within members of the reference class is a statistically typical contribution by it to their individual survival [or] reproduction."[23]

Already, there are some tricky issues here for Boorse's model—perhaps most significantly what counts as a "part" of an organism. The human organism relies on an extensive array of bacteria—to the extent that the gut microbiome is sometimes described as its own ecosystem—in order to maintain digestive function. Are the bacteria that make up the microbiome parts of the organism? How can we, on Boorse's analysis, count such "good" bacteria as parts, and not "bad" bacteria as parts, without begging the question? (The microbiome aids our survival, whereas harmful bacteria often impede it, but you can't count as "parts" all and only the things that fit your analysis.)

But let's leave questions of parthood to the side. Consider the case of BPH, which doesn't actually lower life expectancy or fertility rates in people who have it. Boorse can say that someone with BPH has a prostate that is not making a species-typical contribution to the organism's overall survival and reproduction. So the organism itself could be doing fine, but the prostate is not pulling its weight. (Or, given the particulars of this case, perhaps we should say it's pulling too much weight.)

But this strategy alone is insufficient. Recall that for something to be pathological, it can't *merely* be atypical. This is precisely why normal function accounts define pathological function—as opposed to merely *abnormal* function—as something that detracts from survival and reproduction. So even if a part or process makes an atypical contribution to an organism's overall function, it doesn't follow—by the standards of Boorse's own model—that it's pathological. To say that BPH is pathological, we need to be able to say that the contribution of the prostate negatively impacts the individual's overall ability to survive or reproduce—even though actual survival and reproduction may be in line with species norms.

One way of interpreting Boorse[24]—and a related view defended in detail by Daniel Hausman[25]—is to shift explanatory emphasis on to the idea of *functional*

---

[23] Boorse (1997), p. 7—originally Boorse wrote that it was survival and reproduction, but later revised the view to survival *or* reproduction (Boorse (2014)), possibly causing his account to overgeneralize (see especially Plutynski (2018)).

[24] See especially his (2014) for elaboration on this point.   [25] See his (2012) and (2014).

efficiency. Individual parts and processes have functional goals, which they can accomplish more or less efficiently. If a part or process falls below the species-typical norm for achieving its end goal, it is pathological.[26] Someone with BPH might have normal life expectancy and fertility, but their genitourinary system is functioning less efficiently compared to species norms. Likewise, someone with hypothyroidism might, if they take supplemental thyroid, be functioning normally overall. But their thyroid itself functions less efficiently relative to the functional goal of the thyroid gland (which is to produce thyroid at species-typical levels).

Assessing "functional efficiency" when it comes to human organisms, though, is tricky. The human species is full of examples of "normal anatomical variation"—missing muscles, extra muscles, extra nerve branches, extra bone. These are variations that cause body parts to function somewhat differently, but are not considered pathology. Likewise, a process in the human body can be abnormal without being pathological. Variation in the length or regularity of the menstrual cycle, for example, is a relatively common abnormality in women and is not by itself pathology.[27] Similarly, humans can have a specific part or process that functions far below the typical level of functional efficiency without thereby being considered to have pathology. The most straightforward example of this is missing anatomy. The palmaris longis tendon in the wrist is absent completely in a significant subset of people, and obviously for these people that part does not function at typical efficiency or meet its functional goals (movement of the wrist). But absence of the palmaris longis is considered a normal anatomical variant, and is not pathological.[28]

Cases like these are precisely what push Boorse toward the emphasis on overall goals of survival and reproduction. In characterizing pathology, we need a way to define "subnormal" function for parts and processes that goes beyond statistical atypicality. And for this reason, Boorse maintains that "subnormal function" should be understood in terms of contribution to survival and reproduction—a part or process has diminished functional efficiency if it reduces (compared to species norms) the organism's ability to survive and reproduce. And this lands us right back at our original problem.

---

[26] Boorse (2014) uses the example of the heart's pumping blood vs. the sound it makes when doing so. If a heart pumps blood at a lower rate than is normal for the species, then it is pathological, because pumping blood is the functional goal of the heart; but if it pumps blood a bit more quietly than is normal, that is not pathology, because the sound a heart makes when beating is not its functional goal. And it's not its functional goal because the sound a heart makes plays no role in the organism's survival and reproduction.

[27] Münster et al (1992). Note that if we consider the function of menstruation to be the shedding of the uterine lining, then an unusually long menstrual cycle would be "subnormal" function (although clearly not pathology in anything like the medical sense of the term). Yet a longer menstrual cycle can be advantageous in other ways, given that the woman loses blood less often. "Functional efficiency" is hard to objectively quantify here, as we'll discuss further.

[28] See, e.g., Thompson et al. (2001).

It's worth noting, however, that some defenders of a normal function-type approach—Hausman most especially—resist this push toward overall "functional goals" of survival and reproduction, placing more weight on functional efficiency of parts and processes.[29] A major problem for such views is, as we've said, that it's hard to give a non-evaluative account of the difference between pathology and normal anatomical variation. (Intuitively, we want to say that things are mere anatomical variants when they don't harm us.) In developing his own view, as a departure from Boorse's, Hausman concedes that there will often not be a clear line between normal and pathological efficiency but maintains that "[r]egardless of where the line between pathology and health is drawn, systems or organisms with higher levels of functional efficiency are, with regard to the particular part or process, healthier, and systems or organisms with lower levels of functional efficiency are less healthy" (Hausman (2012), p. 534).

The problem for a view like Hausman's is that some things that are clear cases of pathology are nevertheless difficult to characterize in terms of reduced functional efficiency. Hip dysplasia is a pathological morphology of the hip joint, for example, but mild forms are also overrepresented among dancers, most likely because they allow for increased range of motion.[30] So let's consider a comparison of two hips: one in an elite dancer with hip dysplasia; one in a completely ordinary, average person. The dancer has greater strength, range of motion, proprioception, and movement control, but she also has some chronic hip pain, a torn labrum (more common for dysplastic hips), and a greater risk of developing hip arthritis as she ages. Ms. Average has a normal hip—she can't do most of the things the dancer can do, but she also doesn't have the pain, the intra-articular damage, or the higher risk of arthritis. Whose hip has more *functional efficiency*?

Arguably, there's not a very good—and almost certainly no naturalistic—answer to that question. The dancer's hip is much better for dancing. It's probably also better for immediate feats of jumping, climbing, or other acrobatics—though you probably wouldn't want to run long distances with it. Ms. Average's hip is more stable, somewhat less injury prone, and better for walking around the neighborhood pain-free when you're seventy.

Hausman grants that functional efficiency depends both on particular goals and on environment. He then argues that we should assess comparative functional

---

[29] Although despite his focus on functional efficiency of parts, survival and reproduction are still important for Hausman. He grants that evaluating the functional efficiency of a part of process will often be relative to the functional goals of a part in a particular context, and that the same part can have different functions (and thus different functional goals). Yet we can still use the broad standard of functional efficiency to evaluate health and pathology because, "[t]ypically, what contributes to systems achieving their goals also contributes to survival and reproduction, but when functional efficiency within a system clashes with functional efficiency within the organism, we can adopt different perspectives. Functional efficiency within the organism is usually of more interest and importance" (p. 535).

[30] Turner et al. (2012).

efficiency by comparing the functional capacities of a part or process. We can compare two sets of capacities of relevantly similar parts. A part is functioning more efficiently with respect to some particular functional goal, according to Hausman, if the functional capacities of that part make it more likely that it will achieve that particular functional goal, in a particular context. So the dancer's hip is more efficient for some tasks in some contexts; the average hip is more efficient for other tasks in other contexts—so far, so good.

The problem is that Hausman says we can evaluate what is *pathological* by looking at comparative functional efficiency. To do this, we look at a basic set of goals (the adaptive goals of the part/process and the overall goals of the fitness of the organism) in a set of basic environments ("benchmark environments"; more on these in section 2.1.3). So in assessing whether a hip is pathological, we consider the basic adaptive functional capacities of the hip joint—its role in bipedal movement within our musculoskeletal system—and how those functional capacities impact our overall fitness. And the trouble here is that while the dancer's hip has a clear instance of pathology, it doesn't look like we can explicate this just in terms of functional efficiency (at least in this sense). The basic adaptive role of a hip joint includes a range of features—dynamic movement, stability, range of motion, and so on. The dancer's hip—strong and highly mobile, but unstable—is exceptional for some of these things, but pretty bad for others. It's an exceptionally strong, mobile, painful, injury-prone hip. Whether it would overall be better or worse to have a hip like hers in "benchmark environments" when considering the "goals of the organism" seems like a toss-up. The key point is that the dancer has a hip that functions exceptionally well along some key dimensions, despite—and indeed because of—a clear instance of biomedical pathology.[31]

Perhaps we might say that hip dysplasia is pathology because it tends to, *on average*, produce lower levels of function in the hip joint, even if elite dancers with mild cases sometimes buck the trend. Again, I don't think this solution will work. For defenders of the normal function view, the question of whether something is pathological is a question about the individual (and in this case, the individual hips), not about population averages. That is, it's not a question about how hip dysplasia typically affects the efficiency of human hip joints, but rather a question about how it affects these particular hip joints. And that's because health, as defined by normal function accounts, is had by individuals, and is a matter of the functioning of individuals. This caveat is an important one, since there are features that are in general considered harmful or pathological, but that may not be pathological at all in a particular individual. Bradycardia (unusually slow heart

---

[31] Note that cases like this also pose a problem for comparativist approaches like the one defended by Schroeder (2013), since we don't have a clear answer both to whether the dancer's hip is more or less functional compared to a comparison class of normal hips in her context, or even to future versions of her own hip (a surgical correction, e.g., would reduce some of her long-term risks but would substantially reduce her range of motion).

rate), for example, is often a serious form of arrythmia, but it can also be a harmless side effect of intense cardiovascular training in elite athletes.[32] Similarly, there are things that can, overall, be healthy for populations but that are pathological in individuals,[33] and likewise things that are bad for populations but not health problems for individuals.[34] In assessing function and pathology, we're looking at the function of the individual as compared to the species averages, not just to the averages themselves.

2.1.2 Dispositions and counterfactuals

Setting aside functional efficiency for the moment, how else might a naturalistic account like Boorse's explain the idea that conditions that don't actually kill or sterilize you nevertheless impair your *ability* to survive and/or reproduce? Perhaps the thought is that "ability" here expresses something dispositional. Indeed, Boorse (2014) suggests this explicitly in his discussion of Kingma's (2010) criticisms.

Consider the disposition of fragility—fragile things are more disposed to break, and that's true whether or not they ever do break. And something can make an object more fragile without breaking it. So perhaps ability, for the normal function account, is likewise dispositional, and something can detract from your ability to survive and reproduce without actually making you live a shorter lifespan or have fewer children.

We'd have to say something more specific than just that "ability" is dispositional, however, to make the case that things like alopecia really do detract from the ability to survive and reproduce. A particular fragile thing may not ever break—that is, its disposition to break may never manifest—but in general fragile things break more often than non-fragile things. If we had a class of objects that didn't break often or easily, we'd have no reason to suppose that such objects are fragile.[35] And the point I'm belaboring here is that, in our contemporary social context, a wide range of pathological conditions aren't associated with early death or reduced fertility. What reason do we have, then, for thinking that, in a modern Western social context, the people that have these conditions in fact have a reduced dispositional ability to survive and reproduce?

I think that in order to defend a workable notion of the "ability to survive and reproduce" clause, a defender of the biostatistical theory must say something

---

[32] Stein et al. (2002).

[33] The most famous case of this is sickle cell anemia in populations where malaria is prevalent. Those who are homozygous carriers of the gene have sickle cell disease, while those who are heterozygous do not—but being a heterozygous carrier provides substantial protective effects from malaria. So while the presence of the gene can be harmful to individuals, it can also be healthy for populations. See Aidoo et al. (2002).

[34] A substantial overpopulation of males relative to females, for example, is bad for a population without maleness being pathological for any particular male.

[35] Setting aside, for the moment, complex issues of masked or finkish dispositions.

irreducibly counterfactual. (That is, the solution to the counterfactual problem is a counterfactual reading of "ability.") The thought is something like this: yes, people with BPH or hypothyroidism (or osteoarthritis or eczema etc.) don't actually die earlier or have more trouble reproducing, but their *ability* to survive and reproduce is still reduced because they are more vulnerable to a change in circumstances. Had circumstances been slightly different—had they faced different obstacles or impediments, or had access to different resources—they would be less likely to survive and reproduce than the average human. And it's in this (inherently counterfactual) sense that they have reduced *ability* to survive and reproduce.

It's of course true that people with relatively benign pathologies still face vulnerabilities and threats to their survival that someone without such pathologies won't. Someone with BPH, for example, is more prone to bladder infection and complications of urinary retention. They might also deal with the long-term impacts of chronic pain or of interrupted sleep. So it seems plausible that someone with BPH will face specific vulnerabilities—which might legitimately threaten their survival in some contexts—that those without the condition don't face. What's harder to establish is that these vulnerabilities reduce—*even counterfactually*—their ability to survive and reproduce relative to species norms.

Almost everyone has something a bit wrong with them—mild nearsightedness, allergies, plantar fasciitis, migraines; the list goes on. Having a body, as an adult, usually involves having some conditions that are pathological. It's normal to have a few things that are abnormal. This is especially true when we narrow our comparison class to those of the same sex and a similar age. It's quite common for males of a certain age to have trouble with their prostate. That doesn't mean it's not pathological; it just means it's common. Can we say, then, that the vulnerabilities encountered by people with this kind of (relatively common, relatively benign) pathology make them *less able* to survive and reproduce across a range of counterfactual situations? No, because the species average we're comparing them to is a species average that will always include some pathologies or other (and, as a result, some vulnerabilities or other). That is, we're not asking whether someone with BPH is more vulnerable across a range of counterfactual scenarios than the ideal human form from an anatomy textbook; we're asking whether they're more vulnerable than the average adult human. And there doesn't seem to be a clear or obvious answer to that question. (They might, for example, be more vulnerable in situations that require a solid night's sleep, but less vulnerable in situations where getting up regularly throughout the night could alert you to dangers.)

Merely establishing that people with a pathological condition face counterfactual vulnerabilities is insufficient. What the normal function account would need to establish, for a counterfactual account to succeed, is that people with such conditions are *more* vulnerable—in either a wider or a more salient range of non-actual (but in some sense "normal") situations. It looks very difficult, how-ever, to give a specific and principled account of which counterfactual situations are relevant. After all, there are lots of things that might've killed each of us, but didn't. Which non-actual worlds are closer or more salient in evaluating claims like "The person who has psoriasis could have died more easily than the person without psoriasis," for two people who both lived normal lifespans, is complicated at best. Nor is it obviously a question we can answer in non-evaluative terms, given how much our interests and our conversational context constrain which counterfactual scenarios are salient.

2.1.3 Hausman on benchmark environments

Worries about the inherent evaluative constraints of counterfactual salience notwithstanding, what we need is a principled way of specifying which particular non-actual circumstances are relevant. Daniel Hausman ((2012), (2015)) suggests that we can do this via appeal to "benchmark" environments. Whether a feature is disadvantageous will often depend on the environment in which it occurs. Pale skin, for example, is advantageous in climates with very little direct sunlight but disadvantageous in climates with lots of direct sun and high levels of UV exposure. According to Hausman (2015):

> The determination of what is pathological full-stop, rather than pathological within a specified environment, requires that one specify a "benchmark" envi-ronment. A benchmark environment will be typical of the most common environments in which *Homo sapiens* have lived. Only statistically normal functioning in benchmark environments defines health.   (p. 11)

We can thus read Hausman as offering the idea of benchmark environments as a potential solution to the counterfactual problem. The counterfactual circum-stances that are relevant are the benchmark environments, at least when it comes to evaluating health and pathology.

But I don't think that this solution will work. Consider Hausman's own example of skin color. If we consider what is typical of the "most common environments in which *Homo sapiens* have lived" across the entire history of the species, then (by far) the most common will be those with large amounts of direct sunlight and relatively high levels of UV exposure. Skin with darker pigmentation is advantageous in these contexts, and pale skin is disadvantageous. Very pale (e.g., white) skin is also statistically atypical for the human species. So if we're judging based on function in benchmark environments, pale skin is pathological.

Or consider common health conditions that are direct responses to a person's environment. Allergy to penicillin can be a serious health condition, requiring substantial lifestyle modification and medical care. Yet the most common envir-onments, across the history of our species, do not contain penicillin. And a person with a penicillin allergy would function normally in the most common environments in which *Homo sapiens* have lived. Yet it seems incorrect to say penicillin allergy is not pathology.

Nor does appeal to benchmark environments dissolve the problems of comparing "functional efficiency." Consider again the dancer with hip dysplasia. Which is more advantageous in benchmark environments, flexibility or the absence of pain? There's probably not a good way to answer that question (and almost certainly not a good way to answer it naturalistically). Moreover, it's not clear why the answer would be relevant to whether the dancer's hip dysplasia, as she experiences it in her own environment, is pathological.

2.2 Selected Function Theory

Let's take stock. Naturalistic theories typically try to ground the difference between the normal and the pathological in terms of what compromises our ability to survive and reproduce. But many pathologies don't, in the circumstances we actually find ourselves, bring early death or prevent us from passing on our genes if that's what we're into. So we're faced with the question of why such conditions count as pathological. So far, we've been looking at (and finding fault with) potential answers to this puzzle for the biostatistical theory, which views normality in terms of present statistical norms for a species. But there's another obvious option for understanding "normal function."

This brings us to the second major articulation of function-based theories of health: selected function theory. According to advocates of selected function theory, normal function should be understood as the function of a part or process that proved adaptive—and for which that part or process was thereby passed on via a process of natural selection—in that organism's evolutionary past. So, for example, Karen Neander defines "proper function" in an organism as follows:

> Some effect (Z) is the proper function of some trait (X) in organism (0) iff the genotype responsible for X was selected for doing Z because doing Z was adaptive for O's ancestors.   (Neander (1995), p. 111)

A trait is malfunctioning, according to Neander, if it cannot perform its proper function. (Note that this is not the same as functioning abnormally—a trait is malfunctioning only if it *cannot* perform its proper function, not if it performs extra functions or performs its proper function somewhat unusually, etc.)[36] Note the difference from Boorse; for a view like Neander's, we're no longer construing

---

[36] See also Millikan (1989a) and (1989b).

normal functioning as statistical typicality but instead examining "proper functioning" with reference to a species' evolutionary past.[37] Here, then, we have a specific answer to the counterfactual problem—the question isn't whether a particular state *currently* interferes with survival and reproduction but whether it *would've done so* in the kinds of environments humans inhabited in our evolutionary past.

Similarly, Jerome Wakefield (1992) defines the "natural function" of an organ or mechanism as "an effect of the organ or mechanism that enters into an explanation of the existence, structure, or activity of the organ or mechanism." He then argues that dysfunction is the "failure of a mechanism to perform a natural function" and stipulates that dysfunction is a necessary condition for pathology. Wakefield then argues that pathology should be understood as *harmful* dysfunction—where "harm" is not a purely naturalistic notion but incorporates the values and goals of the society in which an individual lives. Wakefield's theory is thus not a purely naturalistic account, but I'm going to discuss it here because of its emphasis on normal function as a *necessary condition* for pathology. (We'll return to this combinatorial aspect of Wakefield's view when we discuss hybrid views—see section 6.)

It's worth noting that Wakefield's specific definitions of natural function and dysfunction do not work. For example, a common effect of the female menstrual cycle is iron deficiency (and iron deficiency anemia). This effect is part of the explanation for the activity of this particular mechanism—in particular, it's part of the explanation for why menstruation will typically cease if a woman is under extreme physical stress, has become very thin or malnourished, or similar.[38] Likewise, degenerative joint disease in the hips and knees is a common effect of bipedal locomotion. And that particular effect of that particular activity is hypothesized as part of the explanation for the structure of both joints.[39] In both cases, we have effects of an organ or mechanism which played a direct (causal) explanation in the existence, structure, or activity of the organ or mechanism. And yet it would be absurd to label either a natural function, or their absence a dysfunction.

But the specifics of Wakefield's definition aside, can we use something like the selected function approach to provide a solution to the counterfactual problem? I don't think we can. The basic idea is alluring—sure, someone with a hypothyroid condition who is taking thyroid replacement medicine can get by perfectly well in a modern, industrialized society, but they would have been at an obvious

---

[37] Plausibly, then, proper functioning views, unlike views grounded in statistical typicality, are normative in at least some sense—although they don't appeal to notions like good/bad, harmful, etc. but rather appeal only to what was/wasn't adaptive.

[38] Cohen and Gibor (1980).

[39] This includes, e.g., structural attempts to balance mobility with stability and to widen weight-bearing surfaces. See, e.g., Aiello and Dean (1990).

disadvantage in our hunter-gatherer past, and their thyroid is not performing the function it was selected for as our species evolved. Likewise, someone with psoriasis might be fine now—in a world with indoor plumbing and antibacterial soap—but their disrupted skin barrier would have made them far more prone to deadly infections and parasites in our evolutionary past.

The problem is that this "state of nature" counterfactual scenario—one in which we're imagining escaping predators and living a hunter-gatherer lifestyle in a harsh environment—can't give us a proxy for determining what counts as pathology in a contemporary context, which is so radically different. For one thing, our lifespans in that period were shorter. As Anya Plutynski (2018) argues, it's difficult to maintain that many slow-growing cancers of old age were "selected against" in this way, simply because they would've been very unlikely to occur in our hunter-gatherer past (we wouldn't have lived long enough for them to show up), and thus our propensity to develop them wouldn't have been a disadvantage. More strongly, there are arguably some things that might've been advantageous in our evolutionary past that are actively disadvantageous in our current context.[40] Likewise, some things that might've been benefits in that context can rightly be considered pathological in our context.

There's some evidence, for example, that insulin resistance (which is a pre-cipitating factor for type 2 diabetes in the modern Western context) may have been beneficial in our evolutionary past, when food was scarcer and its supply less reliable.[41] Likewise, some researchers have hypothesized that ADHD and similar disorders, although they can be highly disadvantageous in a modern context, might have in fact been evolutionarily advantageous, allowing our ancestors to continually scan for predators, forage for food, and so on.[42] And many researchers hypothesize that the same exaggerated immune response that now causes auto-immune and allergic disorders was a positive adaptation historically, when we encountered more infectious diseases and pathogens than we do in our current environments.[43] Particular examples here will inevitably be somewhat controver-sial and speculative, but the basic point is that the link between what would have been (dis)advantageous in our species' past and what is pathological for us now can be tenuous.

A trait could perform the function it was selected for, yet still rightly be considered pathological. An exaggerated immune response, for example, may

---

[40] Indeed, there's a vast literature on the interplay between evolutionary biology and health that attempts to argue that many modern health problems are caused by a "mismatch" between traits that were favored by natural selection in a hunter-gatherer context and a modern context in which those traits are harmful. We might rightly be skeptical of some of this literature, but it at least makes a compelling case that "state of nature" scenarios aren't a particularly good basis for defining modern pathology. See especially Diamond (2003) for discussion.

[41] For discussion, see Diamond (2003) and Joffe and Zimmet (1998).

[42] See, for example, Jensen et al. (1997).

[43] Sironi and Clerici (2010); Le Souëf et al. (2000), pp. 242–4.

well have been the function for which a particular genotype was selected, and yet you still have a disease if that same exaggerated immune response—placed in a strikingly different environment—manifests as an autoimmune disorder. We evolved as Paleolithic hunter-gatherers, but we get sick in the here and now.

2.3 What Normal Function Views Are Missing

I've been discussing the counterfactual problem at such length because I think it's a way of emphasizing just how difficult it is to give an extensionally adequate theory of pathology (and of health, by proxy) in non-evaluative terms. Function-based accounts—both biostatistical and selected function—struggle to give naturalistic explanatory grounds for what counts as pathology. The naturalistic resources they appeal to in explaining the difference between pathology and mere variation—evolutionary adaptation, efficiency, survival and reproduction, —get us partway to understanding pathology, but always seem to come up short. And to lay my cards on the table: whatever the details, I don't think a fully non-evaluative account can succeed here. That is, I don't think there's a workable naturalistic notion of "normal function" (or "selected function") that is either necessary or sufficient for pathology.[44]

I probably haven't said enough—and won't be able to say enough—to fully establish this claim. But I hope I've said enough to make it plausible. So much of the work that we need a theory of health (and thus a theory of pathology) to do for us lies in explaining its relationship to our lives: how reductions in health cause us distress, interrupt our projects, or matter to our communities. When explaining what health is, we're trying to give an account of something that can explain its biological significance, but also its normative, political, and phenomenological significance. And purely naturalistic theories just don't seem like they have the explanatory resources to do this.[45]

What I'm suggesting is that central to the idea that a condition is pathological— that is, that it reduces health, at least along some dimension—is the idea that it is in some sense *bad for* or *harmful to* the person who has it. Again, this is not to say that all the things we might consider medically pathological are by themselves

[44] Wakefield's view of health, notably, is not strictly naturalistic and includes an evaluative component, but treats departure from species-typical selected function as a necessary condition of disease. Likewise, Neander and Millikan's appeal to "proper function" is arguably normative, but not in the stronger sense that most value-based theories of health have in mind—it doesn't reference what we value and care about, what is good or bad for us, etc.

[45] It's also worth emphasizing that someone like Boorse can't simply claim to be characterizing a biologically interesting kind and leaving the normative or political implications as a separate matter— perhaps then biting the bullet and saying that many of the conditions we care about for normative reasons aren't pathological in the sense he intends. Boorse's stated aim is that he wants to give an account of the notion of pathology that is of use and interest to medical researchers and epidemiologists. But these researchers all treat BPH, osteoarthritis, psoriasis, etc. as instances of pathology.

harmful to our overall wellbeing—deafness might be a biomedical pathology and yet people who are culturally deaf might be no worse off than hearing people, and have good reason to celebrate and value their differences (more on this in Chapters 2 and 4). And yet, part of our understanding of reduction in hearing as biomedical pathology involves distinctive normative significance—if a building company's noise pollution is causing premature deafness, that's harmful, if black Americans are far more likely than white Americans to go deaf from diabetes, that's another woeful consequence of America's racial inequality, and so on.

I don't think we can give a workable account of pathology (or of health more broadly) that doesn't make at least some reference to more normatively robust ideas than function and dysfunction. We need to talk about what harms us, what causes suffering, what makes us feel unwell, what causes distress. And I don't think those dimensions of health—the distinctive normative, political, and phenomenological significance of health—can be fully captured simply by talking about basic naturalistic notions like function and natural selection. So from here, we'll turn our attention to evaluative theories of health.

3. Normative Theories

Making health evaluative, though, doesn't obviously solve the problems encountered by function-based views so much as create different ones. If our understanding of health relies on norms or values, then the obvious question becomes: what values? As we'll see, there are many different answers to that question, each with unique problems of their own.

One option, for evaluative theories of health, is simply to use basic evaluative concepts like good and bad to explain health and disease, respectively. This is perhaps the most straightforward way of giving an evaluative theory of health, and things get substantially more complicated from there. So let's start with the most straightforward approach, and move on from there.

3.1 Cooper's Account of Disease

Rachel Cooper (2002) has given an evaluative theory of disease that starts with the bedrock assumption that a disease is "a bad thing to have." Cooper points to failures of naturalistic theories of health (most especially Boorse's) as motivation for the idea that any successful account of pathology must make direct appeal to normative/or evaluative concepts.

Cooper attempts to give an account of disease, where "disease" broadly refers to "all pathological conditions—whether diseases in the narrow sense, injuries, wounds or disabilities." She gives three criteria, which she argues are jointly necessary and sufficient for something's being a disease. Something is a disease, for Cooper, if and only if it:

(i)   is a bad thing to have
(ii) is such that we consider the afflicted person to have been unlucky
(iii) can potentially be medically treated.

Clause (i) is intended to differentiate pathology from simple biological difference. Clause (ii) is intended to differentiate pathology from the normal limitations of the human body (e.g., it might be bad for humans that our skin is sensitive to UV damage, but it's not a disease). And clause (iii) is intended to differentiate biomedical pathology from the other slings and arrows of outrageous fortune that meet criteria (i) and (ii) (bereavement, social disadvantage, etc.), but to do so in a way that doesn't rely on an underlying naturalism about which things are pathological. Something can count as pathological, on this account, if the collective group of institutions we label medicine can—or can potentially—treat them, but there's no assumption that medicine is tracking some underlying biological kind in the class of things its attempting to treat.

It's worth noting that appeal to evaluative concepts doesn't get Cooper's account out of the same kinds of comparison-class and contextual worries that Boorse's account got into. Whether we consider a person to be unlucky because of a particular pathology is highly context-sensitive. We can feel empathy for a seventy-five-year-old with osteoarthritis, but we probably don't consider them *unlucky*, given the ubiquity of arthritis in later life. And, more strongly, we can sometimes view an instance of pathology as a way in which a person is in fact quite lucky. My great-aunt, for example, lived to be ninety-eight in wonderful health, and then died (instantly and painlessly) in her sleep. It seems obvious that the stroke she died from was biomedical pathology, but I don't consider her at all unlucky because of it. Nor do I consider the friend who got an incredibly mild early case of COVID-19—and the subsequent months of pre-vaccine immune resistance—unlucky; indeed, the ideal scenario during the initial phases of something like the COVID-19 pandemic is to get the virus, but get a very mild case. But that doesn't change the fact that viral infection, even mild infection, is obvious pathology. Our judgments about whether and to what extent a person is unlucky are highly context-dependent, and don't correlate all that well with biomedical pathology. People are often considered unlucky because of pathology, certainly, but they're sometimes considered lucky as well.

Likewise, there are interesting counterfactual difficulties for clause (iii), just as there are counterfactual difficulties for naturalistic accounts. Cooper can't say that all pathology can be medically treated, since some pathologies currently have no available treatment. But when we consider what could *potentially* be treated by the institutions of medicine—not what they actually treat—we cast the net far wider (and arguably too wide). The film *Eternal Sunshine of the Spotless Mind*, for example, imagines a psychiatric treatment in which painful memories are erased in order to make people happier. Such a treatment is obviously not available now, but there's little reason to think it's not *potentially* available in some advanced future. But painful memories—even though they might be bad to have, and unlucky—aren't biomedical pathology.

By far the most substantial and interesting clause in Cooper's account, however, is (i). Central to our understanding of pathology, she argues, is that biomedical pathology (diseases, injuries, illnesses, etc.) are *bad for you*. Cooper explicitly declines to specify in what sense pathology is "bad for" the individuals who have it, instead appealing to an "intuitive sense" of which conditions are bad for individuals. She's clear that by "bad for" she intends something explicitly evaluative. For Boorse, departures from normal function are "negative" just in case they detract from survival and reproduction—regardless of a person's attitude to the resulting condition, the condition's effect on wellbeing, and so on. But Cooper is clear that her intended sense of "bad for" is normative—pathology is "bad for" the individuals who have it in the sense that it is undesired, causes them distress, makes them unhappy, or inhibits their wellbeing.

Cooper further specifies that on her understanding of "bad for," a particular condition can be bad for one individual but not for another, depending on their reactions to it. She thus accepts the conclusion that the same physical condition can be pathology for one person, but not for another. She elaborates this with a gardening metaphor: "A plant is only a weed if it is not wanted. Thus a daisy can be a weed in one garden but a flower in another, depending on whether or not it is a good thing in a particular garden."

Cooper's own account is thus explicitly subjective. Whether a condition is pathology depends, on her view, partly on the particular subjective reactions of the person who has it. Obstruction of the fallopian tubes, for example, might be clear pathology—and deeply distressing—for a woman trying to get pregnant, but the same condition might be actively sought out by a woman undergoing tubal ligation. The very same condition can, on Cooper's view, be pathology ("disease") for one individual, but not for another.

The nature—and plausibility—of Cooper's account thus hinges largely on what she means by conditions that are "bad for" the people who have them. You might hate the shape of your nose, and find it genuinely distressing. Indeed, people might consider the shape of your nose unlucky. And you can seek medical treatment for it. But it doesn't follow that the shape of your nose is biomedical pathology (especially when we consider the ways in which ethnic, racial, and gender stereotypes affect which nose shapes we consider unfortunate, which are more likely to cause distress, and so on).

Conversely, bodybuilders often actively seek a state of dehydration—to make a target weight or to achieve a "chiseled" look for photographs or competitions.

Dehydration, in these cases, is actively sought and desired. But it seems difficult to maintain that it's not biomedical pathology as a result, given how dangerous it can be, and given the widespread effects it can have on the body.

It's difficult to adjudicate such issues for Cooper's account, since she explicitly declines to specify a particular sense of "bad for."[46] But in evaluative theories of health more broadly, there are two main strands of thought about how we should understand the way in which pathology (disease, illness, injury, etc.) is "bad for" the people who have it: teleology-based views and wellbeing-based views. Let's examine these in turn.

3.2 Teleological Views of Health

A hallmark of most normal function theories of health is their emphasis on a broadly naturalistic approach. Normal function, they argue, is a descriptive concept that can be explicated in entirely non-evaluative terms—terms that could be acceptably deployed in the natural sciences. But another class of views is interested in species-normal or species-typical function from a much more intentionally normative perspective.

In a tradition that traces its roots back to Aristotle, some philosophers argue that health can be explicated by a teleological notion of *species form*. On such views, the world includes teleological facts about correct or proper function. The eye is for seeing. An eye that doesn't see is failing to do what it is supposed to do. These are irreducibly normative facts about proper functioning. That is, advocates of neo-Aristotelian teleological views typically resist the idea—favored in more naturalistic accounts—that the proper functioning of a part or process can be reduced to something non-evaluative (such as its contribution to survival or reproduction).[47]

Let's look specifically at the view defended by Philippa Foot (2001), whose articulation of this view is perhaps the most thorough and well known.[48] Foot, it's important to note, isn't primarily engaged in developing a theory of health. Her discussion of teleology is focused on giving an account of what she calls "natural goodness." But she does sketch an interesting—and influential—neo-Aristotelian theory of health in the process.

Foot argues that the notion of proper form can be extended from specific parts and processes to whole organisms, and more generally to species. There is a

---

[46] Her examples and discussion suggest that she means something like "detracts from a person's overall wellbeing" (combined with a somewhat subjectivist take on wellbeing).

[47] See, e.g., Foot (2001), Thompson (1995) and (2004), and Nussbaum (2006). See also especially Glackin (2016) for a helpful overview and incisive criticism.

[48] But see also Thompson (1995) and (2004), whose view developed alongside and in response to Foot's.

proper or intended form, according to Foot, for members of a given species. It's by the standards of this form that we can say, for example, that an oak tree's growth is stunted (even if all the oaks in the area are stunted), or that a weed is especially strong and vibrant (even if we don't want it around), or that dachshund's long spine is unstable (even though it's bred specifically for its length). We're judging, on this view, whether something is an excellent example of its kind—a flourishing instance of the kind of thing it is.

There are, for Foot, objective facts about such conformation to and exemplification of species form. A 50-lb dog is a normal-sized dog even in a society that favors—and selectively breeds for—8-lb "toy breeds." And such facts, according to Foot, are irreducibly normative. Contra the more naturalized teleology of someone like Boorse, species form isn't a simple matter of statistical typicality. A flourishing human *ought to* be able to live into their seventies, for example, even in times and places (medieval Europe, perhaps) where people's actual life expectancy has been far, far shorter. Likewise, strolling through the Olympic village, you'll see many instances of excellence for the human form, but almost no instances of statistical typicality. And Foot similarly denies that species excellence is reducible to enhanced survivability and reproductive capacity. While she argues that species form arises from the traits that were adaptive (conducive to survival and reproduction in the particular environment in which natural selection was occurring), on Foot's view teleological facts about correct species form, once they emerge, are independent of what in fact promotes the actual survival or reproduction or a particular individual in its current environment. The fastest or most impressive-looking deer, she argues, might be the one that falls victim to a hunter—and its swiftness or size might, in such a case, be directly causally related to its death. But on Foot's view, the swiftness or size are still clearly indicative that the deer is an excellent example of its kind. Indeed, that very excellence makes it the target of the hunter.

Foot thus maintains that the notion of species form is inherently normative, as is the corresponding notion of being an excellent or flourishing instance of one's kind (that is, a particularly good instantiation of a particular species form). Armed with these normative concepts, Foot can then give a teleological account of health. Health, for human beings, is the flourishing of the human organism. You're healthy just in case you're an excellent example of the species form of *Homo sapiens*. Something is a dysfunction (a pathology, an illness, a disease, etc.) just in case it impairs the instantiation of species form—that is, it impairs your functioning in the way a thing of your kind *ought* to be able to function. The resulting view of health is objective, normative, and also gives us the first example of a positive characterization of health. Foot is giving an account not just of pathology but also of what it is to be *healthy* (which on her view is to be a flourishing exemplar of one's species form).

Importantly, for Foot, claims about health (species function or dysfunction) are distinct from—and can't be explained in terms of—claims about wellbeing. Put simplistically, health for Foot is the excellence of the human organism, whereas wellbeing is excellence of the person. You can exemplify the human species form and yet not be living a good life—because you're cruel, frustrated, malcontent, or any other range of features that are bad for you as a person but fully compatible with being an excellent human organism. And, likewise, you can be living a good life but not be an excellent example of the human form. More strongly, Foot argues that some things that actively detract from species form can enhance the goodness of a life—sterility, for example, is dysfunction of the human form, but can be actively sought as part of a good life, and can easily be something that makes a person's life go better for them. The excellence of the human person and the excellence of the human organism are, for Foot, two importantly different things, although both are teleological.

There's much that's insightful about Foot's views on health—especially her distinction between the flourishing of the organism and the flourishing of a human life. (And we'll return to this view, in a different context, in Chapter 5, since I think it pinpoints something important about the basic idea we're trying to capture when we're talking about health.) But ultimately I don't think it succeeds as an explanatory account of what health *is*. Consider a dilemma: either the teleological notion of species form directly correlates to the forms and processes that were evolutionarily adaptive, in a process of natural selection, for members of a particular species, or it doesn't. Foot is clear that proper species form is related to—and arises from—such historical adaptivity. But it's less clear whether proper functioning can be reductively explained in terms of those features that were selected for—because of their contribution to survival and reproduction—in a species' evolutionary development. This type of reduction of proper/normal function to facts about natural selection and evolutionary history is what selected function theorists attempt, as already discussed (section 2.2). And if this is Foot's notion of proper species form, then the same objections apply: there are plenty of things that were adaptive in our evolutionary past but that contribute—in systematic ways—to poor health in our current environment.

Yet Foot's view of proper function seems much more strongly normative than the selected function theorists', and while she maintains that teleological facts about species form arise from facts about natural selection, one can read her as arguing that they are nevertheless independent from—and irreducible to—such facts. So let's suppose, for the sake of argument, that teleological facts about species form can't be reduced to historical facts about adaptive traits. Now consider the following question: is red hair unhealthy? If health is a matter of exemplifying proper species function, then this will be a question of whether red hair is a dysfunction or defect of human species form. The trouble is that if facts about proper species form are irreducibly normative, then it's unclear how we should go about answering such a question. We don't answer it based on statistical typicality, or actual facts about survival and reproduction, or historical facts about adaptivity. The normative facts about proper functioning aren't reducible to any of these things.

And so maybe, for all we know, red hair really is a dysfunction—people with red hair are just a little too freckly and pale and altogether *Celtic* to count as flourishing examples of the human species. The worry here is that it's unclear how we could rule such a scenario out. And, more generally, it's unclear how we could come to have knowledge about which things are dysfunctions, which individuals are healthy, and so on.

But we *do* know that red hair isn't dysfunction. A theory of health shouldn't make our knowledge of simple observations like this mysterious or unobtainable. Of course, Foot would agree (I'd assume) that red hair isn't dysfunction, and that we can know this. It's statistically atypical, but all we have to do is observe flourishing redheads to see that one can be a perfectly good instance of the human species form while being atypical with respect to hair color. And this is true even if red hair might've been maladaptive in our evolutionary past—making its bearers easier targets for predators, or more vulnerable to social ostracism, for example. Selection against red hair might account for its extreme rarity, but that wouldn't make it unhealthy in our current social environment.

We can know this, arguably, because we observe redheaded people being perfectly good examples of flourishing human organisms. But here we find the central worry for a strongly normative teleological theory of health: to render normative facts about proper form both non-mysterious and non-reductive, we need to rely on our understanding of health and healthiness. Judgments about which things are flourishing instances of their kind presuppose rather than explain our understanding of which individuals are healthy. It may be *true* that flourishing human organisms are healthy, and that healthy people are flourishing human organisms. But what I'm suggesting here is that if our understanding of "flourishing human organism" doesn't reduce to facts about natural selection, but also remains non-mysterious, it relies on—and is close to co-extensive with—our understanding of "healthy human organism." That is, we can't explain what it is to be healthy in terms of what it is to be a flourishing organism because our judgments about which organisms are flourishing depend on our judgments about which organisms are healthy. The reason we think red hair isn't unhealthy isn't that we have a prior grasp of a flourishing organism, which allows us explain the otherwise perplexing health of some redheads. Rather, our judgment that red hair is compatible with being a flourishing human organism—even if it was at one point maladaptive and even if it is rare—is based at least in part on our judgment that many redheaded people are perfectly healthy.

So we can frame the objection to teleological views of health as a trilemma. If teleological facts reduce to facts about adaptiveness in a process of natural selection, then the same sorts of objections leveled against selected function views (section 2.2) apply. Some things that may have been adaptive in our evolutionary past are no longer healthy, and some things that are perfectly healthy might've been somewhat maladaptive in our evolutionary past. If they don't reduce to such evolutionary-historical facts, then they're either epistemically opaque or they rely on (and presuppose) our understanding of health. Either way, such normative facts don't give us a workable theory of health.

Now, none of this is a problem for Foot herself, who's primarily interested in developing her theory of natural goodness, and discusses health only in passing. Nor is it to say that her account isn't getting at something really important about how we think and speak about health—again, we'll return to this idea of the "flourishing organism" in Chapter 5. But it is a problem for an attempt to use Foot-style normative teleology to provide an informative, explanatorily adequate theory of what health *is*.

3.3 Wellbeing-Based Views

Let's review the dialectic so far. We began with non-evaluative views, with the upshot that at least *some* normative or evaluative component was needed for a successful theory of health. We then looked at Cooper's basic normative view, which relies on the intuitive notion that diseases are "bad things to have." But the upshot there was that we needed to say more about in what way pathology is "bad for" the people who have it. We then turned to one way of locating normative judgments about health and pathology: teleological judgments about species form. Let's now focus on another—and by far the most common—place to locate the normative significance of health: wellbeing.

As Cooper's discussion suggested, it's common to think of the evaluative nature of health in terms of its effect on our wellbeing. Pathology (illness, disease, injury, etc.) is *bad* for us insofar as it detracts from our overall wellbeing or quality of life. We'll discuss this idea in detail in Chapter 2, looking especially at the idea that the distinctive value of health is its correlation with or contribution to wellbeing. But what we're going to examine here are views of health that attempt to give an evaluative account of what health is by saying that health can be understood in terms of (some specific aspect of) wellbeing. Such views of health are clearly evaluative, since wellbeing itself is an evaluative notion—it is not just the descriptive question of what a life contains, it is the normative question of whether and how a life is going well for an individual. But wellbeing-based views can be either subjective or objective, since theories of wellbeing can likewise be subjective or objective.

3.3.1 Health as wellbeing
Let's begin with perhaps the simplest approach in this area: we can say that health *just is* wellbeing. Quite (in)famously, the World Health Organization (WHO) does

exactly this in their definition of health, characterizing health as "a state of complete physical, mental and social well-being."[49] And some philosophers have put forward similar views. Sean Valles (2018), for example, defends a "life course" theory of health according to which health is a state of flourishing—physical, psychological, social, emotional, and cultural—across a lifespan. On a view like Valles', health—because it encompasses so many aspects of a person's life beyond their physical or psychological function—is basically the same thing as their overall wellbeing across their life, and intentionally so.

There are many advantages to construing health in this way—most especially that it allows for a more holistic and less mechanical view of health that, crucially, can encompass the social determinants of health. The more we study population health, the more we come to understand that there is a tight causal connection between social goods like education, socioeconomic security, and community inclusion and health, such that health can't really be understood in isolation from such social factors. Similarly, emotional and social variables (stress, mood, social support, family and relationships, etc.) play a huge role in determining health, and there's often a complex causal network of interaction between our emotional lives and our health.[50]

A straightforward way, then, to acknowledge all this is to simply collapse the divide between our health and other factors that contribute to our wellbeing—to say that health *just is* wellbeing and wellbeing *just is* health. But the problem with this collapse is that it obscures more than it clarifies. Health and wellbeing are importantly related, but they aren't the same thing.

I'll argue for the importance of the health/wellbeing distinction in detail in Chapter 2, but for our purposes here let's look at one important place where health and wellbeing often come apart, and where it's important to have theoretical concepts that allow us to describe this: disability.

Consider the hypothetical case of Juan, who has moderate cerebral palsy. Juan has a family he adores, a job he enjoys and is good at, and leisure time filled with fun and laughter. He's economically secure and doesn't worry about meeting his basic needs. He considers himself a very happy, satisfied person. If health just is wellbeing, then there are two options for what we can say about Juan—either he is not (despite his own appraisal) flourishing, or his disability does not substantially compromise his health. And I think both of these options are wrong. As I've argued at length elsewhere,[51] there's substantial reason to accept that disabled people can and do flourish. That is, there's substantial reason to think that high levels of wellbeing are consistent with having a disability. Whatever wellbeing is, it seems like we should say that someone like Juan has it. His life is filled with love

---

[49] See the WHO Constitution:.   [50] We'll discuss these issues in detail in Chapter 3.
[51] Barnes (2016).

and joy, he is pursuing his projects and his goals, he is happy. He's living a good life.

But if we say that Juan has a high level of wellbeing, and health *just is* wellbeing, then we ought to say he has a high level of health. Yet it seems like we ought to deny that he has a high level of health. Juan's condition shortens his life expectancy, it requires ongoing management of muscle spasticity, it places him at higher risk of osteoporosis, it makes him vulnerable to various gastrointestinal problems. It doesn't follow that he's automatically in worse health than any non-disabled person, or that there aren't some ways in which we can truly say that he's healthy. But cerebral palsy is a complex health condition, with substantial consequences for long-term health outcomes. As we'll discuss in detail in Chapter 4, while it's a mistake to overly medicalize disability—to think that we can understand disability simply via the lens of biomedical pathology—disability very often involves serious reduction in health, even if it doesn't involve serious reduction in wellbeing. What we need to be able to say, for someone like Juan, is that he has high levels of wellbeing but seriously compromised health (and likewise that while his wellbeing might be at or above typical levels, his overall health is substantially below them). And we simply can't say that if health and wellbeing are collapsed into a single entity.

3.3.2 Health as an aspect of wellbeing

A more moderate stance than saying that health just is wellbeing, however, is to say that health is some distinctive aspect of or precondition for wellbeing. And that's exactly the approach taken by several prominent philosophical theories of health. Two of the most influential of these are those developed by Lennart Nordenfelt (2006) and Sridhar Venkatapuram ((2011) and (2013)). We'll examine each in turn, as they offer good examples of subjective and objective approaches, respectively. Both Nordenfelt and Venkatapuram are seeking to give positive theories of health (that is, they want to give an account of what it is to be healthy, not merely an account of pathology). And they both agree that any successful theory of health must be normative—that it must include evaluative as well as descriptive claims. Moreover, they both agree that this normativity should be located in the relationship between health and wellbeing. But they disagree about whether the relevant sense of human wellbeing is subjective or objective and thus differ in whether the account of health they give is subjective or objective.

Nordenfelt defines health in terms of an agent's "vital goals"—aims and projects whose successful pursuit is necessary (and jointly sufficient) for the agent's "minimal happiness" across their lifespan. You are healthy, according to Nordenfelt, if you have the ability, in standard circumstances, to successfully pursue your vital goals. Different agents can have different vital goals, though— different criteria for their minimal happiness—and so something that compromises health in one person might not compromise it another. To use the classic example of infertility, it might be a health problem for the person whose vital goal is to have their biological child, but irrelevant, or even a desired outcome, for someone who doesn't have that goal.

For Nordenfelt, health is not the same thing as wellbeing, but it is a precondition for wellbeing. Wellbeing is meeting your vital goals. Health is being such that you have the ability—in standard circumstances—to meet your vital goals. Suppose a vital goal for you is meaningful relationships. You're unhealthy if for some reason you cannot pursue this goal—you can't leave the house or you can't communicate. But you're perfectly healthy if you have the *ability*, both physically and mentally, but instead just decide to spend all your time scrolling Twitter. Likewise, if you move to a country where you do not speak the language, your vital goals might be temporarily interrupted—you might not be able to communicate well or get a good job. And your wellbeing might thus be at least temporarily compromised. But as long as you have the ability to do these things (by learning the language), you are still healthy, according to Nordenfelt.

Nordenfelt's view is thus highly counterfactual in much the same way that Boorse's is, and arguably encounters many of the same difficulties in specifying the relevant sense of "ability." There are also many questions we might ask about the details of Nordenfelt's approach—most especially what is meant by "in standard circumstances"—especially since this adds a further counterfactual layer to the view. But for the purposes here I am going to focus on Nordenfelt's notion of "minimal happiness."

Importantly, vital goals can't, for Nordenfelt, just be anything that you desire, or anything you desire especially strongly. You might dream of being rockstar, but you're not unhealthy if you can't be one. And we can't fix this just by strength of desire—an athlete might desire to win the Olympics more than they desire anything else in their life, and yet not be unhealthy if they can't succeed in that goal. So Nordenfelt restricts to a notion of "minimal happiness." Your vital goals are the ones that are necessary for you to have a very basic level of enjoyment or subjective quality of life across your lifespan.[52]

At this point, we can frame a dilemma for Nordenfelt's view, once again over the question of whether Juan, the happy and fulfilled person with cerebral palsy, has compromised health. Either "minimal happiness" is relatively expansive or it is more restrictive. Nordenfelt might construe a goal as a "vital goal" if it is necessary for minimal happiness along some important dimension or with respect to some important project—such that something can be a vital goal even if, *overall*, you can be reasonably happy without meeting that goal. In this way, we can say that Juan's health is compromised if his ability to meet this broad set of vital goals is impaired. Even for people who live happy and fulfilled lives with

---

[52] Again, Nordenfelt is committed to the idea that which goals are "vital" is subjective to an individual's wants and desires, but he argues that there will be broad agreement across cultures as to the most common or most central vital goals.

disabilities, there are almost always frustrations—there are things that you want to do that you cannot do, things that are more challenging and difficult for you than they are for others, and so on. Even the simple fact that he has a condition that requires ongoing medical management might be a frustration of a vital goal in this sense.

The problem, if we opt for this expansive construal of vital goals, is that it implies that *everyone* has reduced health. Everyone has frustrations and limitations with respect to some areas of their life or some projects that are important to them. And more strongly, everyone is frustrated by the vagaries of their minds and bodies—whether we wish we could concentrate better or run faster or sleep longer or exercise without getting so sore, we all have impaired ability to do some of the things we want to do and some of the things that are important to us.

So to avoid the absurd result that no one is healthy, or that something universal like being mortal is a disease, we need to restrict to a narrower reading of "minimal happiness." Your vital goals, on this reading, are those aims and projects that are necessary for a very basic level of life satisfaction. But on this reading, there's no reason to say that Juan's health is compromised. Juan quite obviously has a basic level of life satisfaction. He is able to pursue the projects that are most important to him, and he's happy. What we need to be able to say is that Juan has a high level of wellbeing despite the fact that his cerebral palsy involves a significant loss of health. But Nordenfelt's view—given how closely it ties health to wellbeing— doesn't allow us to say this.

Let's shift attention to Venkatapuram's view, which focuses on an agent's "capabilities" rather than on their subjective mental states (such as happiness). The capabilities approach to wellbeing, pioneered by Amartya Sen[53] and Martha Nussbaum,[54] characterizes human flourishing in terms of the capability to achieve certain basic "beings and doings." According to Nussbaum, for example, we should understand wellbeing in terms of our capacity to (successfully) pursue certain basic goods, such as food, shelter, loving relationships, and meaningful projects.[55] Crucially, though, the capabilities approach to wellbeing is objectivist—these basic capabilities are part of a minimally good life whether or not an agent desires them, and whether or not an agent is motivated to pursue them.[56]

Venkatapuram argues that we should understand health in terms of Nussbaum-and-Sen-style capabilities. But he further argues that health is not just one of many capabilities that together comprise basic wellbeing. Rather, on Venkatapuram's view health is a *meta-capability*—it is the physical and mental ability to access the other capabilities (relationships, projects, play, artistic

---

[53] Sen (1985), (1990), and (1993).   [54] See especially her (1993), (2001a), and (2001b).

[55] See especially her (2001b) for elaboration

[56] There are, famously, cases of "adaptive preference" in which, Nussbaum argues, an agent can form preferences that are incompatible with their own wellbeing (Nussbaum (2001b)). Although see especially Khader (2011) for caution over when and how we apply this diagnosis.

endeavor, etc.) that together make up human flourishing. Health, for Venkatapuram, is the ability to have abilities; it is the way your body and mind need to be in order for you to be able to access the capabilities that determine wellbeing. And so, for Venkatapuram, health is not the same thing as wellbeing. Rather, just as with Nordenfelt's view, health is a necessary precondition for wellbeing. But in contrast to Nordenfelt, wellbeing for Venkatapuram consists in objective capacities rather than subjective mental states.

Returning to the case of Juan, Venkatapuram thus has an avenue of response open to him that Nordenfelt does not. Perhaps the right response is that Juan's wellbeing really is substantially compromised because his health is compromised, even if he doesn't consider himself unhappy and even if he judges his life to be going well for him. Yet again, though, we'll encounter a dilemma. If we make the basic capabilities that compromise wellbeing robust and specific enough such that Juan doesn't have them—despite his rich and full life—then it looks difficult to make the case that such capabilities really are necessary, objective, and universal conditions for basic human flourishing.

But if we make the capabilities more basic and generic—you just have to be able to access some basic goods like friendships, meaningful projects, and so on—then it looks like Juan's wellbeing-related capabilities are not impaired. Again, what we need to be able to say is that Juan has a flourishing life—a high level of wellbeing—but that his health is substantially impaired. And any view that ties health very closely to wellbeing (and especially those that make health a necessary precondition for wellbeing) makes this impossible to say.

The broader worry for such views—for which a case like Juan's is merely one salient instance—is the concern of *healthism*.[57] Health can and does have a profound impact on wellbeing—and arguably has a closer connection to wellbeing than many other valuable things in life. But as I'll argue in detail in Chapter 2, health and wellbeing can also come apart in distinctive ways—it's possible to have a good life without being in good health, and it's possible to be in good health without living a good life. More strongly, we can sometimes make rational choices to compromise our health for the sake of our broader wellbeing—to pursue that sport we love even though we know it's why we keep getting injured, to keep going with that job that's stressing us out and keeping us from sleeping and eating right because the work is meaningful and rewarding. Health matters to wellbeing, but it isn't the only thing that matters to wellbeing, and it doesn't perfectly correlate with wellbeing.

Healthism, though, is the tendency to over-value and over-moralize health—to treat it as more central and more primary than it often is, and thus to be too

---

[57] Kukla (2022) presents an informative discussion of healthism in response to Sean Valles that, though targeted specifically at Valles' life-course view, is applicable to many of the wellbeing-based views discussed here.

zealous in its pursuit and too critical of behaviors that don't promote it.[58] Yes, health is valuable, but other things are valuable too—sometimes things that conflict with health. If we conceptualize health as too closely related to wellbeing, we can account for its distinctive normative value, but we risk *over-valuing* it in the process.

3.4 Summary

Let's again take stock. Non-evaluative theories of health struggle with extensional adequacy. Differentiating health from pathology by purely non-evaluative criteria doesn't seem feasible, given how deeply ingrained evaluative notions such as *bad, harmful, or unlucky* are to our understanding of pathology. A stark alternative is to make these normative concepts central to our definition of health. But here we encounter further difficulties. We cannot say simply that pathology is bad or harmful—lots of things are harmful that aren't biomedical pathology. So we have to specify *in what way* health is good for us, and loss of health is bad for us. And that's where things get tricky. If we try to specify a value that is specific to health— such as physical flourishing or proper functioning—we risk smuggling in a pre-theoretic notion of health, and thus not really explaining what health is. (The states that are flourishing are just the states we think of as healthy, the proper functionings are just the ones we tend to judge as non-pathological, etc.) But if instead we say that health is good for us (and pathology harmful to us) because of its relationship to wellbeing, we risk over-simplifying the correlation between health and wellbeing.

4. Phenomenological Theories

An alternative way of developing a normative theory of health emerges from the phenomenological tradition. Such theories focus on how it *feels* to be healthy, arguing that there are distinctive phenomenological states associated with health and illness. They further argue that these experiential states are how we should understand both what health and illness are and why they matter. Such theories are, of course, subjective insofar as they make health a matter of personal felt experience rather than an objective biological condition.

---

[58] See, e.g., Crawford (1980). Crawford also argues that tying health too directly to wellbeing—which is typically conceptualized as individual flourishing—makes health too much of an individual good rather than a social and public good.

In a tradition that was spearheaded by Merleau-Ponty[59] but is most thoroughly explored by Havi Carel,[60] the phenomenological approach argues that we understand health as a type of comfortable, carefree embodiment. You are healthy, roughly speaking, if you can more or less take your physical and mental functioning for granted. If you feel "at home" in your mind and body—if they don't feel like a struggle or something you are having to worry about or work against—then you are experiencing health. And that's both *what it is* to be healthy and why we care about health. To put it a bit too simply: people want to feel good. We want to be able to go about our daily life without constantly asking ourselves whether we have the physical stamina to do normal tasks, whether we need to take medicine or whether a standard activity is unusually risky or difficult, whether we will have the emotional stability and energy to cope with a normal day's activities, and so on. And so, in the famous phrasing of the surgeon René Leriche, health is "life lived in the silence of the organs"[61]—an everyday experience of embodiment that is comfortable and that can take for granted physical and mental stability.

Havi Carel ((2008), (2018)) then characterizes *illness* (which she uses as a broad category to include felt experiences of physical distress,[62] injury, limitation etc. which might not always correlate with biomedical disease) in contrast to this sense of carefree embodiment. The distinctive experience of illness, for Carel, is of a sudden awareness of vulnerability and instability. You are constantly aware of how you feel, constantly interrogating whether you can accomplish normal and everyday tasks, constantly alive to sensations (respiration, heart rate, the exact position of your limbs, etc.) that were previously white noise. And it is this distinctive experience, according to Carel, that both explains why illness is bad for us and, more strongly, characterizes the nature of illness. Health, in contrast, is the relative absence of such negative experience—the comfortable embodiment we experience when illness doesn't intrude on us.

A prima facie problem for phenomenological accounts, of course, is that plenty of things that substantially compromise our health have no symptoms—and thus are fully compatible with comfortable embodiment (and don't produce "illness" in Carel's sense). You might have a growing aortic aneurysm or advanced ovarian cancer and yet experience no signs of illness at all—your sense of embodiment might be just as comfortable and carefree as it's ever been. (For the hypochondriacs who immediately stopped reading in order to google aortic aneurysms and ovarian cancer, welcome back.) And while experience is clearly *relevant* to our

---

[59] Merleau-Ponty (2012).   [60] See especially her (2010), (2008), and (2018).
[61] Quoted in Canguilhem (1991).
[62] In characterizing illness, Carel explicitly does not take herself to be giving a theory of disease or pathology per se. She's giving a theory of what it's often like to *experience* pathology. Someone can have significant biomedical pathology, on Carel's view, and yet not currently be experiencing *illness*. Likewise, someone can arguably experience illness without the obvious presence of biomedical pathology. (More on this shortly.)

assessment of health, I'm going to assume that any successful account of health needs to be able to say that someone with a massive aortic aneurysm is not in good health, their comfortable embodiment notwithstanding.

But the phenomenological account has an available response to such worries. Measurements of health sometimes distinguish between an individual's "health state" at a particular time and the presence of various diseases, defects, and so on. Your health state at a particular time is how you feel and what you are able to do at that time. You are considered to be in a good health state if you are not in pain or distress, not overly fatigued, and can carry out "normal activities of daily living." And thus a good *health state* is compatible with a serious disease because the presence of that disease might as yet not be impacting how you feel or your basic everyday functioning—you have biomedical pathology, but you are not experiencing *illness* in Carel's sense. Pathology, on this understanding, is still related to reduction in health states—and to the development of illness—because pathology will affect your health state over time, either by reducing your future health states, or by (if it kills you suddenly) reducing the amount of time in which you can experience positive health states, or both.

We can thus interpret Carel—and the phenomenological tradition more broadly—as giving an account of health states. Likewise, Carel's notion of *illness* is arguably an account of reduction in health states—she's not equating illness with pathology, or giving a theory of pathology, but rather giving a theory of which kinds of experiences are distinctively (on her view) related to reduction in health states. Carel is thus, to some extent, engaged in a slightly different project than someone like Boorse or Cooper, who are primarily concerned with giving a theory of pathology (and then explaining health in terms of the relative absence of pathology).

For Carel, a distinctive type of experience is central to understanding both what health is and why it matters. That's what makes Carel's account distinctive, and also the distinctive place in which I think the account fails. She's identified a set of experiences that are evocative of many people's overall experience of illness, and overall experience of health. But I don't think those experiences are specific or universal enough to tell us what's characteristic of health.

Firstly, not everyone experiences health as comfortable embodiment. In his essay "Walking While Black," Garnette Cadogan describes his experience of navigating pedestrian sidewalks at night as a black man—his constant awareness of his height and his posture, his attempts to make himself seem less threatening and more approachable, his jitteriness. The experiences Cadogan is describing resonate strikingly with Carel's description of illness. He's aware of his body constantly, he can't easily do things most people take for granted (e.g., walk down the sidewalk at night without being perceived as a threat), and he feels physically on edge. He's certainly not describing comfortable embodiment. And yet Cadogan isn't ill. Just as with wellbeing, it's important that in discussing the relationship between health and experience our account of health doesn't over-reach. Cadogan isn't suffering from an illness; he's suffering from systemic racism. (And, of course, illness and systemic social inequalities are deeply entangled—but it's important that we have a theory of health that can explain how they are closely connected but not the same.)

Similarly, women experience their bodies through the lens of sexist norms about women's appearance. There are plenty of women, especially young women, who are constantly aware of their bodies, who constantly feel uncomfortable and awkward, who do not feel at all "at home" in their bodies. Some of these women really are ill—they have eating disorders or body dysmorphia. But many more have the simple angst and discomfort of feeling like they don't quite measure up to the smooth, spotless, hairless, weightless ideal that they are told determines their acceptability. Again, these women aren't ill, and it's important that we say they aren't ill. They're having a very normal, understandable reaction to the situation they find themselves in. And yet they don't seem to have the comfortable, "at home" embodiment that is meant to characterize health.

On the flip side, many people with serious and substantial illnesses don't experience the kind of alienation or discombobulation that Carel describes. Carel gives a striking portrait of many people's experience of sudden-onset *acquired* illness. But, to take perhaps the clearest case, consider the contrast of congenital illness. Some conditions that substantially impair health are present from birth. One compelling thing about bodies, though, is that you only get one, and you only know what it's like to live in the one that you've got. Someone with a congenital illness likely won't experience their condition as intrusive, alien, or otherwise disorientating—even if they experiencing it as limiting and at times distressing—simply because it's the only type of embodiment they've ever experienced. Carel's notion of illness is characterized in terms of an experience of contrast, but not all illness is experienced as contrast.[63]

To sum up, then: it seems like people can be healthy and yet not feel "at home" in their bodies, and likewise that people can be ill and yet not experience the distinctive sense of alienation that Carel describes. And so, even construed as a theory of health states, I don't think the phenomenological account succeeds.

5. Social Constructionist Theories

Yet another place we might locate the evaluative content for a theory of health is in our collective judgments, norms, and practices. Although the term "social construction" means many different things to many different people, theories of

---

[63] Although Carel's discussion focuses primarily on physical illness and the experience of the body, it's also worth noting that some serious mental illnesses aren't universally experienced as intrusive, alien, or otherwise distressing in the way that Carel describes. Sometimes the hallucinations of schizophrenia, for example, are experienced as positive and welcoming.

health that pursue this approach are generally referred to as social constructionist. The main idea is this: what determines whether a particular condition is pathological is how that condition is collectively viewed and treated within a particular society. Such views tend to be objective—there are objective facts about which things a society considers illnesses and diseases that are independent of the felt reactions of any particular individual. But they're also normative—whether something is pathology is determined by collective normative judgments about which things we should consider as disease, should consider as appropriate for medical intervention, or similar.

Within our society, we have complex institutions of medicine and healthcare—doctors, hospitals, nurses, physical therapists. And it's generally accepted that these institutions exist in order to treat pathology (and thus, perhaps instrumentally, to promote health). Social constructions argue that whether something is pathology has to do with its relationship to these social institutions.

We can all agree that poverty is bad for you, but it doesn't follow that you can go to your doctor and ask for a prescription for money. In contrast, if your poverty contributes to your high blood pressure, your doctor can prescribe medications to treat your high blood pressure. The difference between diseases and other harms or ills, on social constructionist approaches, consists in precisely this discrepancy. Disease should be understood, on this model, as a category constructed from the social norms and practices of medical institutions (and perhaps our social norms and practices in responding to these institutions). And so, in its most straightforward form (as in Englehardt (1996)), the social constructionist view maintains that something is a disease if and only if it is considered treatable by the institutions of medicine.

The familiar problem for such views, though, is simply that they don't leave room for a society to be wrong about which things are diseases.[64] We have, historically, considered things as appropriately treated by the institutions of medicine that aren't, in fact, pathological. Likewise, we've failed to consider things that are in fact pathological as appropriate targets for medical intervention. A classic example of the former is homosexuality, which in the mid-twentieth century was widely accepted to be a disease, was listed in the *Diagnostic and Statistical Manual of Mental Disorders*, and was regularly treated by psychiatrists. In the reverse direction, epilepsy has, in various historical contexts, been seen as a sign of witchcraft, demonic possession, the malign influence of a full moon, and so on—an appropriate target for spiritual rather than medical intervention.

---

[64] Although there may be some equivocating on what is meant by "wrong" here. Suppose a society treats homosexuality as a disease. Standard social constructivist views leave open the possibility that a society has adopted a *morally wrong* standard of evaluating diseases, but not that they're making a *factually incorrect* judgment in classifying homosexuality as a disease, given the standards of that society. See especially Glackin (2019) for interesting discussion on this point.

And yet, arguably, we should be able to say that homosexuality never was a disease and epilepsy always was a disease.

But if something is a disease if and only if social norms and collective social practice consider it a disease (that is, they consider it an appropriate target for treatment by the institutions of medicine), then it's impossible for a society to be wrong about which things are diseases. We can have inter-society disagreement. That is, we can say from our perspective—based on our norms and judgments—that people in the 1950s were wrong to say that homosexuality was a disease. But at that time and in that place, if the basic social construction analysis is correct, they weren't wrong. Likewise, people in a society that treated epilepsy as a sign of witchcraft would be correct to say—within their particular social context—that epilepsy is not a disease.

We begin to lose our bearings on what we're trying to do with a theory of health and pathology, I'd argue, if we cannot say that a society makes a mistake if they don't classify epilepsy as a disease. (Epilepsy is a well-defined and potentially fatal neurological disorder, so I'm willing to hold as a fixed point of analysis that it is a disease.)

Social constructionist theories, however, can be more nuanced than this basic approach. Perhaps the most detailed and well-developed such theory is the one defended by Quill Kukla (2014):

> **The Institutional Definition of Health:** A condition or state counts as a health condition if and only if, given our resources and situation, it would be best for our "collective" wellbeing if it were medicalized—that is, if health professionals and institutions played a substantial role in understanding, identifying, managing, and/or mitigating it. In turn, health is a relative absence of health conditions (and concomitantly a relative lack of dependence upon the institutions of medicine).

Unlike the basic social constructionist model, Kukla's model allows that a society can be wrong—entirely wrong—about whether something is a health condition (and, more specifically, whether something is a disease; it's important to note that Kukla is specifically giving an account of health, and a definition of health conditions, and not attempting to give a theory of "disease"). Suppose everyone collectively agrees that epilepsy is not a health condition and collectively agrees that it should not be medicalized—we all say that you should see your priest and not your doctor if you have seizures. We'd all, despite this agreement, arguably be better off if epilepsy *were* medicalized. That is, even if we send people with epilepsy to a priest instead of a doctor, it would—presumably—be in everyone's best interest if we sent them to a doctor instead of a priest. And so we can say that epilepsy *really is* a health condition (and perhaps more specifically that it *really is* pathology) within a particular society, even if that society doesn't consider it to be.

There are questions about the details here, of course—especially when it comes to how to interpret what it means for something to be best for "our collective wellbeing." The medicalization of homosexuality is an apt case. In the early twentieth century, homosexuality was criminalized and widely considered to be a distinctive kind of moral failure. Very often, medicalizing something is a way of removing associated blame or moral condemnation—if someone has a disease, on the common understanding, it is not their fault and they shouldn't be blamed for it. And so, perhaps surprisingly, gay rights campaigners were a major part of the effort to medicalize homosexuality in the mid-twentieth century.[65] And, although it was clearly a limited strategy, the medicalization of homosexuality did have many substantial—though perhaps short-term—benefits for gay people. Being sent to the doctor for being gay is insulting and offensive, but it's not as bad as being sent to jail.[66]

It also seems important to distinguish between conditions that compromise your health and conditions that are perfectly compatible with good health and that are nevertheless appropriate targets for medical intervention. As Kukla acknowledges, a woman might see her doctor for birth control pills when she doesn't want to get pregnant, receive regular prenatal exams during a pregnancy, and consider various medical strategies to ease her experience of menopause. And yet fertility, pregnancy, and menopause—though all appropriate targets for medical intervention—don't seem well classified as the same kinds of health conditions she might ordinarily see her doctor for. Kukla is not offering a model of disease, and so their account isn't committed to the idea that fertility, pregnancy, and menopause are diseases simply because they are medicalized. The model is, however, committed to the view that they are "health conditions"—at least on the assumption that their medicalization promotes collective wellbeing. The model then further defines health as "the relative absence" of health conditions and lack of "dependence" on the institutions of medicine. Much will hang on what is meant by "relative absence" and "dependence," but at least arguably many women are dependent on the institutions of medicine to manage their fertility and reproduction, and this management is crucial to their wellbeing.[67] Women can often be thus somewhat more dependent on the institutions of medicine—for birth control, for gynecological exams, for management of the issues related to the menstrual cycle, for pre- and postnatal care—than men of a similar age.

---

[65] For a brief overview of this history, see Gooren and Gijs (2015).

[66] Kukla's view explicitly allows that which things are health conditions might change over time, as both social arrangements and the tools of medicine change. So it's open for their model to say that homosexuality was once a health condition but isn't any longer—and that it's appropriate for us, from our perspective, to view it as the kind of thing that it would be silly to call a health condition. But it's arguably uncomfortable—and not quite enough distance from the major objections to basic social constructivist views—to bite the bullet and grant that homosexuality was ever a health condition.

[67] This is, of course, not true of *all* women—and is true of some people with female anatomy who don't identify as women.

And yet it's not at all obvious that, as a result, women are less healthy, as Kukla's model seems to imply (given that they have more health conditions and more dependence on the institutions of medicine). A woman can, arguably, be perfectly healthy while experiencing fertility, pregnancy, and menopause—even if she has to regularly visit her doctor to manage these conditions.

Whether regular visits to a doctor for birth control or prenatal care render you less healthy might, admittedly, be a debate that boils down to a clash of intuitions. The larger issue for Kukla's model, though, lies in the *explanation* for whether something is a health condition. Take the example of chiropractic medicine— which at least in the United States is well established and credentialed enough to count as an institution of medicine. Many of the diagnostic categories of traditional chiropractic medicine—especially those involving "vertebral subluxation"— have been largely scientifically discredited. And yet, people often *feel better* when they visit chiropractors, and chiropractic medicine is a relatively effective treatment modality for non-specific back pain.[68] It seems as though it's an entirely open question whether it's best for our collective wellbeing to treat "vertebral subluxation" as a health condition. And yet, whether or not it might promote people's wellbeing to treat "vertebral subluxation" as a health condition seems largely orthogonal to the question of whether "vertebral subluxation" *really is* a health condition.

It might of course be objected here that what is being treated as a health condition is back pain, not "vertebral subluxation." Kukla argues that medicalizing something makes it a health condition but not that all the anatomical beliefs involved in medicalizing something are thereby correct. It's not clear to what extent this response works, however, given that explanations of "vertebral subluxation" are arguably part of why the treatment is effective—patients are reassured by being given a mechanistic reason for their pain. But the broader worry is that for a model like Kukla's, whether or not "vertebral subluxation" is a genuine health condition is determined by the instrumental impact of treating it as such and not by any underlying biological reality.

There are all sorts of reasons why going to a healthcare provider might increase wellbeing—it might provide reassurance, it might provide the benefits of a therapeutic relationship, it might provide the power of the placebo effect, or it might deliver medically effective treatment by accident (or at least without a good understanding of the underlying condition being treated). And so it seems like it ought to at least be possible for there to be useful fictions—for there to be things such that we benefit collective wellbeing if we treat them *as though* they reduce your health, even though they aren't *in fact* things that reduce your health.[69]

---

[68] See, e.g., Walker et al. (2011).
[69] Some psychiatrists and historians of psychiatry make the case that this is exactly the scenario we find ourselves in with extremely broad diagnostic categories like "depression." Arguably, some people

But Kukla's model doesn't allow for this, precisely because *what it is* for something to be a health condition is for our collective wellbeing to be increased if we medicalize it.

Similarly, there are all sorts of reasons why going to a healthcare provider might fail to increase wellbeing in a given context—your current medical system might have false beliefs about treatment, or might be especially biased against the people most likely to have a particular health condition. Consider, again, the case of epilepsy. Growing awareness of epilepsy as an organic disease in eighteenth-century Europe coincided with a range of terrifyingly counterproductive medical treatments—people with epilepsy were confined to sanitoriums, stuffed full of sedative and barbiturates, and subjected to invasive experimental procedures.[70] Overall, it's far from obvious that it was best for our collective wellbeing—given the state of medical knowledge and medical practice at that time—for epilepsy to be medicalized. Everyone, and most especially people with epilepsy, might've been better off just getting a blessing and a good dousing of holy water from the local priest. And yet it seems we ought to be able to say that they had a condition that objectively compromised their health—just one that wasn't well understood or well treated by the institutions of medicine at that time. If we define health conditions as those that it would be in our best interest to medicalize, however—and health as the absence of health conditions—we cannot allow for this.

A potential fix for such worries, on Kukla's account, would be to add in a further layer of idealization. On their model, we already say that health conditions are not merely those we *in fact* say are pathological, but rather those that we would all be better off saying are pathological. We could complicate this further: something is a health condition just in case it would be best for our collective wellbeing if it was treated by an ideal version of the institutions of medicine. But the problem with this solution is that it's hard to fix on what an ideal version of the institutions of medicine would be without specifying that they accurately diagnose and treat the things that are, in fact, health conditions (that is, conditions that do in fact reduce our health).

We've previously (section 2) discussed the shortcomings of purely naturalistic analyses of health. And yet, it seems crucial to our understanding of health and

---

who meet the diagnostic criteria for depression have substantially compromised mental health, but others have the bad feelings that are a normal reaction to all sorts of social ills and stressors. Medicalizing all of these people under the broad label "depression" has played an important role in the destigmatization of such experiences, but making progress on our understanding of mental illness plausibly requires that we can at least entertain the idea that some people, although they have benefited from a medical diagnosis, and have negative feelings that are very real, don't obviously have a condition that compromises their mental health—they don't, more specifically, have a *mental illness* in the way that others, including others in the same diagnostic category, do. For discussion, see Horowitz and Wakefield (2007) and Harrington (2019).

[70] See especially Gross (1992) and Harrington (2019).

pathology that there is at least *some* objective biological component to it. Health, as we've discussed, has distinctive biological significance, and that distinctive significance is often independent of cultural norms and judgments. Multiple sclerosis (MS) is a disease which reduces health. More specifically, MS is a condition in which the immune system attacks the central nervous system and causes widespread demyelination, which in turn impairs neurological function. Healthy immune systems fight pathogens that invade your body, not your own central nervous system. And this basic biological malfunction is at least a large part of what makes MS a disease which reduces health—*really* a disease and *really* something that reduces health, whether or not we recognize it as such, and even if it wasn't in our collective best interest to treat it as such. Imagine, for example, a society in which the most common treatments for MS were in fact—unbeknownst to everyone involved—quite harmful. It might not, in that society, actually promote wellbeing to medicalize MS. (This was, arguably, the case for many diseases during the era of leeches and mercury.) And yet MS is a disease—and something which compromises health—whether or not it's currently and actually in our best interest to treat it as such. And that's because at least *part* of what determines and explains whether MS detracts from health is the basic biological reality of the condition.

And so the key problem, I'd suggest, with Kukla's position is not a failure of extensional adequacy but rather a failure of explanation. Even if the view got the correct result for all cases—even if it said of all diseases that they are health conditions and of all healthy people that they are healthy—it seems as though it's locating the *explanation* for why this is true in the wrong place. At least part of the reason why something like MS is a health condition, and why having something like MS reduces your health, should be the basic facts of the biological process involved.

6. Hybrid Theories

Again, let's take stock of where we are. Prospects for giving a purely non-normative account of disease look dim. It seems plausible that in order to separate pathology from mere anomalies or oddities, we need to appeal to something evaluative—something about the way in which disease is harmful, bad, or unwanted. And yet attempts at locating this normativity prove equally frustrating. If we try to say that the badness of pathology comes from its relationship to wellbeing, we oversimplify the relationship between health and wellbeing. If we try to say it comes from how pathology makes us feel, we struggle to account for all the varying forms of pathology (and to differentiate biomedical pathology from other socially marginalized traits). And if we try to say that pathology is determined by collective social judgments, we miss out the importance of an objective, physiological component to disease.

Part of the difficulty here is that all of these theories seem to be getting *something* right. Biological dysfunction is a central part of how we understand pathology. The way that pathology makes us feel and the ways it can negatively affect our wellbeing are a major part of how we conceptualize the difference between the pathological and the merely atypical. And, likewise, the role that the social institutions of medicine play in defining and treating a condition seems like a major factor in how we understand medical pathology, in contrast both to mere physical difference and to other forms of disadvantage. Yet none of these factors, by themselves, is sufficient to give us a full account of health. If we try to understand health merely, or even primarily, in terms of biological dysfunction, wellbeing, felt experience, or social norms, we miss major pieces of the puzzle.

At this juncture, then, we find hybrid accounts. Hybrid accounts are typically conjunctive in nature—they attempt to give an account of health by saying that health has both a non-evaluative and an evaluative component. Typically, these accounts give both a normative and a non-normative condition for pathology, with the claim that each component is necessary and together they are jointly sufficient. The main idea is that non-evaluative and evaluative views each get part of the story on health right, so by combining them we can get the complete picture.

The most well-known such account is Wakefield's "harmful dysfunction" model.[71] According to Wakefield, a condition is pathological if it is a biological dysfunction that causes harm to the individual who has it. Jacob Stegenga (2018) offers a similar account to Wakefield's, but with importantly different details. According to Stegenga, a condition is a disease (is pathological, in my terminology) just in case it meets two criteria. The first is the Causal Basis of Disease: something is pathological only if it involves a failure of specific physiological or psychological mechanisms to perform their functions at typical efficiency. The second is the Normative Basis of Disease: something is pathological only if it is devalued or considered a bad thing to have. According to Stegenga, for any condition to be a disease, it must satisfy both criteria. And satisfying both criteria is what it is to be a disease—anything that satisfies both criteria is thereby a disease.

The most salient feature of hybrid accounts like Wakefield's and Stegenga's, however, is that they tend to inherit the problems of the views they combine. If normal function accounts simply had some slight problems of extensional adequacy—a few cases where it seems like they over- or under-generalize—then it might be promising to look for a condition to add that would address these problems at the boundary. Likewise, if some of our extant normative accounts of the harms of pathology, or the connection between pathology and wellbeing, got most cases right but just had some minor difficulties, then they would be plausible

---

[71] Wakefield (1992), (1995), and (2014).

candidates to combine with a normal function account. But, as I hope I've shown in the previous sections, the explanatory problems for both function-based accounts and normative accounts run far deeper than problems at the edges. It's not just that these accounts have difficulties at the boundaries and need some tinkering. It's that they have major, fundamental issues in the explanations they attempt to give for what health is and what differentiates health from pathology. Combining extant theories might yield promising results if extant theories were simply leaving something out. But I've argued that extant theories have more fundamental problems than this—problems that can't simply be fixed by linking them together.

To illustrate this, let's look at the details of Wakefield and Stegenga's accounts. As we've already discussed in section 2.2, there are significant problems with Wakefield's natural selection-based account of dysfunction, and thus significant problems with treating it as a necessary condition for pathology. But, as discussed in section 3, it also looks too strong to say that pathology always causes harm—at least without much more discussion about what is meant by "harm." If, in mid-2020, I had an extremely mild case of COVID-19 (without noticeable symptoms but with the subsequent months of increased immunity before the availability of vaccines or effective treatments), it is not at all obvious that it caused me harm. Indeed, it seems like it would be a dramatic benefit to me. And yet being infected with the COVID-19 virus is clearly pathology.

Stegenga's account encounters similar problems to Wakefield's. As we've discussed in detail, the idea of normal functional efficiency is riddled with problems, at least when used as a basis for an analysis of pathology. Because his account is conjunctive, Stegenga treats failure to perform a function at species-typical efficiency as a necessary condition for disease, but not a sufficient condition. But it is not at all obvious that it can do even this. Someone with intense allergies, for example, clearly has pathology. But it's not that their immune system doesn't target pathogens at the levels of efficiency we'd expect. It's just that it also targets peanuts, for reasons all its own. Whether there is non-ad hoc way of explaining this difference in terms of reduced functional efficiency is unclear (see section 2.1.1). More significantly for Stegenga's account, however, it doesn't look like the combination of his two criteria—decreased functional efficiency plus social disvalue—are jointly necessary or sufficient for pathology.

Let's take a simple example: skin aging. Normal aging of the skin, including "fine lines" and wrinkling, is partly due to the skin's failure to perform at species-typical efficiency[72]—our skin loses collagen and becomes less elastic as we age.

---

[72] Perhaps, like Boorse, Stegenga could relativize to an age-controlled reference class in order to determine typical efficiency. But we would then run into the familiar problem that some diseases are quite common in old age. Pathological-level function might, for some age groups, also be relatively typical-level function.

And it is absolutely disvalued socially. But there's a difference between disease and the normal processes of aging, including the aging of the epidermis. Our skin may be slightly less efficient as we age, and society may tell us that this is horrible (indeed, they may even medicalize it), but that doesn't mean that normal skin aging is pathological.

And regardless of what you make of functional efficiency, it doesn't look like Stegenga's normative criteria are necessary for pathology. As we've discussed at length, something can easily be pathological without being devalued or considered bad for the individual that has it. Again, much of this will hang on what is meant by "bad," but there are clearly cases in which something that is pathological is not socially devalued[73] and clearly cases in which something that is pathological doesn't cause net harm to the individual who has it. Being socially devaluated (or considered a bad thing to have) is neither sufficient *nor necessary* for being pathology. And so combining the idea of social disvalue with the idea of functional efficiency doesn't generate jointly necessary and sufficient conditions for pathology or fix the underlying problems with a functional efficiency-based account.

What we have here is a basic situation in which two wrongs don't make a right. Extant theories of health often get at something that we care about or that is significant in our discussions of health—something about its biological, normative, political, or phenomenological significance—but that doesn't adequately explain what health (or pathology) is. Hybrid theories can't, simply by combining such theories, arrive at necessary and jointly sufficient criteria for pathology (and for health, by proxy).

7. Summing Up

Health is complicated. The most prominent theories of health enjoy their philosophical popularity for a reason—they all manage to get at something significant that we care about when we talk about health. Part of what matters to health is the basic functional efficiency of the human organism. Part of what matters is our overall wellbeing—what we can do, what projects we can pursue, our distress and our suffering. Part of what matters is how we feel—good or bad, tired or energized, relaxed or in pain. And part of what matters is where we fit in a broader social network—our relationship to others and to the institutions of medicine.

What we see, in philosophical theories of health, however, is that it's difficult to characterize all these aspects of health simultaneously. Normal function accounts emphasize the biological significance of health but struggle to account for its

---

[73] Ranging from the Hmong's traditional view of epilepsy as a sign of divine favor, as chronicled in Anne Fadiman's *The Spirit Catches You and You Fall Down*, to the extreme thinness of "heroin chic" in 1990s fashion magazines.

moral or phenomenological significance. Sometimes, for example, part of what determines the differences between normal variation and pathology seems to be how a condition makes us feel, or the impact it has on our lives, in a way that isn't obviously reducible to a naturalistic account of species functioning. Wellbeing-based accounts, in contrast, emphasize the normative significance of health, but do so at the risk of both losing the biological significance and of *over*-weighting the normative impact of health (that is, of lapsing into healthism). A social constructionist model like Kukla's captures the political significance but again struggles to give a full account of the biological significance—we're given an explanation for why it matters to us socially that epilepsy, for example, be classified as a health condition, but not an explanation of why it's *true* that epilepsy is a health condition, regardless of how it's viewed by society. And phenomenological accounts, though they capture something distinctive about the phenomenological importance of health, struggle to explain both the normative and the biological significance of health, since many things other than health conditions can give rise to interruptions in comfortable embodiment, and many health conditions (especially those had from birth or early childhood) can fail to give rise to such interruptions. The pattern that emerges here is that our most influential theories of health often capture something important and distinctive about the nature of health, and fail to give full, extensionally adequate accounts of health. And often they don't just fail at the margins—they miss in central and substantial ways.

Nor can we fix the problem simply by cobbling extant theories together, as hybrid theories attempt to do. Part of the problem is that many of the things we care about, when it comes to health, can at times actively compete against each other. Sometimes things that make us feel bad promote our biological functioning. Chemotherapy can be taken for previously asymptomatic cancer, for example. Sometimes things that are perfectly biologically normal nevertheless cause us distress, are socially devalued, and are things we might seek medical intervention for. Hot flashes and mood swings during menopause can be perfectly normal, but they can also be uncomfortable, and stigmatized, and something we might treat medically. Some things that represent poor physiological functioning can nevertheless make us feel good or be socially desirable as signs of fitness—the "ripped" look of dehydration in bodybuilding competitions, for example, or excessively low levels of body fat. Sometimes things that make us feel the worst, and interrupt our daily lives the most, lack obvious biological correlates of "disease" or "dysfunction" in the typical sense—as in many cases of depression, anxiety, or chronic pain. And so on.

We can't fully explain health—what matters to us when we talk about health, the things we value when we value health—simply by combining different theories or layering together different concepts (normal function, harm, social value, etc.). The various dimensions of health seem to be more shifty and more complicated than that. In the following chapters, I'll explore in more detail what I take to be some of the most complex aspects of health that any successful theory of health needs to grapple with. We'll look first at the relationship between health and wellbeing (Chapter 2), then at health and the subjectivity of personal experience (Chapter 3), then at health and disability (Chapter 4).

Through all these cases, I'm going to argue that health exhibits a strange kind of shiftiness—we're pulled in different directions, we have compelling reasons to say things that feel somewhat contradictory, and the only way of adequately capturing everything we want to say is to aim for a "nuanced middle ground" that might not exist.

It's this inherent shiftiness that I'll then use as the basis to defend my own preferred account of health, based on the idea of *ameliorative skepticism*. Nothing, I'll argue, can do all the work we want a theory of health to do for us (Chapter 5). But this doesn't mean we need to abandon or replace the idea of health (Chapter 6). Rather, the inherent instability and shiftiness of health—although ineliminable and confusing—is part of what makes it especially useful for much of the work we need it to do.

2

Health and Wellbeing

In Chapter 1, I made the case that none of our extant theories of health can do the work that we want a theory of health to do. Although they each capture some important aspect of health, they fail to fully account for the distinctive biological, normative, political, and phenomenological roles we ask health to play. In the following chapters—this chapter and Chapters 3 and 4—I'm going to explore some of the reasons why health is so difficult to theorize.

When we're trying to give a theory of health, we often find ourself pulled in different—at times competing or inconsistent—directions. Crucially, though, many of these competing axes of emphasis for a theory of health are themselves interconnected and interdependent. Explaining health is like trying to navigate through a series of Scylla and Charybdis-style traps, with the added wrinkle that Scylla and Charybdis are in some kind of weird entangled state.

The upshot of all of this, I'll argue, is that health displays a distinctive kind of shiftiness—a shiftiness that needs to be built into our understanding of what health is (Chapter 5) and what work a theory of health can do for us (Chapter 6). But to lead up to that, I want to first examine three key areas where theorizing health displays this "pulled in competing directions" phenomenon—wellbeing (this chapter), subjectivity (Chapter 3), and disability (Chapter 4). Let's begin, then, with the vexed relationship between health and wellbeing.

Many of the things we care about have a somewhat "love it or leave it" relationship to our wellbeing. Sports are great if you enjoy them, but if you're clumsy and not a fan of sweating, that's fine too—you can do something else with your time. Dogs are pure joy for some people, but others are allergic and others prefer cats. Even the things most central and valuable in many of our lives—kids, romantic relationships—are wonderful if they're wanted and cho-sen, but aren't for everyone. They're the center of some people's lives, but others lead rich lives without them. Building a flourishing life is very much a mix-and-match process—there are many valuable things in the world and many ways to combine them, and different people can build equally good lives by picking different things to value or combining what they value in different ways. Ruth Chang likens comparing how different values "hang together" to building a jigsaw puzzle—but a puzzle where we don't have a completed image and we just have to piece things together into a good shape as

Health Problems: Philosophical Puzzles about the Nature of Health. Elizabeth Barnes, Oxford University Press.
© Elizabeth Barnes 2023. DOI: 10.1093/oso/9780192883476.003.0003

we go along.[1] We can extend this metaphor to overall wellbeing by thinking of flourishing as a mosaic—the human life is the completed artwork, the things we value are the colored pieces of glass. You can build different but equally rich mosaics both by picking different pieces and by blending the pieces together in different ways.

With health, however, the situation is more complicated. To state the obvious, you can't choose to *completely* disregard or disvalue your own health in order to pursue other goods, yet still have an overall high quality of life. You can completely ignore some things of value—music, art—and still live a great life, so long as you're prioritizing other good things. But you can't similarly disregard health, simply because if you *completely* disregard your health, you'll probably end up dead. And you can't have a good life if you're dead.[2] Likewise, if you completely disregard your health—if you place no value on it at all—you'll probably end up *feeling bad* fairly quickly—your body will be sore and painful and tired. And while it is very easy to flourish while ignoring opera or sports or the outdoors if those don't interest you, it's a far different thing to flourish if your body is in a constant state of physical distress.

Loss of health can also have a profound impact on your ability to pursue other things of value in your life. A catastrophic loss of health—cancer, amyotrophic lateral sclerosis (ALS)—can impair your ability to do the job you love, to pursue the hobbies you value, to be the parent or the partner you want to be. And some mental illnesses—anorexia, addiction—can distort your thinking and your goals so pervasively that every aspect of your life is affected. Health thus has a much closer and more direct connection to wellbeing than many other things we value.

And yet, health and wellbeing are not the same, and they can come apart. The relationship between health and wellbeing is a vexed "it's complicated" tangle. And sorting through this tangle is vital to understanding both what health is and why it matters.

In this chapter, I'll first (section 1) argue for the importance of the distinction between health and wellbeing. I'll then (section 2) argue against the view that, even if health and wellbeing are different things, the value of health is its *contribution* to wellbeing. Let's call views that collapse the distinction between health and wellbeing "constitution views" and views that say that health is valuable (only) insofar as it contributes to wellbeing "contribution" views. Both types of views on the relationship between health and wellbeing, I'll argue, end up being overly simplistic and fail to allow for the complex, non-linear ways in which health and wellbeing can be related. I'll then argue (section 3) that the relationship between

---

[1] See Chang (2004). See also especially Chang (2012) and (2017) for discussion of a view of values according to which values can be good in different ways, such that they are "on a par."

[2] This is, importantly, not to say that you can't make choices to promote your wellbeing in ways that would lead to a shorter, but better, life.

grief and wellbeing gives us a useful model for thinking about the connection between health and wellbeing.

1. Health and Wellbeing: The Importance of the Distinction

Both philosophers and public health researchers have become increasingly interested in the intimate connection between health and wellbeing. Alongside this, there has been a growing emphasis on the social determinants of health; income, housing, work, interpersonal relationships, social stressors such as discrimination and harassment—they all affect health in profound ways. Quite obviously, things that affect our health can have a dramatic impact on our wellbeing, and likewise things that impact our wellbeing can have a dramatic impact on our health. Yet traditional biomedical models of health—which focus directly on the functioning or impairment of the organism but in many cases don't consider the broader context—often struggle to capture this aspect of health.

As a corrective, it's now common for health to be defined specifically in terms of wellbeing. The World Health Organization (WHO), for example, currently defines health as "a state of complete physical, mental and social well-being."[3] In this definition, health and wellbeing are directly equated—what it is to have wellbeing and what it is to have health are one and the same.

In a similar vein, many influential contemporary philosophical theories of health—including those defended by Venkatapuram (2011), Valles (2018), and Nordenfelt (2006)—explain health in terms of (some basic component of) wellbeing. As we discussed in Chapter 1 (section 3), such theories view the normative significance of health as its connection to wellbeing and construe health itself either as wellbeing or as a necessary precondition for (adequate) wellbeing. Venkatapuram, for example, understands health as a "metacapability" required for achieving the basic capabilities necessary for human flourishing. On his view, then, health and wellbeing are not the same, but health is a necessary (though not sufficient) condition for wellbeing. And Valles construes health as a type of flourishing across one's entire "life course," making health roughly the same as wellbeing across one's lifespan. We've already discussed some of the obstacles for views like these—see Chapter 1 (section 3.3)—and I won't go back into that discussion in any depth here. The point is simply that it's increasingly common, in philosophical literature, to see the lines between health and wellbeing blurred.

Let's call such views "constitution views" for the purposes here. They are views that understand the relationship between health and wellbeing as at least partly constitutive—health either just is wellbeing or is some necessary constituent of

---

[3] www.who.int/about/governance/constitution.

basic wellbeing. Constitution views, although they aim to correct an overly mechanistic view of health, are, I suggest, an *over*correction. To fully understand health, we need to understand not only the ways it's intimately connected to wellbeing but also the ways in which the two can come apart.[4]

1.1 You Can Sacrifice Your Health to Increase Your Wellbeing

One major reason for insisting on a distinction between health and wellbeing is that we can—and indeed we often do—compromise our health for the sake of our wellbeing. And those kinds of tradeoffs can only be intelligible if we understand health and wellbeing as different things.

You have gastritis (an inflammation of your stomach lining that can cause discomfort and impair digestion). Your doctor tells you to stop drinking coffee, as coffee can exacerbate the problem. But coffee is delicious. And coffee is an essential part of your daily ritual and your working life. You imagine drinking tea while you read the news in the morning, or sipping kombucha at your favorite coffee shop while you try to write. The loss doesn't seem worth it, so you let your stomach take one for the team. You'd rather drink coffee and have an inflamed stomach lining than not drink coffee and have a healed stomach.

You're into soccer, and you play weekly with a local pick-up league. You love the adrenaline rush of the game, the camaraderie with your teammates, the challenge. But you're getting older, and you've had some pretty tough injuries. Your doctor is frank—soccer is a rough game, and it's taking a toll on your body. If you keep it up, you're going to keep getting injured and those injuries will likely become more substantial. What about switching to a gentler form of exercise? But you love soccer—both the game itself and the culture surrounding it are a huge part of your life, and have been since you were a kid. You decide, all things considered, that the injuries and the risks are worth it in order to keep going.

In cases like these—scenarios that are incredibly commonplace and normal—you are making a choice to compromise some aspect of your health for the sake of other things you value because you think it is what's *all things considered* best for your life. This is importantly different from places where we choose to compromise our health, but that choice is akratic. In akratic choices, we keep smoking or we don't get enough sleep even though we judge that it would be, overall, best for our lives if we made a different choice. But in choices like the ones discussed above, we aren't acting against what we judge to be our own best interests. Rather, we are making the conscious (and rational) choice to prioritize things other than

---

[4] See Hausman (2017) and (2015) for a related but somewhat distinct set of arguments for the same conclusion. Hausman has been perhaps the most vocal critic (within this literature) of collapsing the distinction between health and wellbeing.

our health. Health is one of the things we value, and we value it a great deal. But it isn't the *only* thing we value. And it can make perfect sense to accept some degree of loss in health in order to gain other things we care about.

A defender of a health-as-wellbeing view might want to re-describe the cases above as ones in which you compromise your *physical* health for the sake of your *mental* health. But this is too quick. For one thing, it's important that we don't understand "mental health" so expansively that anything that promotes positive mood or good feeling is thereby contributing to our mental health. (This also begins to look question-begging—if "mental health" can encompass anything that promotes our wellbeing, then there won't be a difference between health and wellbeing so broadly construed, simply in virtue of how we're using the term "mental health.") In the coffee case, let's assume the person in question could cultivate a taste for some other drink. They just love the taste of coffee and love coffeeshop culture. It's a stretch, I think, to say that they are thus promoting their mental health by persisting in their coffee drinking—they're simply choosing to value taste and preference over digestive health. Likewise, in the soccer case, let's assume the person in question could get the fitness and stress-relief benefits from some other, gentler form of exercise, and could replace the social activity of team soccer with something less damaging to their body but equally social. They don't *need* soccer for their mental health in this case. But they love soccer, and being a person who plays soccer and hangs out with other people who play soccer is a big part of what they care about. If they choose to accept injury to persist in this, it again doesn't look like a choice that's well described as prioritizing mental health—they could promote their mental health in other, less physically damaging ways. It's just a choice to value soccer, even though soccer comes at a cost to health.

More strongly, it also seems possible—and again, not uncommon—to compromise mental health for the sake of overall wellbeing. Many parents, for example, report that parenting young children is a fairly intense strain on their mental health, and yet they also value being parents. And it's not that they simply assume that the costs to their mental health while caring for an infant will be compensated by benefits to mental health once the child is older—maybe they will, maybe they won't. (Indeed, one of the stressful things about parenting is that so many aspects of it are unpredictable and outside your control.) But it can make sense to think that parenting makes your life go better, overall, even while granting that it can be a strain on mental health.

Likewise, many people value careers that are stressful, anxiety-inducing, or otherwise a strain on mental health. Suppose you work in a demanding job for a community-focused non-profit organization. You work long hours, your finances are tight, the situations that you deal with on a daily basis are extremely stressful and sometimes border on traumatizing. This is all a drain on your mental health, and plenty of career options would've led to less stress, less anxiety, and more peace of mind. And yet you love your job and wouldn't trade it for anything. Quite obviously, you might value the meaningfulness of your work, your sense of place in the community, your feeling that you're trying to make a difference. You might even value the adrenaline rush of constantly putting out fires, or the camaraderie you have with your coworkers who struggle alongside you. These are all things that can contribute to your wellbeing—the overall sense in which you're flourishing or living a good life—even while incurring costs to your mental health.

1.2 Things Can Harm Your Wellbeing without Harming Your Health

Conversely, it is not only possible, but again quite common, to experience a detraction in wellbeing that isn't well construed as a loss of health. Plenty of things can reduce your quality of life without thereby harming your health. Suppose, for example, that you don't like your job. It's not that your job is particularly stressful or morally questionable—but it's very boring, and you find yourself trudging through it. I think it's fair to say that a job like this could negatively impact your wellbeing. You'd be better off if you loved your work, if you found it meaningful or were excited by it. But it doesn't follow that the job is harmful to your health, even if health is quite broadly construed. You can be healthy while also being bored at work.

Or, to take a more substantial example, consider a painful life event like a breakup. While going through something like a breakup definitely can harm your health—both physical and mental—I don't think it's right to say that all breakups, even all painful breakups, are damaging to your health. Indeed, in many ways the negative emotions you experience during a breakup—disappointment, frustration, a sense of loss—are very healthy. They're the kind of thing you need to be able to experience during sad or difficult times.[5] This is another key reason for differentiating mental health from positive mood—sometimes feeling negative emotions is an important part of mental health, and an inability to feel those unpleasant emotions would itself be an impairment in your mental health. But while the sadness and frustration involved in processing a breakup may not harm your mental (or physical health), they can definitely detract from your wellbeing. Going through a tough breakup can diminish your quality of life while you're in the midst of it, even in cases where that reduction in quality of life isn't explainable as a loss of health.

---

[5] See Horowitz and Wakefield (2007) for an extended defense of the idea of "normal sadness"—bad feelings that can be intense and debilitating but are not at all pathological and shouldn't be understood as "mental health conditions."

1.3 You Can Have Poor Health but High Levels of Wellbeing

It is also entirely possible—and again, not at all uncommon—for people with serious and substantial health problems to have high levels of wellbeing. As discussed in Chapter 1 (section 3.3), perhaps the clearest case of this is people with medically complex disabilities. We'll discuss the relationship between disability and health in detail in Chapter 4, but for the purposes here the simple point is that people seem to be able to lead rich, valuable lives with disabilities that involve serious costs to their health. Take, for example, Harriet McBryde Johnson's account of her own life in her book *Too Late to Die Young*. Johnson describes her career as an attorney and activist, her immersion in the disability rights community, her everyday joys of friendship and family—all in ways that paint a picture of a rich and flourishing life. And yet Johnson had a congenital neuromuscular disease that profoundly affected her health, and eventually took her life.

If we collapse the distinction between health and wellbeing, it seems impossible to accurately describe Johnson's experience. We must either say that her health was not, in fact, severely compromised or we must say that she did not, in fact, have a high quality of life. And both these options look wrong. If we construe health as flourishing, or as an overall positive life course (or as a necessary precondition for either) then there's pressure to say that if someone like Johnson is clearly flourishing—clearly living a good and vibrant life—she is thereby healthy. But Johnson had profound health issues, and to deny this would misdescribe her life in important ways (more on this in Chapter 4). Indeed, much of what she tried to convince the public of is that it's possible for reduced health and a good life to exist alongside each other. A discussion of health starts to look like magical thinking if we instead say that, if you are happy enough or have enough other good things in your life, health problems can be rendered non-problems. On the flip side, if we simply say that Johnson—despite her happiness and her success and the richness of her life—could not have had high levels of wellbeing because of her health issues, it looks like we've overinflated the role of health in determining wellbeing.

1.4 The Dangers of Healthism

This brings us back to the more general worry—raised in Chapter 1 (section 3.3.2)—of collapsing the distinction between health and wellbeing: the threat of *healthism*. Part of what makes health distinctive is its *complex* relationship to wellbeing. Health and wellbeing aren't directly correlated, yet they are intimately connected to each other in a way that many other things we value aren't. But if we say that health and wellbeing are the same (or that health is a necessary condition of wellbeing), we lose this distinct complexity.

According to the WHO, for example, to be healthy *just is* to have a high level of wellbeing. But despite such definitions, organizations like the WHO don't then shift to theorizing wellbeing—they still recognize that there's an underlying biomedical reality to health, that many aspects of health can be measured in epidemiological terms, that if you have a condition like multiple sclerosis (MS) your health is harmed regardless of your attitude toward it. That is, though they take themselves to be investigating wellbeing (because health just is wellbeing), they then rely heavily on the investigative tools of the biomedical sciences, assume that the biomedical sciences have special expertise in saying what health is and who is healthy, and so on. In doing this, such views risk overstating the role of health in determining wellbeing; likewise, they risk drastically overestimating what the biomedical sciences can tell us about what a good life is or what it means to thrive. Yes, health is important and health matters to wellbeing. But it's not the only thing that matters, and it doesn't have a perfect correlation with wellbeing. Nor do the health sciences have special epistemic authority in telling us which people are flourishing or what kinds of lives are good lives. They can, at best, only tell us about health. But you can be very healthy without having a good life, and you can have a good life without being very healthy.

"Healthism" denotes the trend—promoted by an ever-increasing wellness industry—to equate being healthy with having a good life and to treat being healthy as a morally valuable end in itself.[6] Healthism makes sense if being healthy really is just the same thing as flourishing. But there are obvious reasons to deny this connection. And as we'll see in subsequent discussion, the trend toward healthism gets a lot wrong about the importance of health and why it matters to our lives.

2. The Value of Health

Another common way of thinking about the connection between health and wellbeing focuses not on what health *is*, but on why health *matters*. According to a standard way of thinking about the value of health—especially popular in both philosophy of health and health economics—the value of health is determined by its contribution to wellbeing. Health and wellbeing are different things, on this view, but health is valuable to us (only) insofar as it contributes to wellbeing.[7] Let's call such views "contribution views."

---

[6] Again, see Kukla (2022) for insightful discussion.

[7] This view is perhaps most pervasive within health economics, where it undergirds measuring practices like the QALYs (quality-adjusted life years). (The thought is that we can measure the utility or disutility of various health states via their contribution to wellbeing, and further that wellbeing is a matter of preference satisfaction. For extensive critical discussion of this set of assumptions, see Hausman (2015).) For notable articulations of this view within philosophy, see especially Broome (1988), (2002), (2004a), and (2004b), Bognar and Hirose (2014), and MacKay (2017).

A few caveats are in order. First, we must distinguish between the value of health *to individuals* and the *public* value of health. The public value of health includes questions of why health matters to societies more broadly, and encompasses lots of issues—the economic impact of disease outbreaks, public safety, and so on—that go beyond the value of health to any particular individual. During a pandemic, you, as an individual, might be primarily concerned with not getting sick and with not getting others (including your loved ones) sick. But governments are also concerned with things like whether people feel safe enough to engage in civic behavior like voting, whether essential services will continue to run if enough people become ill, or whether hospitals will max out their capacities.

The claim that the value of health is its contribution to wellbeing is a claim about the value of health to *particular individuals*. Your health, the thought goes, is valuable to you insofar as it promotes your wellbeing. Likewise, a loss of health is bad for you insofar as it detracts from your wellbeing. If you get cancer, for example, that is bad for you both insofar as it detracts from your wellbeing at a particular time—by making you feel unwell, by requiring you to spend a lot of time in the hospital, by interrupting your goals and affecting what you can do. And it's also bad for you insofar as it prevents subsequent wellbeing (by ending your life or by impairing your future health in a way that also detracts from your wellbeing). Cancer *by itself* is not bad or good, the thought goes. It's just a collection of cells with an unusually high cell turnover rate. It's bad for the people who get it precisely because of—and only because of—the impact that it has on their wellbeing.[8]

To be clear, this is not a claim about what health *is* (it's consistent with many different views of health). Rather, it's a view about the relationship between health and wellbeing—and more generally a view about why we should care (as individuals) about our health. Health, at least in part, consists of biological processes, and loss of health consists in alteration of those biological processes. It's a matter of how our bodies and minds are. But, as Hume famously tells us, you can't get an ought from an is. And so the question becomes: why is health valuable to us? That is, why should we care about health and why is health something that matters?

Again, separate out the social value of health—the costs of medical treatment, the need for a healthy workforce, and the like. The question at issue is why an individual's health is valuable to her as an individual. And the thought, as Daniel Hausman (2015) succinctly puts it, is that when it comes to individuals, "wellbeing is what matters." Health is valuable to us (only) insofar as it promotes our wellbeing. And loss of health harms us (only) insofar as it detracts from our wellbeing.

---

[8] Again, separate this out from questions about why societies might disvalue cancer—which will include things like the cost of hospitalization and chemotherapy, etc.

The genesis of such views—and their enduring popularity within bioethics and health economics—comes from a desire to measure the comparative severity of different health conditions, to prioritize health conditions for treatment, to assess the success of health interventions, and so on. When we're evaluating specific health states, we're typically evaluating something descriptive—a person has a demyelinating neurological disease, a person has breast cancer, a person has type 2 diabetes. We can look at basic information about "mortality and morbidity," but typically what we want to know is how *bad* the loss of health is or how much *better* a treatment makes it. That is, what we want to know is normative. Amounts of health are notoriously difficult to quantify and compare.[9] But, the thought goes, the reason why health *matters* to us is its impact on our wellbeing.[10] "We are interested," as John Broome puts it, not in measuring specific amounts of health or disease themselves but "in the harm done by disease, and the benefit caused by preventing disease."[11] And disease is bad for us or harms us, on this view, insofar as it detracts from our wellbeing.[12]

There's a lot that's attractive about the contribution view. To begin with, it's fairly intuitive. We should care about health and value health, the thought goes, because and to the extent that being healthy makes our lives go better and being unhealthy makes our lives go worse. And this makes sense. People value being healthy because they want to feel good, they want to pursue their desired goals and activities, they want to avoid things they dislike such as hospitals and uncomfortable medical treatments, they don't want to cause their loved ones distress. They value health, that is, because it increases their quality of life. Likewise, people disvalue disease because of its impact on quality of life. If you catch a virus but have no symptoms (and so your projects aren't impacted, you don't feel ill, your family doesn't worry, etc.), then the virus doesn't harm you, as an individual, even though it's a clear case of your body being in a diseased state.

In taking this approach, the view also avoids the unnecessary moralizing associated with healthism. There's nothing special or shiny about being healthy over and above the fact that it contributes to your quality of life. You should exercise or eat kale or take vitamins or do whatever other health-related thing

---

[9] See Hausman (2015) for a particularly effective discussion of this point.

[10] It's for this reason that major health surveys such as the Global Burden of Disease study use utility functions rather than simple epidemiological surveys. For further discussion, see Broome (2002).

[11] Broome (2002).

[12] It's sometimes suggested that there is a restricted notion of "health-related quality of life," and that the value of health is its contribution to the distinctively health-related aspects of wellbeing. (See Bognar and Hirose (2014), chapter 2 for discussion.) There might be other contributors to wellbeing, on this view, which outweigh or mask the distinctively "health-related" components of wellbeing when considering overall wellbeing. Plenty of people are skeptical that we can parcel out wellbeing in this way (see Hausman (2015) and Broome (2004a)). But even if we could, I intend the arguments provided in what follows to apply even to the view that the value of health is its contribution to a narrowly defined or restricted aspect of wellbeing.

you're advised to do if doing so will in fact improve your quality of life (both now and long term), but there's nothing special or morally superior about acting in a way that promotes your health. As John Broome (2002) puts it:

> Measuring health is one thing and valuing health is another. Which should we aim at? I assume we are interested in a person's health as a component of her wellbeing. Wellbeing consists of various different sorts of goods, such as health, comfortable living, freedom from oppression, and so on. So we are interested in health as one good thing to be set alongside others.

This approach to valuing health also lets us make useful comparisons about the relative value of various health states. Osteoarthritis of the hip and osteoarthritis of the fingers are both common forms of age-related joint changes. It is hard to give any biomedical account of which is "worse" as they are essentially the same process of joint erosion, just in two different places in the body. And yet hip arthritis is *typically* much more life-altering than hand arthritis. The hip is a weight-bearing joint, so when you have pain and limited mobility in your hip, many of your daily activities become altered or constrained. A hip replacement— the standard treatment for severe hip arthritis—is a major operation with a substantial recovery period. And so, typically, hip arthritis tends to have a much more profound impact on quality of life than arthritis in smaller joints. If we say that the value of health is its contribution to wellbeing, we then have a ready explanation for why hip arthritis is "worse" than many other forms of arthritis without needing to provide any explanation of how hip arthritis is somehow biomedically more serious than other forms of the same disease; it is worse because it tends, on average, to detract from wellbeing more. But, at the same time, we can say that hand arthritis might be worse for a particular individual— a pianist, say—than hip arthritis. That is, we can differentiate between the typical or average contribution to wellbeing of a health state type, and the contribution to wellbeing of a particular token of that type. Since the value of a particular health state for an individual, on this view, is its contribution to that individual's wellbeing, this might be different from its *average* contribution to wellbeing.[13]

These advantages notwithstanding, I think this approach to the value of health is misguided. The relationship between health and wellbeing is complicated. And the complexities that arise in the interaction between health and wellbeing show

---

[13] It is for reasons such as these that Hausman (2015) reluctantly concludes that assessing the private or personal value of health via its contribution to wellbeing is the best of a range of relatively unsatisfactory options. I agree with much of what Hausman says about the relationship between health and wellbeing, and the difficulty of measuring either, but am more skeptical that contribution to wellbeing can tell us what we want to know about the private value of health for individuals, for the reasons we're about to get into.

that the individual value of health can't be merely its contribution to wellbeing. In what follows, I'll discuss a few ways in which the impact of health on wellbeing can be mediated by other factors—factors that don't seem to alter the *value* of health, even if they affect its interaction with wellbeing. The assumption behind the contribution view is that things are valuable to individuals insofar as they contribute to an individual's wellbeing or quality of life. Likewise, things are harmful to individuals insofar as they detract from an individual's wellbeing. I'm going to argue that these twin claims about value and harm are too simplistic. The extent to which something—including loss of health—harms you can go beyond its impact on your wellbeing. And, likewise, the extent to which health is valuable to you (a matter of care and concern to you personally, something that is an enriching or good thing in your life) can go beyond its impact on your wellbeing.

In making these arguments, I will try as much as possible to stay neutral across various theories of wellbeing. There is a wide debate, within philosophy, about whether wellbeing is objective or subjective, and about whether it is a matter of having certain capabilities, achieving certain specific ends, having certain goods, satisfying preferences, being happy, and so on.[14] While some of what I say here will doubtless be more amenable to some views than to others, I think a good case can be made that regardless of what your specific take on wellbeing is, the connection between wellbeing and health is far more complicated than either the contribution or the constitution view allows.

I'm also going to help myself to cross-condition and cross-person comparisons of wellbeing, since that's the work that the contribution view is typically put to. But if you're worried that these introduce too much noise, a lot of the arguments I'm presenting here can be reformulated without them. That being said, it's worth noting that some potential ways of reading the contribution view are, I think, not available given the complex nature of health. You might be tempted to read the view as something like an individual difference-making principle, for example: the value of health is the difference that a change in health makes to an individual's wellbeing if we hold everything else in their life (social circumstances, hobbies, relationships, etc.) fixed. The trouble with this kind of view applied to health, though, is that many health conditions are such that you can't have them while holding everything else fixed. (You can't develop schizophrenia or alcoholism, for example, while holding your social and personal circumstances fixed.) So while more precise or counterfactual readings of the contribution view might be ways for it to avoid some of the problems I raise below, I'm skeptical that they're actually live options in the case of health.

---

[14] For an excellent introductory overview, see especially Fletcher (2016) and Heathwood (2021).

2.1 Resilience

A large amount of empirical data suggests that psychological and personality traits can impact how we react to loss of health.[15] And few such traits have generated more interest, in recent years, than *resilience*. Resilience, roughly speaking, refers to a person's adaptability in the face of setbacks or losses. People with high levels of resilience tend to focus on positive aspects of their life—on what they have and what they can do rather than on what they don't have and what they can't do— which seems to promote active strategies for dealing with difficult situations. Perhaps unsurprisingly, then, evidence suggests that people with high levels of resilience tend to cope better with serious illness.[16] In the context of substantial chronic illness, for example, highly resilient people will tend to adapt their lifestyle and hobbies to suit the different needs of a changing body. They'll prioritize connections with other people and restructure the ways they communicate with others if necessary, for example, or they'll put more emphasis on the good things in their life that are unaffected by their illness.[17] In short, resilient people will tend to be fairly flexible about how they live their lives, and will tend to focus on the things they can control rather than the things they can't. All this leads to resilient people being quite adaptable in the context of long-term illness.

And so it's not all that surprising that people who have high levels of resilience tend, in the context of chronic illness, to report higher levels of wellbeing than people with lower levels of resilience. And this makes sense regardless of whether you accept a broadly subjectivist or broadly objectivist approach to wellbeing—the resilient person will have fewer of their desires frustrated because they tend to shift their desires more, for example, but they'll also tend to have a life filled with more objective-list-type goods because the kinds of goods they seek out and promote in their lives are more flexible, and they adapt the "mosaic" they're building to suit the pieces they have available to them.

But if we say that the value of health (for the individual) is its contribution to wellbeing, and resilient people are able to maintain relatively high levels of wellbeing even in the face of significant loss of health, then we end up being forced to say that health is less valuable the more resilient you are. And this seems to entirely misdescribe the phenomenon of resilience. People with strongly resilient tendencies tend to find ways to adapt to hardship—to things like the loss of a loved one or a major career setback. But the very idea of their adaptability is predicated on the assumption that what they lose in these setbacks is of significant value to them. It's not that the resilient person just loved their family member less and so grieves less when they die. It's not that they were less invested in their career and

---

[15] For an overview of some relevant studies, see, e.g., Pavot and Diener (2008). See also Boyce and Wood (2011).

[16] For an overview of findings, see Stewart and Yuen (2011).   [17] See, e.g., Kralik et al. (2006).

so care less when they get fired. It's that they can pick up and move on *even when they lose things of great value to them.*

Suppose that—for whatever reasons of luck, social support, community integration, or similar—you're a relatively resilient person. You develop a serious long-term illness—Crohn's disease, say. You're devastated. But you decide to make the best of a difficult situation. You have to change your diet, so you use this as an impetus to finally take those cooking classes you've been thinking about for a while, so that you can have the skills to cook well for yourself. Cooking becomes a relaxing hobby, and strangely enough, you're more into food post-Crohn's than you were pre-Crohn's. You have to spend a lot of time in hospital waiting rooms, so you become an avid iPhone gamer to pass the time. And so on. You miss not having Crohn's disease. You miss what it's taken from you. But you're doing pretty well, all things considered.

It's not that resilient people who get sick care about being healthy less than non-resilient people, or don't feel as bad when they are ill, or have their lives disrupted to a lesser degree. Rather, what seems to happen is that they are somewhat more flexible in adapting their lives and their attitudes to accommodate illness. But if the value of health is simply its contribution to wellbeing, we don't have a good way of capturing this. If, in fact, as evidence suggests, resilient people have significantly higher levels of wellbeing when they face major health challenges than non-resilient people, then their health would simply be less valuable to them.

2.2 Social Stigma

As discussed, one of the benefits of construing the value of health as its contribution to wellbeing is that it allows us to weight the relative "seriousness" of different health problems in a way that goes beyond their basic biomedical manifestations. Hip arthritis is more serious than hand arthritis (at least for most people) in part because of complex social factors about how we live our lives. Some of the outsized impact that hip arthritis has is biological (we're bipedal, and the hip is a weight-bearing joint whereas the hand is not), but some of it is more directly tied to our social context—it's easier to "work around" limitations in the use of a hand than a hip, and most people's daily lives are structured in a way that makes limitations in hip mobility much more significant than limitations in hand mobility. The contribution view can handle all this complexity—without trying to differentiate the social and biomedical aspects of a health condition—simply by saying that a health condition is in general worse if it in general tends to have a greater impact on wellbeing.

And yet there are some cases in which precisely this feature of the contribution view causes problems. Some health problems affect wellbeing in dramatic ways, for example, because of how they interact with social norms and social stigma.

Perhaps the clearest example is acne, especially acne in teenagers (the population most commonly affected). Acne is a health condition—it is an infection of the hair follicle in the skin. Even mild to moderate acne has been shown to have a severe impact on wellbeing for those affected, including negative impacts on self-esteem, increased rates of depression, and social isolation.[18] The impact on acne on quality of life appears to be similar to severe asthma, epilepsy, and arthritis.[19] And unsurprisingly, the impact of acne appears to be more profound for women, who experience greater reduction in self-esteem and more appearance-associated anxiety.[20]

What it seems like we ought to be able to say, in such a case, is that acne is itself a relatively mild health condition, and that it has the impact on wellbeing that it does because of the stigma we place on individuals affected, and on the social value we attach to blemish-free skin. It's not that acne is a severe health condition or that having clear skin is as important to your health as having working lungs or a brain that doesn't spontaneously lapse into seizures. Rather, the impact that acne can have on quality of life has relatively little to do with health and so much more to do with cultural norms about appearance.

But on the contribution view, But on the contribution view, the importance of a distinction like this is hard to maintain. If health matters (only) insofar as it affects wellbeing, then it looks like we're forced to say that if I have breakouts and you have severe asthma, and both conditions affect our overall wellbeing to the same extent, we've each sustained an equally valuable loss of health. Likewise, we're forced to say that clear skin is just more valuable to the health of women than the health of men because the pressure we place on women to be beautiful makes acne more stigmatized.

Again, this seems like a place where the contribution view is forcing us to misdescribe what's going on. What we ought to be able to say is that skin blemishes like acne are of relatively minor value qua health condition—that the impact they make is relatively unimportant to our health. And yet we often care about them a great deal, and they impact our lives a great deal, for reasons that have nothing to do with our health. But the contribution view struggles to fully articulate this type of distinction. If the value of health just is its contribution to wellbeing, then if a condition like acne has a very negative impact on wellbeing—even if

[18] See, e.g., Gieler et al. (2015).
[19] Thomas (2004) and Cresce et al. (2014). Note that Cresce et al. looked specifically at measures of "health-related quality of life," so defenders of a view like Bognar and Hirose's can't protest that this case shows why we need to distinguish between overall wellbeing and the health-related components of wellbeing. Precisely because of their health condition, people with acne often avoid social interactions, restrict their activities, etc. It's their health condition that is causing their reduced wellbeing, but the reason it has the dramatic impact that it does is because of broader social arrangements—and evaluating the impact of health conditions in their social context is precisely what contribution views are intended to do.
[20] Hassan et al. (2009).

that impact is mediated by factors like social stigma—we have to say that in losing clear skin you've lost an aspect of your health that's just as valuable as your lung function or your joint mobility.[21]

2.3 Previous Health Experiences

But perhaps the place where the contribution view encounters the most difficulty is in cases of long-term chronic illness and disability. Just as with the constitution view, the contribution view struggles to accommodate the ways in which people with serious, long-term health challenges often have relatively high levels of wellbeing. But the issue faced by the contribution view is slightly different. If we allow that people with substantial long-term health challenges can have high levels of wellbeing, then the constitution view is committed to saying that health is less valuable to (many) sick people than it is to healthy people.

As we'll discuss in Chapter 4, many disabled people whose disabilities involve substantial loss of health nevertheless report high levels of wellbeing. And, more generally, substantial and serious loss of health seems to have a complex relationship with wellbeing, at least when that loss of health is extended over time. Perhaps nowhere is this clearer than in the phenomenon known as *response shift*. Response shift encompasses many different phenomena, all related to the ways in which major life changes can impact a person's assessment of their own quality of life.[22]

The two main aspects of response shift I want to focus on here are *recalibration* and *reprioritization*.[23] Recalibration refers to ways in which experiences— including experiences of illness—can affect the standards and the metrics by which an individual judges their quality of life. Reprioritization refers to ways in which experiences—including experiences of illness—can shape which goals and values a person emphasizes in their life. Like so many adaptive processes, recalibration and reprioritization can encompass both positive and negative ways of coping with difficulty. Sometimes, we deal with hard times by lowering our standards and expectations in a way that's harmful to us. And sometimes we deal with setbacks by giving up on our goals in way that's harmful to us.

---

[21] Note that the contribution doesn't need to thereby commit to the claim that acne and asthma represent an equal amount of health lost, or equally serious (in the medical sense of "serious") health conditions. It's just that the view struggles to allow for an important distinction between things which are health condition, but matter to us for reasons other than their status as health conditions, and things which matter to us precisely because of the loss of health (and the distinctive loss of health-related value) they involve.

[22] For an overview, see especially Schwartz and Sprangers (2000). For more detailed discussion, see Sprangers and Schwartz (1999).

[23] Again see Schwartz and Strangers (2000) for a helpful overview and discussion of theoretical constructs.

But I want to focus here on ways in which recalibration and reprioritization can be positive (and quite common) coping mechanisms.

Loss of health can be shocking. As Havi Carel (2008) eloquently discusses, so many of the things a person took for granted about their body and their daily lives suddenly become difficult or impossible. But for many people, this shock wears off over time.[24] The fatigue that was overwhelming at the beginning of chemotherapy treatment feels manageable six months into the process;[25] people with lupus can go about their daily activities about as well a decade into their disease process as they could ten years earlier, even though in most cases their disease has progressed in severity;[26] and so on. For many people who face long-term loss of health, adapting to their "new normal" involves a shift in standards—how much fatigue will ruin their day, how much pain will distress them, how bothered they are by needles or tubes. This process of recalibration is something we all engage in, and it's utterly familiar.

Perhaps pain tolerance is the most familiar case. If you're an exercise novice and start a new jogging routine, having never run in your life, you'll probably be sore. A veteran marathon runner might be equally sore after a day's training. But what feels like unbearable soreness to you is probably barely noticeable to her—her standards for what counts as running-related discomfort have become, through long years of experience, very different than yours. Part of this adaptation is skill. The marathon runner knows how to train more effectively than you do. She knows how to use ice packs and foam rollers and physical therapy and whatever else to manage her discomfort. But part of it is simply recalibration in standards. The kind of soreness you can't stop thinking about and focusing on is pure white noise to her because by this point she's experienced that level of discomfort—and much worse—as a regular part of her life for years.

When people with long-term illnesses undergo a process of recalibration, it's often a similar phenomenon. Things that seem like a *very big deal* to the fully

---

[24] Carel (2008) and (2018) discusses the changing shape of this experience as the process of discovering "health within illness".

[25] Salmon et al. (2017), Andrykowski et al. (2009)

[26] Following a group of lupus patients for 8 years, for example, Kuriya et al. (2008) found little overall change in self-assessed quality of life (based on the measures used in the SF-36, which purports to assess "health related quality of life"), even though physical functioning appeared to decline. In their systematic review, Mcelhone et al. (2006) found that disease activity and objective markers of damage are not correlated with self-assessments of quality of life, and that there is limited evidence for correlation between disease duration and self-assessment of quality of life (including self-assessment of health and health-related quality of life), even though disease course often worsens over time. Findings like these aren't, of course, unique to lupus—McCabe et al. (2009), e.g., followed a cohort of 382 people with MS over a two-year period and found that, although MS is typically a degenerative disease process, on average their overall self-assessed quality of life (including measures of health-related quality of life) improved over the measurement period, with the biggest gains due to development of enhanced coping strategies and greater community inclusion. (The MS group actually reported higher levels of psychological quality of life than healthy controls, and placed more emphasis on positive coping strategies.)

healthy person—a trip to the hospital, a significant medical procedure, a wide range of physical discomforts—become commonplace. They're not pleasant, but they're *normal*. And because they're normal, they're less distressing than they would be to the fully healthy person.[27]

This process of recalibration often goes hand in hand with a process of reprioritization. Someone dealing with the fatigue of lupus, for example, may in some ways simply be less bothered by fatigue several years into her disease than she was in the first few months. But she also may change her activities and values in a way that accommodates her fatigue. Her love of going to concerts is slowly replaced by working on that novel she'd always thought about writing. Nights in crowded bars are superseded by chats over coffee with close friends. She gradually restructures the things she chooses to focus on and prioritize in a way that better fits her health condition, meaning that even if the health condition doesn't improve (or, indeed, even if it gets worse), its ability to interfere with what she wants and what she cares about is much more limited than it used to be.

Again, this process of reprioritization is not unique to illness—it's a common way we respond to changes in our circumstances. Your job forces you to move from a big city to a small town. So much of how you lived your life, when you lived in the city, was centered around what city life made available to you—you loved eating all sorts of different cuisines at a huge range of restaurants, you loved going to exhibits at the art galleries and museums, you loved the bars and clubs. Your new town doesn't have any of that. But rather than pine miserably over how much you miss the city, you try to focus on what your town does have. It's a pretty place, and you notice how people who live here are really into the outdoor life—hiking, kayaking, climbing. So you throw yourself into that. There's not a great range of restaurants but because of all the nearby farms there's a big local food movement and a great farmers' market. So you throw yourself into that too. You don't necessarily end up loving the small town the way you loved the city—you might still prefer to move back if you could—but you shift your habits and your lifestyle in a way that makes small-town life much more enjoyable a few years in than it was when you first moved.

But when reprioritization occurs in the context of illness, it's one way that people adapt—they adapt both what they pursue and what they care about in ways that better accommodate their loss of health. Together with this recalibration, this means that significant challenges to health are often less subjectively distressing and less objectively impairing (in the sense that they interfere less with what a

---

[27] It's worth noting that this is a common response to changes in physical function even the absence of major adverse health events. Barclay and Tate (2014) compared a cohort of aging men who survived strokes to a control group of similar age who had not had strokes. *Both groups* showed substantial evidence of recalibration over time as they aged. Spuling et al. (2017) likewise found evidence for recalibration in assessment of health status in aging regardless of major health events, but found that reprioritization was more associated with a significant change in health status.

person does in their day-to-day life) over time. And this, of course, has compli-cated implications for the relationship between health and wellbeing.

Here are two health-related anecdotes based on slightly fictionalized versions of two of my students. Let's call the first Tim. Tim fractures his ankle in the fall semester and is *devastated*. He's young and healthy, and this is the first significant injury he's had in his life. He plays intramural rugby, but the injury means he has to sit out the rest of the season. He has to be on crutches, and he finds them unbearably frustrating. He can't do so many of his favorite activities, and so much of his lifestyle—his tendency to get up just before class and then jog to make it on time, his habit of meeting up with friends to "just go wherever"—is impeded. He finds it hard to concentrate on his classes and he begins to struggle through a period of depression. His grades, unsurprisingly, suffer as a result.

Tim's classmate is Jana. Jana has had active Crohn's disease since she was twelve. She mentions this to me at the beginning of class, but assures me it's not a big deal. Around the same time Tim breaks his ankle, Jana is hospitalized for an intestinal obstruction. She emails to let me know, but casually says that she'll only be in the hospital for a day or two, that this has happened several times before, and that she's still able to do the reading while she's hospitalized. A cheerful Jana shows up in class the next week, obviously weakened but ready to go. Her semester—filled with online political organizing, small-group hangouts with her friends, and the various student coalitions she's involved in—proceeds largely unimpeded.

Jana has obviously both recalibrated and reprioritized in the process of having Crohn's disease. A trip to the hospital for a major illness event—something that for most people would be all they could focus on—is for her something that is relatively undramatic, and something she can happily read textbooks while enduring. And her hobbies and pursuits are all structured in a way that accommodates fluctuations in her health and makes an illness event something that's manageable.

I think it's fair to say that Tim's wellbeing is significantly more impacted by his loss of health than Jana's. And because of this, comparing the difference between the two cases can help illustrate why the value of health can't simply be its contribution to wellbeing. A small loss of health affects Tim's daily life much more than a major loss of health affects Jana's, but it shouldn't follow that Tim's health is simply more valuable to him, or that Jana's health matters less to her. Likewise, it shouldn't follow that in hurting his ankle, Tim has lost something of greater comparative value than Jana has lost in enduring an intestinal obstruction. Rather, what I think it makes most sense to say for a case like this is that Jana has adapted her life in a way that accommodates a loss of significant value.[28]

---

[28] There is some empirical support for the view that people with substantial loss of health still value health (even if they have outsourced their overall wellbeing to other places). Lenert et al. (1999), for example, found that individuals with substantially impaired health valued moderate gains in health more than healthier individuals.

2.4 Where Contribution Goes Systematically Wrong

The contribution view has, prima facie, a lot going for it. It allows us to compare the relative badness of health conditions or the effectiveness of treatments without engaging in the often-impossible task of comparing amounts of health. It lets us explain why health—and health interventions—are so important to us without placing any intrinsic value on health, or lapsing into healthism. Health is valuable to us as individuals because—and only insofar as—it promotes our wellbeing. This takes the focus off health as a goal to pursue for its own sake, and instead emphasizes the broader context of people's lives.

And sometimes, this focus is helpful. A shift to focus on wellbeing and quality of life has allowed us to emphasize the seriousness of many mental health or chronic pain conditions, for example, where objective changes in physiology or function are often lacking but the impact on quality of life is extreme. Likewise, it allows us to highlight that in treating medical conditions we want to make people's lives better, not simply alter their physiology. Insulin pumps and insulin injection therapy might have similar overall effects on blood sugar in people with diabetes, but research shoes that insulin pumps have a far more positive effect on people's experience of their disease, and thus on their overall quality of life.[29] Insulin pumps are thus a better treatment for diabetes—given that what we care about is the impact that diabetes has on a person's overall wellbeing, not their blood sugar values for their own sake.

What goes wrong, then, if we use something like the contribution view as at least a rough proxy to understand the value of health? This, I take it, is something like the view defended by Daniel Hausman (2015). There might be places where we can protest the details of the contribution view, but as a general rule focusing on wellbeing is a good way of doing the things we need to do, like making decisions about health policy.

The problem is that while there's something appealing about the contribution view—and undeniably an intimate connection between health and wellbeing—using the contribution view as a guide or proxy will get things *systematically* wrong in key places. Let's look then, at some specific cases that can illustrate this phenomenon. I don't want to hang anything too heavy on any of these particular cases—empirical data in this area should be treated with due caution, and there's a lot we don't know, especially when it comes to evaluating studies of self-assessed wellbeing. My main contention, specific examples aside, is that we have good reason to think that using something like the contribution view as a "close enough to true" proxy will go wrong in systematic ways.

---

[29] See Pouwer et al. (2009).

Let's start with some macro examples and then zoom in to more specific cases. To begin with, a large amount of data suggests that when we're comparing both the self-assessed quality of life/life satisfaction and the health satisfaction of different countries, they don't track health outcomes in quite the way we'd expect.[30] Better health outcomes such as increased life expectancy and decreased infant mortality have a strong positive correlation with increased life satisfaction and happiness (even controlling for other variables like political stability and income.) And yet, there are curious outliers—people in Pacific Rim countries like Japan and Hong Kong, for example, have extremely good health outcomes but relatively lower life satisfaction compared to other wealthy countries.[31] And at least based on data from the World Gallup Poll, there are some very striking places where health and life satisfaction fail (and fail repeatedly and systematically) to correlate—former Soviet countries, for example, report both very low life satisfaction and very low health satisfaction compared to similar counties. They also report far lower health satisfaction, for example, than much poorer countries that have much less access to medical care.[32]

But it would be odd to describe situations like these as ones in which there are regional differences in the *value of health*. If we can run the pilot program for our innovative new stroke therapy in either Norway or Japan, it doesn't seem right to say that we should run it in Norway because Norwegians, since they tend to report higher levels of wellbeing, will get more value out of it. Nor does it seem right, conversely, to say that we should run it in Japan, because people in Japan, although in objectively similar health to people in Norway, have greater need for the added value of health interventions because of their lower levels of (self-assessed) wellbeing. People from Japan and Norway may well differ in systematic ways in the relationship between health and self-assessed wellbeing (although findings like these should always be treated with a certain amount of caution, given the range of variables involved.) But supposing this were true, it shouldn't license a conclusion about how comparatively valuable health inventions are for people in Japan or Norway.

Similar, systematic divergence between health and wellbeing crops up at finer levels of grain as well. Some research suggests, for example, that Asian Americans tend to be more stoic, and have more robust coping skills, when they experience pain (including severe cancer pain) than other groups.[33] Again, let's suppose this is true, for the sake of argument. (What's very plausible, at least, is that there might be some significant cultural differences in pain coping and pain tolerance.) It would be absurd to conclude from this that we offer something of less value if we

---

[30] There is, of course, plenty of reason to be skeptical about some of this data, and about what if anything it tells us about wellbeing, but given that it's the kind of data that's often appealed to by defenders of contribution-style views, it seems relevant to the debate over their plausibility.

[31] https://worldhappiness.report/ed/2017/.   [32] See, e.g., Deaton (2008).

[33] Im (2007), Kwok et al. (2014).

give analgesics to an Asian person than to someone from a different ethnic background. Likewise, we shouldn't deprioritize the treatment of the Asian person's pain because "she can handle it better" or because "pain isn't as bad for her". The fact that she might cope very well with pain—that it really might impact her overall wellbeing less because of a broad range of cultural influences in how people cope with and interpret pain—doesn't diminish the badness of her being in pain. (We'll return to this issue in Chapter 3.)

A substantial body of research also suggests, for example, that the presence of depression can mediate the effects of medical interventions on quality of life. People with pre-existing depression seem to have lower overall improvements to their quality of life from stroke rehabilitation programs, and report poorer outcomes from orthopedic surgeries (even when biomedical factors are controlled for).[34] It's tricky to know what to make of such findings, since part of the phenomenology of depression is a tendency to devalue good things that happen to you. But it wouldn't be surprising if people with pre-existing depression—precisely because of the difficulties created by that pre-existing depression—struggle to rebuild their quality of life in the wake of something like a stroke, even as rehab programs help them regain physical functioning. And yet surely the upshot of this shouldn't be that quality rehabilitation after stroke is simply less valuable to people struggling with depression, or that hip replacements are worth less if given to people with depression. Depression might mediate the impact such treatments have on overall wellbeing, but the right conclusion shouldn't be that it diminishes their value (or mitigates the disvalue of what they address).

Likewise, as discussed above, studies on quality of life in the context of chronic illness suggest that many things other than health status mediate the relationship between such conditions and overall quality of life. And some of this mediation highlights the good features of the contribution view. People who are vulnerable in various ways—people who are poor, who are subject to major social stressors or social discrimination, who lack access to work or stable housing, who have major caregiving responsibilities—all seem to have their wellbeing substantially more impacted by loss of health.[35] (More on this in the Appendix to Chapter 4.) And it makes sense that we would want to say that getting a chronic illness really is worse if you're in such a socially precarious position, and we should prioritize the treatment and management of people in such conditions, since loss of health has an outsized impact on their lives.

On the other hand, individual variables such as personality factors also play a major role in the relationship between chronic illness and overall wellbeing. Studies of people with rheumatoid arthritis, for example, have suggested that individual attitudes to coping, sense of coherence, and positivity play a major

---

[34] Gillen et al. (2001), Rosenberger et al. (2006).   [35] See especially Smith et al. (2005).

role in how a person adapts to their disease (and that individual disease activity was a poor predictor of overall quality of life).[36] Again, though, it doesn't seem right to say that of the cohort that is coping well that the loss of health simply matters less to them, or that they aren't in as much need of treatment since their wellbeing hasn't suffered as much as their health has.

Similarly, many serious longterm illnesses seem to show relatively similar impacts on overall wellbeing as their (objectively) milder counterparts. (We'll return to this thorny issue in Chapter 3, Section 6.4.) Ulcerative colitis, to pick one of many examples, is a serious inflammatory bowel disease that can often involve regular hospitalizations and invasive treatments, frequently leads to removal of part or all of the colon, and substantially raises the risk of colon cancer. Yet studies suggest that people with irritable bowel syndrome—a much more common and less medically serious condition—report similar (or worse) impacts on their wellbeing.[37] There could be many reasons for these similarities, but on salient factor at play is that people with ulcerative colitis have often had their disease for longer, and it often requires more substantial and pervasive adaptive processes. They will often have fundamentally changed their lives over an extended period of time, and will typically have changed their expectations as well. Many people with ulcerative colitis adapt well, and report high quality of life. But we don't want to conclude from this that the health they've lost is simply of less value to them because of that adaptation, or that people dealing with a milder—though distressing—disease have lost something of greater value simply if it impairs their own sense of wellbeing more.

Exactly this juncture is, indeed, where many disabled people describe a type of double-bind. To combat stigma against disability, they often want to (rightly) emphasize the ways in which they are flourishing, the ways in which their well-being is not compromised, the ways in which they wouldn't automatically be better off if they were non-disabled. And yet most disabilities unequivocally involve substantial loss of health. And, partly due to the influence of the contribution view, in order to have that loss of health taken seriously—to see it as something that matters to them and deserves to be ameliorated—disabled people are then pressured to emphasize how much their quality of life is compromised.[38] In many contexts, though, the right account may simply be that their overall wellbeing is not compromised, and yet the loss of health is still, in an important sense, bad for them. (We'll come back to this in Chapter 4.)

[36] Hamilton et al. (2005), Goulia et al. (2015).
[37] Seres et al. (2008), Pace et al. (2003). Seres et al. (2008) found that people with IBS reported worse health-related quality of life and health-related distress when compared to people with ulcerative colitis, and that people with ulcerative colitis seemed to have developed more adaptive ways of coping.
[38] For discussion of this double-bind in the context of legal damages, see especially Bagenstos and Schlanger (2007).

3. Grief, Loss, and Wellbeing

To get a better handle on the complex relationship between health and wellbeing, at this point I want to draw an analogy to grief. The main reason for this is that I think grief usefully illustrates another common way in which value and wellbeing come apart. But grief can also be a helpful model for understanding adaptation to loss of health, and thus can further show why the value of health to individuals isn't reducible to its contribution to wellbeing.

3.1 A Small Story about Grief and the Good Life

To begin with, here is an anecdote from my own life. Willow was the first dog that was *our dog*. I'd had dogs growing up, but as much as I loved them, they were family dogs. Willow was the first dog I had as an adult, and the first dog I had together with my husband. We adopted her from an abusive situation, and in the process we learned how to care for a dog as adults and learned how to rehabilitate a fearful (but gentle and sweet) dog. We both shared a bond with that beautiful girl that I don't think we'll ever share with another living creature. When she died several years ago, we were both devastated.

As much as we missed her, though, our life moved on. We adopted another dog—Breccan, a happy-go-lucky muppet as different from Willow as any two neurotic rescue border collies could be from one another. And then we adopted yet another—sweet little Rowan, who was taken out of a terrible hoarding situation and greets every new day like she's won the lottery. We also started fostering, so a strange parade of hard-luck dogs make their way through our household. We're madly in love with each of our dogs, and we also love the pack our family has become.

The life we have now, with these dogs, is different in many ways than the life we had with Willow. It isn't better; it isn't worse. Both versions are filled with incommensurable and incompatible joys. We would never have adopted Breccan or Rowan (much less taken on foster dogs) had we still had Willow. And because of Willow's fear issues, we couldn't have had multiple dogs while we had her. Our wellbeing suffered severely in the immediate wake of Willow's death, but our lives now, years later, are as full and rich—though different—as they were when she was with us.

And yet with Willow's death, we lost something precious. And we will never get it back. This loss—and the ongoing impact of our grief for Willow—can't be measured by its impact on our wellbeing. Our wellbeing has adapted: our lives are different without Willow, but they are still very good. But the adaptation in our wellbeing doesn't change or mitigate the loss. Willow is gone, and our lives will always be missing something because of that, in a way that can't be captured simply by evaluating our quality of life.

3.2 Grief and Wellbeing

Grief can take a devastating toll on wellbeing, especially in its acute stages. And there are few things that can affect nearly every part of your life as profoundly as grief can. Yet the long-term impact of grief—and the value of the loss over which we grieve—can't be adequately understood via effects on wellbeing.

To begin with, grief doesn't end when wellbeing recovers. You can still mourn and miss what you've lost, and ache for its absence, even while living a good and happy life. Any account of wellbeing has to allow that human flourishing is consistent with some degree of sadness—feelings of sadness and loss are part of the human condition, and they can exist alongside (and mingle together with) the things we value. Likewise, it would be absurd to think that joy and happiness can't co-exist with feelings of loss and grief. When we report on our wellbeing, we're typically talking about how our lives are going all things considered. And your life can be going *all things considered* very well, even while containing elements of struggle and pain—including grief.

In her podcast *Terrible, Thanks for Asking*, writer Nora McInerny discusses her experience of falling in love in the wake of her husband's death. And she describes an experience in which deep joy and deep sorrow existed at the same time. Her grief for her husband, she emphasizes, didn't go away—it wasn't fixed or "healed"—when she fell in love with someone new. Instead, she says, her grief just "scooted over a little" to make room for a new range of emotions. As she tells it, her quality of life (her overall wellbeing) improved, even while her grief remained profound. And that's because other things in her life changed, some related to her grief (her loneliness, her sense of isolation) and others related to a new and different love—the strength of which didn't in any way diminish the strength of her love for the partner that died.

Secondly, though, the impact of grief on long-term wellbeing is mediated by all sorts of factors—personality, social support, methods of coping, and so on—such that how and to what extent grief alters a person's wellbeing over time tells us nothing about how much they valued what they lost. Suppose Arya and Bran both lose their father at a young age. Over time, Arya adapts—she throws herself into new experiences and strives to be the person her father would've wanted her to be. She ends up flourishing, although she never stops missing her father. Bran, on the other hand, withdraws into himself. He becomes more and more isolated, and spirals into a depression. His long-term wellbeing is dramatically impacted by the loss of his father in a way he never fully recovers from. It would be perverse to conclude, from this, that Bran loved his father more than Arya did, or that his relationship with his father was more valuable to him than Arya's relationship to her father. Nor can we conclude that Bran grieved more than Arya did. The extent to which grief affects long-term wellbeing is influenced by many things other than the grief itself or the value of the thing that's lost. Arya and Bran are simply two different people processing a loss in two very different ways.

Perhaps missing this point about the connection between grief and wellbeing, some philosophers have argued that if a person's wellbeing recovers relatively quickly in the wake of significant grief, it implies that what was lost has turned out to be relatively unimportant[39], or that such recovery is something to be regretted because it in some way diminishes the value of what was lost.[40] But as Ryan and Erica Preston-Roedder (2017) persuasively argue, such arguments focus far too much on a person's overall wellbeing. Unsurprisingly, a person's overall wellbeing is influenced by a wide variety of factors, and a person can live a good life by combining many different (often incommensurable) good things. And so, in focusing on the extent to which a loved one—a partner, a friend, a parent—contributes to our overall wellbeing, it makes sense to say that they are *in that minimal respect* replaceable. But we also don't love people simply as means of achieving our own wellbeing. Part of loving someone is, as the Preston-Roedders argue, about their particularity—it's about loving them for the particular individual they are. The role they play in allowing you to live a good life might be taken on by other people or other things. But they themselves—and the particular value they bring to your life in virtue of that uniqueness—are irreplaceable. When we grieve, we grieve the loss of the par- ticular individual—the loss of the unique thing we valued, the thing that is now gone. And we can't measure the extent of that loss just by thinking about its effect on our overall quality of life.

3.3 The Relevance of Grief to Thinking about Health

But what does any of this have to do with health? A very common assumption— the assumption of the contribution view—is that health is valuable to us insofar as it contributes to our wellbeing, and that we can understand the value of health to an individual in terms of the extent to which it affects that individual's wellbeing. And often this view is assumed to be common sense. How else could health be valuable to us except by contributing to our wellbeing? And what would it be for an individual's health to be valuable to her except that her life goes better for her if she has health and worse for her if she doesn't?

Grief gives us an interesting parallel case in which this kind of reasoning is mistaken. We can't measure how valuable a loved one is to us by how badly our wellbeing is affected long term if we lose them. The value of interpersonal relationships—and the badness of grief—isn't the same as their contributions to wellbeing.

---

[39] Moller (2007).   [40] Cholbi (2019).

The comparison is particularly apt because for both loss of health and loss of loved ones, the greatest complexities in the interaction between that loss and overall wellbeing arise over time. Arya doesn't love her father any less than Bran simply because she adapts to the loss differently over time. Likewise, the resilient person doesn't value her health any less than the passive person, even though they adapt quite differently, over time, to a long-term loss of health.

Thinking about the analogy to grief is relevant because long-term reduction in health is—like grief for loved ones—fundamentally a matter of *loss*. And like grief, adapting to change in health status is very often about adapting to the loss of something that was valuable, and that will never be recovered. How people adapt to this type of loss will depend on many factors—their personality, their network of social and financial support, their lifestyle, the length of time they've spent adapting, and so on. And thus the overall effect on their wellbeing will be a complex and multi-dimensional thing. That a person adapts to illness well doesn't mean that their health wasn't as valuable to them or that the loss has been less significant for them.

Grief also presents a useful model of an "it's complicated" relationship to wellbeing. Grief and wellbeing are closely tied together. Grief—especially acute grief—has a tremendous impact on wellbeing, and there can be experiences of grief from which a person's wellbeing never truly recovers. Likewise, a flourishing and happy life might be incompatible with truly overwhelming amounts of grief— the loss of one's entire family or community, for example. But at the same time, grief and a flourishing, rich life are obviously compatible. As painful as grief is, it is a consequence of loving and being loved under the basic constraints of mortality. The only way to shield yourself from grief is to not love things that will die, and that's not a path toward a rich and flourishing life. Nor are amounts of grief and reduction in wellbeing directly correlated—sometimes we grieve the most because we have loved and been loved in the best ways, such that the experience of grief is directly tied to some of the best things in our life. Perhaps the most we can say about the relationship between grief and wellbeing is that they have a close connection—it's not as if whether grief negatively affects your wellbeing is simply a matter of your attitude toward it or your other plans and desires—but that this connection isn't as strong as a perfect correlation.

And that, I think, is the best way to think about the connection between health and wellbeing. There can be losses of health from which wellbeing does not recover. And some reductions in health might simply be incompatible with high quality of life—not only because some reductions in health will kill us but also because some reductions in health can have such a devastating effect on how we feel. But there isn't a linear correlation between loss of health and reduction in wellbeing. The relationship between health and wellbeing, like many intimate relationships, is a complicated one. It's mitigated by many factors—both social and personal—and it often plays out in surprising ways.

4. Conclusion

The relationship between health and wellbeing is weird. In the above discussion, I don't take myself to have offered a positive account of this relationship, but rather to have outlined the complexities that any such account would need to navigate—the complexities that most standard accounts run aground on.

Both constitution and contribution views are right to emphasize that a major part of why we care about health is its role in our overall wellbeing—our happiness, our ability to pursue our goals, and our capacity to live the lives we want. And yet the straightforward relationship between health and wellbeing offered by both such views obscures more than it clarifies. To truly understand the relationship between health and wellbeing, we have to appreciate both the closeness of the connection and the many ways in which the two can come apart.

Daniel Hausman (2015) has argued that health and wellbeing must be understood as distinct, and that standard presentations of the contribution view are oversimplified. But he further argues that, because of the undeniably close connection between health and wellbeing, the contribution view is the best of a limited set of options, at least for assessing the personal value of health. Although he doesn't endorse it fully, Hausman in effect argues that the contribution view is a reasonable proxy for what we're trying to get at. Here, I'm arguing for something stronger. The contribution view is not just occasionally incorrect, it's *systematically* incorrect. People with high health expectations often suffer substantial reductions in their quality of life after relatively mild health events.[41] People with serious health conditions often rate their quality of life as no worse, sometimes even better, over time, even if their health declines, and even while they clearly value their health.[42] People's self-assessed wellbeing often improves with age (at least in many parts of the world), even as health declines in age—but health status remains an important component of subjective wellbeing in older age.[43] And so on.

There is an undeniable correlation between health and wellbeing. But the relationship is complicated—complicated to an extent that both constitution and contribution views obscure more than they clarify. To capture the relationship between health and wellbeing, we need to be able to offer an account—closer the picture of grief I suggest above—that allows for a trickier connection. There is correlation, but correlation that's often weird and unexpected. And, of course, just as with grief, there's the reality that a full, rich life—lived well—inevitably involves the loss of health.

---

[41] Carr and Robinson (2001).
[42] Again see Schwartz and Spranger (2000) for discussion. See also, e.g., Sinclair et al. (2008).
[43] Steptoe et al. (2015).

3

Health, Subjectivity, and Capability

We explored, in Chapter 2, the complicated relationship between health and wellbeing. Health and wellbeing aren't the same thing. At times, they aren't even all that well correlated—someone can be in great health but have very poor wellbeing, and someone can have very high levels of wellbeing while being in relatively poor health. And yet health *matters* to our wellbeing in inextricable and basic ways. While it's important that we don't conflate health and wellbeing, or overemphasize the connections between health and wellbeing, it's likewise important (but difficult) to capture the ways in which they are intimately connected.

In this chapter, we'll explore another axis along which our understanding of health seems pulled in conflicting directions: the relationship between subjective experience and objective physiological condition. If we overemphasize the objective aspects of health, we miss some of the most crucial aspects of why health matters to us and why loss of health is harmful to us. In contrast, if we overemphasize the subjective, experiential dimensions of health, we miss both the biological importance of health and part of the explanation for the normative significance of health conditions. Strikingly, though, when it comes to understanding what health is, the subjective and objective dimensions aren't just sometimes in tension with each other; they're also often mutually dependent on each other, such that you can't fully separate them.

Some things that matter to our lives matter for objective reasons. And some things that matter to our lives are objective matters of fact. Whether you have money to pay your rent, whether you have access to housing and food, whether you're free from immediate danger—these are simple matters of fact and they don't depend on your attitude to them. Likewise, their importance to your life doesn't depend (entirely) on your opinion about them. No amount of cheerful optimism will make a situation of domestic abuse any less dangerous for you. No amount of repeating "affordability is relative" will make a price-gouged apartment in a crowded city into affordable housing for a minimum-wage salary.

Controversially, I'd also add (at least some) moral and political matters to the list of objective things that matter to our lives objectively. You might pursue libertarian ideals out of the sincere belief that these ideals will make the world a fairer and more just place for everyone. You might also just be wrong about what fairness and justice are, or wrong about what will promote fairness and justice.

Health Problems: Philosophical Puzzles about the Nature of Health. Elizabeth Barnes, Oxford University Press.
© Elizabeth Barnes 2023. DOI: 10.1093/oso/9780192883476.003.0004

Many things that matter to our lives, though, depend at least in part on our subjective reaction to them for their value to us as individuals. And many things that matter to our lives are likewise partly *constituted* or *determined* by our subjective reactions to them. Regardless of whether you accept a subjectivist or objectivist account of wellbeing, a person's feelings and attitudes matter to how well their life is going. The romance with the attractive partner is only good for them if it isn't secretly making them miserable (and part of whether it's a romance or a charade is determined by how they feel about it). The afternoons spent strolling through galleries are good for them if they're motivated by a genuine enjoyment of art, but not if they're motivated by cultural snobbishness and a creeping fear of being outclassed by the neighbors. The career they've worked so hard for might be good for them or it might be slowly grinding down whatever enthusiasm they have left in their life. If the era of social media has taught us anything, it's that a life can *look* happy and fulfilled from the outside and yet be hiding all manner of pain.

One of the most perplexing—and interesting—things about health, however, is that it doesn't fit neatly into either category. Many aspects of health are objective. There is an underlying biological reality to health that doesn't depend on our attitudes or desires. Likewise, many aspects of health matter to our lives objectively—independently of how we feel about or react to them. And yet, health isn't entirely objective—part of what matters to and determines health is a matter of subjective feeling. If you are in pain, or suffering extreme fatigue, or dealing with ongoing depression, then your health is compromised in significant ways, even if there isn't any objective criteria by which we can measure this loss of health. Balancing these strikingly different aspects of health in trying to give an overall explanation of what health is, however, is an incredibly fraught task. And it becomes even more difficult when we consider the ways in which the subjective and objective aspects of health are often interdependent.

1. Objectivity and Subjectivity

Whether something is "objective" or "subjective" can mean many different things in different contexts, especially when we're talking about things we value. For the purposes here, I want to roughly distinguish two senses of objectivity and two senses of subjectivity.

First, consider *epistemic* objectivity and subjectivity. As I'm using this distinction, when x is epistemically objective there is a fact of the matter as to whether x, that fact doesn't depend entirely on individuals' attitude to x or emotive reaction to x, and knowing whether x doesn't involve special first-person authority. This leaves open, among other things, the enticing prospect of being *wrong* or *mistaken* about whether x is the case.

Let's consider Jeff Bezos. Whether Jeff Bezos is human is epistemically objective in the sense that I am using it here. There is a fact of the matter, and that fact doesn't depend on how Jeff Bezos feels about himself or on how any of us react to him. It would still be true that Jeff Bezos is human even if we collectively agreed that billionaires are a separate species—we'd all just be wrong about that.

Importantly, though, in the way that I am using the term, something can be epistemically objective without being a claim about the "natural" or "mind-independent" world. Take, as an example, being wealthy. There are epistemically objective facts about which people are wealthy, given the way that money works in our society. There may be borderline cases, there may be different ways of precisifying the term, but it is not a matter of mere opinion or emotive reaction that Jeff Bezos is wealthy. This doesn't mean that there are society-independent or natural facts about what it is to be wealthy. *Wealth* is plausibly socially constructed, insofar as we collective agree on units of money, on a money-based economy, and so on. It just means that which people are wealthy, within a social context that has an established economy, is not a matter of opinion, attitude, feeling, or individual stance. Which people are wealthy is determined by facts about distribution of economic resources, and those facts are objective (in the sense that I am using "objective")—though non-natural.[1]

Famously, many people in the highest-income brackets do not consider themselves wealthy—they don't "feel" wealthy because they can't afford everything they want, still struggle to pay their bills, or they compare themselves to higher-earning peers. But on this view, whether you are wealthy is a simple matter of your wealth relative to the average wealth in your social community. It's not determined by your attitude or other people's attitude to you. And so you (and your neighbors) can easily be wrong about whether you are wealthy.

Consider, in contrast, whether Jeff Bezos is happy. This is epistemically subjective. One person, and one person only, has special first-person access to whether Jeff Bezos is happy. Likewise, whether Jeff Bezos is happy depends on the specific emotional reactions of one person. Importantly, Jeff Bezos might still be wrong about whether he's happy (he might be kidding himself). The difference is that, in the case of happiness, Jeff Bezos is the primary authority (the best source of information) over whether he's happy, and nothing but his own subjective reactions determine whether he's happy. "Tests determine that Jeff Bezos, despite feeling cheerful and optimistic, is not actually happy" is a headline that doesn't make sense. Jeff Bezos is happy if he's in a particular subjective emotional state (about which he's the best, though not a perfect, authority). There are no further facts to be discovered.

---

[1] To give a similar example: it's an objective matter of fact that Richmond is the capital city of the Commonwealth of Virginia, even though it's obviously a fact that's determined by collective social agreement.

So epistemic objectivity, as I'm using it, is objectivity in what the facts are and what the evidence for those facts is. Normative objectivity is objectivity in whether and to what extent those facts have normative significance in a context. On many views, if Jeff Bezos is human, that matters normatively—the fact that he is human is objectively important to what we owe him and how we should treat him. And it matters, on such views, regardless of how Jeff Bezos regards his own humanity. He could be a critic of "species parochialism" or a devotee of biohacking who wants to transcend the paltry limitations of *Homo sapiens*, and it would still be normatively significant that he's human. Likewise for wealth. There are objective facts about how much money you have. But on some views of distributive justice, there are also at least *some* objective facts about the normative significance of how much money you have. If Jeff Bezos is extremely wealthy, then some views hold that there are objective moral and political consequences—how Jeff Bezos ought to use his wealth, how societies ought to handle the fact that Jeff Bezos has so much more money than other people. And on some normative views, these consequences go beyond how we collective view or react to Jeff Bezos' wealth. Perhaps a society that allows extreme levels of wealth inequality is simply *unjust*, even if most people in that society are fine with it. Resources like wealth, at least on many views, have objective normative significance: they matter, and they matter independently of our (or Jeff Bezos') emotive reaction to them. These particular examples are controversial, obviously, but hopefully they show the basics of the distinction.

Note that normative objectivity, as I'm using it here, isn't the same as moral objectivity. There are lots of normative notions that aren't obviously moral—rationality, truth, aesthetic value, wellbeing. As I'm using it, something is normatively objective just in case it matters *in the contextually relevant normative sense* for reasons independent of our subjective feelings or reactions. I'm hoping readers will go along with me with this rough and ready distinction. What mattes for the purposes here is that sometimes things like health are normatively significant for reasons independent of how you feel about them or how they make you feel.

In contrast, some things—including some very important things—are partly constituted by our reactions them. Imagine that a government, learning about the detrimental effects of loneliness to long-term wellbeing, assigns each person four loving relationships. In the Loving Relationships Program, you're matched with four other people in your community, and these are now your loving relationships, and you're no longer lonely. Problem solved! Obviously, this is absurd—and it's absurd because whether a relationship is loving—whether it's something you cherish and whether it fills a need for companionship—is determined partly by the subjective reactions of the people involved. They have to care about each other.

Likewise, whether something matters is sometimes determined at least in part by how we feel about it. This doesn't diminish the importance or the significance—many of the things we value most matter in this way—but it makes a differences to how we understand the ways in which these things matter to us. Again, consider relationships. Two people might be single, and one might be desperately lonely, the other extremely content. The desire for a romantic partner on behalf of the former is no less important because the latter is content to be single. But, likewise, we make a serious mistake if we try to diagnose the latter as secretly pining for love or descending into unconscious spinsterhood. Different people want different things, and their subjective reactions are part of what determines what's important *for them*.

Having made these (very rough) distinctions, I'm now going, in what follows, to discuss many of the ways in which they intersect and blend together. And that's because one of the interesting—and perplexing—things about health is the way in which it is a confusing mixture of both forms of objectivity and both forms of subjectivity. There are aspects of health that are epistemically objective and that are normatively objective. But they are closely intertwined with aspects of health that are epistemically and normatively subjective.

2. The Objective Nature of Health

Many of the most central and important aspects of health are objectively verifiable matters of fact—that is, they are epistemically objective. Your heart rate, your blood oxygen and blood sugar levels, the level at which your kidneys are filtering waste, the number of white blood cells your body is producing—these are all crucial components of your health, and they are all matters of objective fact.

They are also important aspects of your health *regardless of how you feel about them*—that is, they are normatively objective. The religious ascetic may interpret the shakiness and palpitations of hypoglycemia as a transcendental experience brought on by fasting, but that perception doesn't make a hypoglycemic state any less real or any less damaging to their health. Whether it's damaging to their wellbeing is, of course, a separate question, one that might depend on decidedly more subjective factors. As discussed in Chapter 2, the very same thing can be good for your wellbeing and bad for your health. The point here is simply that whether a state like hypoglycemia is harmful *to your health* doesn't depend on your attitude to or feelings about that hypoglycemic state.

Appreciating these objective aspects of health is important for several key reasons. Firstly, and perhaps mostly straightforwardly, it's important that we understand the ways in which *good health* and *feeling good* can come apart. You might feel great—feel completely in your prime—but have an aneurysm or valvular heart disease that, in fact, seriously compromises your health. Not all damage to health is associated with "symptoms" or with feeling ill. Women with ovarian cancer, for example, often feel completely normal—without any idea that something is going wrong—until the disease is in a relatively advanced state.

An important aspect of health is that people can be mistaken about their own health, for the simple reason that we can feel good while nevertheless being in poor health. And this is, of course, part of why objective medical testing is an important component of healthcare. We use testing not only to diagnose but also, crucially, to screen. Women in their fifties go to the doctor to find out whether the lump in their breast is anything to be concerned about, but they also go to the doctor to get that scheduled mammogram, even if they haven't noticed anything unusual. And that's because they know that they might *feel* perfectly well and yet have breast cancer.

But the importance of this potential to be mistaken about health—our own and others'—extends beyond things like hidden or symptomless illness. It's also crucial that our understanding of health incorporates our capacity, both as individuals and as societies, to be wrong about what promotes and detracts from health. Let's consider the individual case first.

People can be—and often are—simply wrong about their own health. Perhaps the most straightforward cases are those involving mental health, where part of the difficulty of grappling with a mental health condition is the way in which it distorts patterns of thinking. The young woman with anorexia insists that her weight is not a problem, that she is eating enough, that food is not an issue for her. And in many cases, she sincerely believes all of this—she believes that it is really other people that have the problem with food and other people that have the distorted body image, and that she is perfectly healthy and showing admirable self-control with her food choices. And yet, her eating patterns and her weight are still harmful to her health. Similarly, the alcoholic might insist that his drinking is not a problem, that he does not drink to excess, that he doesn't consume alcohol at a level that's bad for him. And again, in many cases this is a sincere belief—he thinks other people are worrying excessively and that he has everything under control. But that belief—or his desire to keep drinking and for others to leave him alone about his drinking—doesn't change the fact that he has a problem.[2]

Leaving particular mental health conditions aside, however, it's entirely possible for people to feel good—indeed, to feel *healthy*—doing things that in fact are harmful to their health. And to capture this, we need a distinction between a how a person feels (their own subjective experience of their body and mind) and the biological reality of their health. A particularly interesting example comes from extreme endurance sports (ultramarathon, endurance cycling, etc.). Many people who pursue these sports claim the increased sense of physical health and vitality they get from their training as a primary motivation, and yet at least some research suggests that repeatedly taxing the body to such an extent can, in fact, lead to

---

[2] Cases like these pose similar problems for phenomenological accounts of illness to the ones discussed in Chapter 1 (section 4). Not all illness involves a sense of disruption—and some illnesses are pernicious precisely because of the way in which they can feel so comfortable.

long-term cardiovascular damage.[3] This particular body of evidence is very controversial—the point is not that the finding is accurate, but rather that such a finding easily *could be* accurate, for all we know, and we need more investigation to find out. What *feels* like the paradigm of health to the extreme athlete could, potentially, be objectively harmful to their health—how we feel is part of what determines how healthy we are, but we can also sometimes be wrong about how healthy we are and about what things promote our health.

This capacity to be wrong about what promotes or detracts from health is magnified writ large at the level of social norms. Societies can—and often are—simply wrong about which things are diseases, about which things are harmful to health, and so on. Homosexuality was never a disease, even when it was widely considered to be a psychological disorder and was listed in the *Diagnostic and Statistical Manual of Mental Disorders*. Epilepsy was always a disease, even when it was widely considered to be a spiritual condition. Draining blood with leeches never cured infectious illnesses, even if physicians and patients alike would swear the treatment worked. It's not harmful for women to swim while they're menstruating, even if collective belief insists otherwise.

There are objective facts about what constitutes, detracts from, and promotes health—facts that go beyond (and can be in contrast with) subjective feeling or widespread norms and opinions. And this is because there is an objective biological reality to health. Part of what we are tracking when we talk about health are physical and psychological facts about the human organism. And those facts obtain regardless of how we feel or think about them.

3. The Subjective Nature of Health

And yet, while it's important to emphasize the objective aspects of health, it's also important not to ignore health's more subjective aspects. *Part* of what we care about when we value health is objective. But not all of it. There are distinctive and central aspects of health that are subjective, and that subjectivity is an important part both of how we experience health and what we value about health.

To state the obvious, most people want to feel good. And part of what we care about when we care about our health is feeling as good as we can for as long as we can. MRIs, blood tests, heart rate monitors, and CT scans can give us so much vital information about health. But they can't tell us whether or to what extent we feel good. Nor is this a simple limitation of our current medical knowledge. Part of the subjectivity involved is that "feeling good" means different things to different people. Some people think the perfect way to relax and feel their best is to sweat in

---

[3] See O'Keefe et al. (2012).

a scalding hot sauna and then dive into an icy plunge pool. For others, that sounds like torture. Some people think the muscle ache of a deep-tissue massage or post-workout soreness is exquisite. Other people just think it hurts. Some people train for marathons but take herbal supplements because they have "low energy." Other people think having the energy to run recreationally sounds like a beautiful, impossible dream. People have different preferences, different standards, and different experiences—all of which render the subjectivity of "feeling good" something that can't be circumvented via better scans or more medical knowledge.

But moving beyond the generic desire to feel good, many of our most distressing experiences of illness—and many of the things we most wish to avoid when we think about avoiding illness and promoting our health—are highly subjective in nature. Feelings like nausea, pain, and fatigue can be overwhelming and incapacitating, but they also can't be objectively quantified the way that blood pressure or heart rate can.

To see this point in detail, let's examine a particularly salient test case: pain.

3.1 Pain and Subjectivity

Pain is an important indicator of injury and disease—it's often the first indicator of the stress fracture in your foot, the flare of your rheumatoid arthritis, the tumor in your brain. It's also one of the things people find most distressing about illness, and one of the things they most want to avoid. And yet pain is also inherently subjective. We can x-ray your foot to determine the extent of the fracture; we can look at your blood work to monitor the markers of inflammation during an autoimmune flare; we can measure the size of a tumor on a CT scan and biopsy it to determine if its malignant. But we don't have a test that tells us whether or how much any of it *hurts*.

To find out how much it hurts, we have to ask you. Suppose that Ama and Brynn both have a migraine. Visiting the doctor, they're both asked to rate their pain on a scale of 1–10. Both rate their pain as a 6. Can we conclude that Ama and Brynn are in the same (or even a similar) amount of pain? Of course not. They may be interpreting the scale differently, for one thing. But they also may be experiencing the pain in different contexts. Maybe this is Brynn's first migraine, whereas Ama has gotten migraines every week for a decade. Maybe Brynn is worried the headache means she has brain cancer. Maybe Ama has had this migraine for five hours already, and that's on top of a truly terrible day. All of these variables can affect how they each experience their headache.

The major impediment here is not just the 1–10 scale and its lack of specificity—it's the amorphous thing we're trying to measure. Nearly everyone can agree that pain comes in different amounts or different degrees. That is, some experiences of pain hurt more than others. And we also tend to think that in many cases we can make meaningful comparisons about differing amounts or degrees of pain. A stubbed toe hurts, but it hurts less than a broken ankle. And yet there's no simple linear ordering of the "badness" of pain experiences. Pain comes in a wide variety of hues and textures. The experience of it is moderated by its duration, our mood when we experience it, the factors that precipitate it, and countless other variables. It's a beast with many faces.

Does the dull throb of a bruise hurt more or less than a mild electric shock? Does your toothache hurt worse than that time you burned your hand? Does the pain of a massage therapist digging their fingers into your knotted muscles hurt less because it's paired with pleasure? Or is it somehow a sensation that's simultaneously very painful and very pleasant? Has that stabbing headache at your temples gotten worse over the course of the day, or has it just become harder to deal with as you become more tired?

Our experience of pain is inevitably bound up with our emotions, our preferences, and our attitudes to what has caused the pain, in a way that makes talking about "how much" pain a person is in both elusive and subjective. A key part of the difficulty here is that we implicitly treat the question "how much does it hurt?" as synonymous with the question "how *badly* does it hurt?". The first is a question about the intensity of particular bodily sensation you are experiencing. Some things itch. Some things itch more than others. The question "how much does it itch?" is a question about to what degree or to what intensity you're experiencing the sensation of itching. Again, this question isn't entirely straightforward for pain, even while we grant that pain comes in different amounts or intensities, and that these can sometimes be meaningfully compared. Different pains have different feels and qualities—burning feels different from aching and from shocking. And these differences can make quantitative comparisons difficult at best. But the question "how badly does it hurt?" is a yet more complicated question.

"How badly does it hurt?" is not merely a quantitative question—it's also both a normative and an emotive question. We are asking a person not only how much pain they are experiencing but how bad or unpleasant this experience is for them. The question of how bad or unpleasant an experience of pain is, however, can be contingent on many factors other than the basic bodily sensation of pain, if there even is such a thing as "the basic bodily sensation of pain."[4] Nor is this merely a problem of introspection. It's not simply that people have trouble distinguishing the question "how much does it hurt?" from the question "how badly does it hurt?"—but rather that these two questions can often be entangled in surprisingly complex ways.

---

[4] There is a basic biological process of *nociception*, by which our peripheral nervous system communicates noxious sensations. Painful sensations are often associated with and caused by nociception, but as we'll discuss, pain isn't the same thing as nociception.

3.2 Subjectivity and the Biological Function of Pain

Pain is your body's alarm system. Pain's primary function is protective, and that protective mechanism is the main reason we experience it. You touch a hot stove, you feel a jolt of pain, you immediately jerk your hand back. You trip and twist your ankle, you feel a sharp, aching pain when you try to walk, you keep weight off the ankle for a bit. Pain is *correlated* with tissue damage (what we need protection from), but it's not simply the perception of tissue damage—especially because the two can easily come apart. If you've ever cut yourself and not felt a thing until you saw blood, you've experienced tissue damage without pain. If you've ever felt a sting in anticipation of an injection, you've experienced pain without tissue damage.

Our best contemporary understanding of pain suggests that it differs, in this respect, from basic sensory mechanisms like vision or hearing. The primary role of pain is not to perceive tissue damage but rather to protect the body from harm.[5] This is one reason, for example, why chronic pain often persists long after an acute injury—the direct tissue damage—has healed. Pain is a threat response, but it's a threat response that can persist even in the absence of damage. You fall, you twist your ankle, your ankle hurts. The pain is, effectively, your body shouting "protect your ankle!" But your body might want to protect your ankle—your ankle might *feel* vulnerable—even after the damage you did in the fall has subsided.[6]

Pain can play this protective role effectively at least in part because of its emotive aspect. Pain isn't just a random sensation—a tingle, a bit of pressure, a gentle warmth. Pain works well as a threat response because we typically find it unpleasant and distressing. You wouldn't jerk your hand back so quickly when touching the hot stove if pain was just a strange tickle. You jerk your hand back because it *hurts*. To be an effective protection mechanism, pains need to be motivating.[7] And pains are motivating primarily because we don't like them, and because they distress us.

Thus the emotive aspect of pain—the sense in which the physical sensation is blended together with a sense of distress, distraction, or unpleasantness—is built into the very nature of pain. The International Association for the Study of Pain defines pain as "an unpleasant sensory *and emotional* experience associated with, or resembling that associated with, actual or potential tissue damage" (emphasis added).[8] The emotional content of pain, at least on many standard ways of

---

[5] For an excellent philosophical overview, see Klein (2015). See also, e.g., Raja et al. (2020) for an introduction to some of the relevant empirical background. For an excellent introductory overview to some key empirical issues, see especially Moseley (2007).

[6] See Moseley (2007).   [7] See Klein (2015) for an excellent extended discussion on this point.

[8] The definition is further elaborated by six explanatory notes: "(1) Pain is always a personal experience that is influenced to varying degrees by biological, psychological, and social factors.

thinking about it, is part of what pain *is* and an essential part of pain's biological function.

But the emotional content of pain is also, of course, highly subjective. How much a particular experience of pain distresses you can be influenced by many factors, including factors that are largely independent of your health status. Pain that's desired or meaningful is often a very different experience than pain that's unwanted. Pain that's attached to other emotions—worry that it won't go away, worry that it's a sign of a serious illness, frustration that it keeps happening—is often a different experience from pain that's annoying but ordinary (a stubbed toe, a paper cut). Pain that comes when you're tired, stressed, or otherwise not at your best is often a different experience than pain that hits you when you're relaxed and happy.

But its subjectivity notwithstanding, pain is objectively important to health. Its role as a protective mechanism is vital—indeed, the rare inability to feel it (congenital insensitivity to pain) carries with it a massively reduced life expectancy and major long-term health consequences.[9] Likewise, pain is an objectively important indicator for many disease processes, both acute (the crushing chest pain of a heart attack) and chronic (the lingering joint pain that is often the first indication of autoimmune disease). And while it doesn't correlate perfectly with tissue damage, it's still strongly correlated with tissue damage, and an objective part of your body's response to such damage. But pain is nevertheless, by its nature, a subjective experience.

3.3 Pain, Subjectivity, and Measurement

Suppose that Ama and Brynn both have rheumatoid arthritis, and they both have pain in their knees. Their blood work shows the same markers of inflammation and their scans show the same degenerative changes to their knee joints. They're both asked to rate their pain on a scale from 1–10. Ama says 4, Brynn says 7. Is Brynn in worse health than Ama?

There may, I suggest, be no good answer to this question. To begin with, we don't know whether Brynn is in fact experiencing worse or more intense pain than Ama. Perhaps Brynn just interprets the scale differently than Ama does, or perhaps Ama has had worse previous pain experiences, which makes her shift

(2) Pain and nociception are different phenomena. Pain cannot be inferred solely from activity in sensory neurons. (3) Through their life experiences, individuals learn the concept of pain. (4) A person's report of an experience as pain should be respected. (5) Although pain usually serves an adaptive role, it may have adverse effects on function and social and psychological well-being. (6) Verbal description is only one of several behaviors to express pain; inability to communicate does not negate the possibility that a human or a nonhuman animal experiences pain" (2020).
[9] For an overview, see Cox (2017).

her current experience further from a 10 ("the worst pain you can imagine") simply because she now has more vivid acquaintance with severe pain. But let's assume, for the sake of argument, that they're interpreting the scale in similar terms, and that Brynn judges her pain to be *worse* than Ama does. What does this mean for comparative judgments about their health?

The inherent subjectivity—and social complexity—of pain may simply create incommensurability here. Brynn judges her pain to be worse than Ama does, and her pain is more distressing to her. It doesn't immediately follow, though, that Brynn is in "more" pain, or that she is in fact suffering more, or that her health is more compromised. Because of the protective nature of pain, our subjective reactions are part of the pain experience, and we can't always disentangle the severity of the pain from the severity of the effect it has on a person, even though these aren't the same thing.

We've all seen cases where someone is, quite obviously, *over-reacting* to pain. Chris and Dev both have sprained ankles, but Chris is acting like he's dying and Dev just shrugs and says "my ankle is a little sore." Our impulse is to say that Chris needs to calm down—to separate his emotional reaction from the (quite mundane) pain sensation, and deal with that pain sensation a little better. But there are also cases where the reaction and the pain are far harder to tease apart.

Chris and Dev both have an ordinary headache, but Dev's hits him early on a relaxing Saturday morning and Chris' hits him at the end of a very frustrating meeting during a long workday. Dev shrugs off the headache and takes some ibuprofen, while Chris furiously rubs his temples, says he thinks his head might explode, and briefly considers bashing it against his desk. Is Chris just reacting poorly to an ordinary pain? Or does the stress and exhaustion—which probably brought the headache on, and are so closely tied with the pain experience itself— actually make the pain more intense? There might, I suggest, simply be no fact of the matter, given the nature of pain.

So let's return to Ama and Brynn. The objective biological markers of their health are the same, but Brynn judges her pain to be worse. Is Brynn in worse health than Ama? It may seem like the obvious answer is yes—the objective features of the disease are the same, and so if the subjective dimensions of the disease are worse for Brynn, then Brynn is in worse health. And, depending on how we fill out the details, sometimes this does seem like the right answer. Suppose Ama is a well-off college student who has few responsibilities and very little stress in her life, whereas Brynn is single mother working overtime to make ends meet. *Of course* the same amount of corrosive joint damage will end up causing someone in Brynn's situation more pain, and of course that should factor into how we comparatively assess her health.

But suppose, instead, that Brynn is a wealthy woman from a highly privileged background whose experience of illness is the first true hardship she's ever endured in her life, whereas Ama is an economically disadvantaged immigrant who has worked demanding jobs since she was young to pay her way through school—something that's required a great deal of emotional and physical perseverance. Again, it wouldn't be surprising if Brynn experiences her knee pain as worse than Ama does, given those circumstances. But it no longer seems as straightforward to say that Brynn is actually in worse health than Ama.

Just as a person's life experience—both health-related and more broadly—can profoundly influence the impact of health on wellbeing, they can also profoundly impact the experience of pain. How *badly* something hurts is partly a function of how distressing a pain experience is. Again, that's just how pain works. But that distress can be influenced by all sorts of factors, including overall stress load, expectations, and previous experiences. And the net effect is that we can't straightforwardly say that worse pain *automatically* means worse health. And yet, quite obviously, pain matters to health.

4. Subjectivity, Equality, and Capability

The complex interplay between the subjective and objective aspects of health becomes especially vivid in the relationship between health and *social capability*. Quite obviously, reductions in our health can limit how we function in our social environment, and these limitations are often clear and objective. If you are in a hospital receiving chemotherapy for cancer, you cannot work a typical 9-to-5 job. If you have advanced congestive heart failure, you cannot physically do many ordinary things that most people take for granted, like walking up a flight of stairs or walking around the block. If you have a severe peanut allergy, you can't eat peanuts without going into anaphylaxis. And so on.

And sometimes, the *subjective* experience of illness can place *objective* limitations on capability or functioning. Severe enough pain will cause you to lose consciousness, for example. Likewise, no amount of grit or positive mental attitude will allow someone with Addison's disease to "push through" the extreme fatigue of an illness flare. If they try, they'll simply collapse, because their body isn't producing enough cortisol to maintain activity.

As we saw in Chapter 1 (section 3), some philosophers place a great deal of emphasis on this link between health and capability. Sridhar Venkatapuram (2011), for example, argues that an ability to access other social capabilities and functionings is in fact *what health is*. And just as with wellbeing (see Chapter 2), other philosophers have argued that its relationship to capability is the source of health's normative significance. For example, while he doesn't endorse the constitutive claim, Norman Daniels (1985) argues that the *distinctive value* of health is the role it plays in opportunity and ability. If two people are in the same socioeconomic circumstances and apply the same level of effort, according to Daniels, they should be capable of the same range of opportunities *provided they have an equal level of health.* And so, to ensure equality of opportunity, we must ensure equality of health.

Such views focus on the way in which reductions in health affect what a person is capable of doing, and likewise emphasize that equality in health is part of social equality more generally. And these basic ideas are, no doubt, correct. Widespread health inequalities are a major form of social inequality, and reductions in health can radically limit what a person can be or do in society. And yet spelling out the relationship between health and capability is complicated—perhaps more complicated than many theorists acknowledge—precisely because of the subjectivity that pervades so many aspects of our health.

It's true that something like pain, for example, can (objectively) limit capability. But the extent to which pain itself is limiting—and the correlation between severity of pain and reduced capability—is complicated by multiple subjective factors, including interpretation of pain's context and individual pain tolerance. And that's true even for the *objective* limitations that pain can place on functioning, such as losing consciousness. One person might faint from the pain of a tattoo and another remain fully consciousness through amputation without sedation, simply because of differing tolerance for, experience with, or interpretation of pain. We can't immediately say that the pain of the tattoo-fainter was more severe than the pain of the conscious-amputee, even though obviously the former is more functionally limited by their pain response.

Likewise, fatigue from the adrenal insufficiency of Addison's disease is objectively limiting, and can't simply be "pushed through." And yet John F. Kennedy was able to serve in the military, the senate, and presidency, all while dealing with Addison's disease. His capability in the face of illness was extraordinary. It also shouldn't be seen as setting a benchmark for the kinds of social functioning a person with Addison's disease not well controlled by medication (and JFK's regular health crises indicate his was not) can achieve.[10]

In explaining the link between health and capability, we are caught between the Scylla and Charybdis of over-reliance on objective measures of health and over-reliance on subjective measures of health. Just as with the connection between health and wellbeing, we cannot simply say that the people whose capability is in fact most impaired are those who perceive it to be. The biological reality of health makes it the case that you can sometimes perceive yourself to be limited by your health in a way that is mistaken—an over-reaction, a hyper-caution, an

---

[10] Daniels emphasizes that equality of opportunity (and equal share in the range of available opportunities) doesn't in any way imply equality in achievement—different people will have different levels of talent, drive, and interest, and so will do different things with the same range of opportunities. So perhaps JFK was just especially talented, and making the most of a limited set of opportunities. As we'll discuss, though, when it comes to health, this distinction becomes difficult to maintain, and it's hard to establish the extent to which two people both affected by fatigue are equally limited in the "range of opportunities" available to them.

overconfidence. But that doesn't mean that we can determine who is most limited simply by looking at objective measures of health. Two people can have the same test results or the same disease process, and yet have very different subjective experiences of illness, in a way that directly affects their ability to function in their social environment. And navigating between these two over-corrections is essential for an adequate understanding of the relationship between health and capability.

4.1 Equality and Subjectivity

Let's examine, for example, Daniels' claim that equality in health will create equality of social opportunity in cases where other social factors (income, age, race, gender, etc.) are held fixed. On a "level playing field," the thought goes, we should all have access to an equal range of opportunities (an equal range of options for forming "life plans").[11] Of course, the playing field is often not level—factors like race, gender, and class all present barriers to the range of opportunities we can reasonably access. But if we compare people of similar social demographics—or if we consider a better society in which those demographics didn't limit us—we can see that reductions in health will limit the range of opportunities that a person has access to. The person in poor health has less choice and less freedom in what she is able to do and how she is able to function in her social environment. So the special moral importance of health, for Daniels, is that it is required for equality of opportunity. Two people who are in roughly equal health and who occupy a roughly similar position in society will have access to the same range of social opportunities. That doesn't, of course, mean that they'll be equal in what they can do with those opportunities—different people have different levels of skill, drive, and interest. But they'll be equal in the range of "life plans" that they can reasonably pursue.[12]

Crucially, for Daniels, the relationship between health and opportunity is objective.[13] People want lots of things for their bodies—to be thinner, better looking, more athletic—and getting some of these things might genuinely increase

---

[11] Daniels (1985), pp. 27–8. Daniels doesn't explicitly state that equal *loss* of health leads to equal loss of opportunity (at least within a given social context, holding other factors fixed), but his arguments seem to rely on something like this principle.

[12] Importantly for Daniels, an equal loss of health might in fact impair the opportunities of one person over another—if two people are affected with arthritis in their hands but one is a cyclist and the other is a pianist, it will in fact have more of an impact on the pianist. Contra the contribution view, however, Daniels maintains that this difference in the impact on quality of life, function, and preference shouldn't affect how we view the value of what is lost—at least if we're considering what we owe to the individuals in question. And that's because, for Daniels, each has lost an equal share of the *range of opportunities* available to them. See Daniels (1985), pp. 34–5.

[13] This is why it becomes crucial to Daniels that he defends a normal function theory of health—his arguments rely on an objective, non-evaluative determination of health status.

some people's wellbeing. But health, for Daniels, is of special value and special moral importance, in a way that athleticism or beauty is not, because of the objective ways in which it is linked to capability and functioning. Beauty or athleticism might *in fact* make people happier or give them more social opportunity, given the contingent ways in which our society is arranged. But for Daniels there is a stronger and more objective link between health and capability. If you are in poor health, you are simply *incapable* of accessing the normal range of life plans, regardless of how society is arranged or what the norms are.[14]

Likewise, Daniels needs to be able to say that how much a health condition (objectively) limits functioning determines how bad that health condition is. Lots of things might affect our attitudes toward a particular health condition—stigma, social desirability, and so on. But in conditions of limited resources, we need to prioritize those whose health most impairs the range of opportunities available to them, and try to improve their health in order to improve that range of opportunity. And to say all this, Daniels needs to be able to say that people who are in objectively similar levels of health will have objectively similar levels of opportunity (other factors being equal).[15]

This is a complicated claim before we even consider the subjectivity of health. A well-rehearsed problem for discussing equality of health is that comparative judgments of health are difficult at best. As Daniel Hausman (2015) compellingly argues, health doesn't come in easily quantified amounts that can then be compared between persons. Who is in worse health: the person with multiple sclerosis relatively well controlled by medication or the person debilitated by hip arthritis who needs a total hip replacement? Emphasizing the link between health and social capability, however, gives us a framework from which we can attempt to grapple with such comparisons. The worse health condition is the one that more substantially restricts a full range of opportunities. There might not always be a clear fact of the matter—there might not always be *any* fact of the matter—in pairwise comparisons, but we at least know what we're trying to compare.

The problem of equality in health becomes more pervasive, however, when we consider the inherent subjectivity of health. It's not just that we often lack

---

[14] Daniels grants that "[f]or many of us, some of our goals, perhaps even those we feel are most important to us, are not necessarily undermined by failing health or disability. Moreover, we can often adjust our goals, and presumably our levels of satisfaction, to fit better with our dysfunction or disability. Coping in this way does not necessarily diminish happiness or satisfaction in life to a level below that achievable with normal functioning." This is part of why he wants to emphasize "this basic fact: impairments of normal species functioning reduce the range of opportunity open to the individual in which he may construct his 'plan of life'" (Daniels (1985), p. 27).

[15] Daniels maintains, however, that the seriousness of a reduction in normal functioning will partly be determined by contingent social factors—nearsightedness might be a much bigger restriction on opportunity in a society that requires lots of reading of small print than in a society that requires mostly manual labor, for example. So social norms and arrangement constrain the range of normal life plans, and then the extent to which functional limitation restrictions access to those life plans determines the relative seriousness of a health condition.

meaningful comparisons between different health conditions, it's that we often lack meaningful comparisons between *the very same health condition*. Many of the most common, and most significant, health problems in modern industrialized societies include a significant degree of subjectivity. The most common causes of application for disability benefits and protections (and for work leave) in America, for example, include non-specific low back pain, depression, migraine headache, and musculoskeletal pain conditions (arthritis, fibromyalgia, etc.).[16] For some of these conditions, there are objective biological markers (joint degeneration, etc.), but these objective findings are poorly correlated with pain severity or with the daily limitations people experience.[17] For other conditions, especially mental health conditions such as depression and anxiety, there are (at least currently) no objective markers.

Suppose our old friends Chris and Dev both have knee arthritis. All the objective measures of disease suggest that Dev's knee is in worse biological shape than Chris'—Dev has more erosion of the joint space, more restriction in range of motion, larger effusion.[18] But Chris reports worse pain, and his daily activities are substantially more limited than Dev's. Whose opportunities are more limited? Who is in greater need of healthcare resources such as ongoing physical therapy or a knee replacement procedure? The objective markers of disease matter here—we can't just assume that the person worse off is the person who shouts the loudest.[19] But the subjective components matter too. Pain is limiting. Pain is debilitating. Pain is also, of course, inherently subjective. And the extent to which a person is limited or debilitated by something like pain is itself a complex and often deeply subjective matter.

The difficulty this type of subjectivity creates in directly correlating health to capability is only heightened when we consider health conditions that have no objective disease markers at all—conditions like depression or generalized anxiety.[20] They aren't any less real or less significant simply because we can't "see" them on an MRI

---

[16] Social Security Administration (2021).

[17] In arthritis, the generation of pain seems to be multifactorial, and may include complex social factors (see Dieppe and Lohmander (2005)). For non-specific low back pain, findings on imaging are extremely poorly correlated with levels of pain and functional limitation (see Chou et al. (2009)).

[18] The knee is a complicated joint, and knee pain is poorly understood. But current research suggests a weaker than expected correlation between objective markers of joint disease and reported pain severity and functional limitation, especially the longer the condition persists. See Sharma (2021) for an overview.

[19] Especially because, for example, research suggests that people with more educational attainment tend to report a larger impact on their self-rated health than those with less education when dealing with the same health problem, perhaps because the former have higher health expectations (see Delpierre et al. (2009)). Likewise, individuals from more disadvantaged socioeconomic backgrounds seem to be more likely to normalize experiences like chest pain, leading to overall worse health outcomes (Richards et al. (2002)). We'll return to this issue in section 6.

[20] For a fascinating overview of the largely unsuccessful search of objective measures or biological markets for our most common mental health conditions, such as depression and generalized anxiety, see Ann Harrington's (2019) *Mind Fixers*.

or CT scan. And yet it's difficult (to say the least) to give interpersonal comparisons of limitation on capability for subjective mental states like anxiety, or to meaningfully compare those limitations to the kind created by things like paralysis or ataxia. Daniels is surely right that we should care about health at least in part because of how it affects opportunity—because of the way that limitations on health constrain what people can be and do. But it's a much stronger claim to say that, if other things are held fixed, health affects opportunity equally or constraints capability in the same ways.

4.2 The Problem of the High Achiever

In her book *An Unquiet Mind*, distinguished psychologist Kay Redfield Jamison discusses her own experience of bipolar disorder. Although profoundly affected by the disease, Jamison has had an incredibly successful career, and she chronicles how bouts of mania helped her to write prolifically (and wonders whether she would've gotten tenure without them). As Jamison herself documents, this link between mania and creativity is not isolated to her own experience—many highly successful, creative people are affected by bipolar disorder, and experience a connection between their creativity and their mania.

But many people don't. For some, mania can involve epic bouts of writing or painting. For others, it can involve out-of-control spirals of substance abuse, or career-ending lapses in judgment. For some, it can involve combinations of both, such that the substance abuse or lapses in judgment prevent any creative output from coming to fruition. You could take two people from Jamison's socioeconomic background, both with bipolar disorder, and one of them could end up a successful tenured professor and the other could end up destitute, in jail, or dead. And it's not because the tenured professor tried harder. It's not because she expended more effort. Nor is it simply because the tenured professor has more "talent" in Daniels' sense—plenty of brilliant, talented people end up completely debilitated by bipolar disorder. Rather, the subjective experience of something like mania is highly variable, and can affect people in profoundly different ways.

It is easy to point to highly successful people like Jamison—or Steven Hawking, or John Nash, or Franklin D. Roosevelt—who manage to maintain an extraordinary level of social functioning and accomplishment despite severe health conditions. It is also easy to use such examples to blame people who are limited by illness for their own limitations, especially if we assume that equal health means equal capability or equal opportunity, provided socioeconomic factors are held fixed. If Nash can deal with schizophrenia and still win a Nobel prize, or if FDR can live with what was likely Guillain-Barré syndrome and still become president, then surely others in similar circumstances are complaining too much if they say that they can't work full time or can't participate in a "normal" range of daily activities. (Or perhaps we're simply left to conclude they just aren't *talented* enough.)

Emphasizing these stories of great success combined with serious illness is sometimes called the *supercripple* trope. We put the focus on the success of a particular individual and then use that success to say that others with similar challenges should be capable of just as much. Or at least insofar as they are not, it is because of a lack of drive or talent or interest rather than anything to do with justice. Blind people have climbed Mount Everest! The only true disability is a bad attitude!

Obviously, those who argue for a direct correlation between health and capability or opportunity aren't claiming that anyone with debilitating neurological disease should be capable of becoming president, or that being president is a reasonable expectation for them. But what they do maintain is that roughly equal health states translate—other social factors being equal—to roughly equal capability or *social opportunity*. This leads to a trilemma: in order to account for their remarkable social capability, we need to say that high achievers like Jamison are either less ill, more talented, or harder working than those whose similar experience of illness is a greater impediment. But none of these answers seems right.

Compare someone like Jamison to another middle-class American white woman whose bipolar disorder leaves her unable to work. We can't simply say that Jamison's high level of functioning means her condition is less severe. As Jamison documents, her illness has involved multiple hospitalizations, ongoing complex treatment, and a tremendous amount of suffering. But nor can we say that the person who cannot work is simply not trying as hard or is not as talented. The other woman might be just as talented and just as motivated, but experience mania in a way that leads to crushing self-destruction or delusion, rather than bouts of writing that might ultimately aid a bid for tenure. Because of both the subjectivity in how something like mania is experienced and the subjectivity in how illness combines with other features of a person's life—their personality, their hopes and desires—two people who experience similar illness in similar circumstances can nevertheless have very different levels of social capability.

4.3 Pain Tolerance and Coping

To further illustrate the complex interplay between subjectivity and objectivity in how we evaluate health-related capability, let's return to the case of pain. Pain, quite obviously, has an effect on capability—on what you can be and do and how you can function socially. But the relationship between pain and capability is mediated by how individuals *cope* with pain. Different people have different levels of tolerance to pain and differing ways of dealing with it. Moreover, pain coping does not merely have a subjective influence on the perceived "badness" of pan; individuals who cope especially well often won't have the same degree of cascading biological responses (increased heart rate, increased levels of stress hormones) that can prolong and exacerbate a pain response.

Pain coping is not an objective biological process—it's a complex biopsychosocial process that's influenced by many things, including a person's previous health experiences and cultural background. Sickle cell disease, for example, is an extraordinarily painful condition that often first presents in childhood. And perhaps unsurprisingly given their life experiences, adults with sickle cell disease often exhibit very high levels of pain tolerance and cope very well with bouts of extreme pain.[21] An adult in the midst of a sickle cell crisis might easily be less distressed and less impacted by their pain than someone with a bout of low back pain—especially if the low back pain is a first major pain experience and the sickle cell crisis is one in a long series of pain events present since childhood.

And just as previous health experiences can influence pain coping, so can various social and cultural factors. As mentioned in Chapter 2, some research suggests that Asian Americans, for example, often complain comparatively little about pain and present very stoic responses, even to severe cancer pain. Beliefs about and reactions to pain are part of a complex cultural network, and such beliefs and expectations modulate both pain tolerance and the subjective experience of pain itself.[22]

Individuals who cope very stoically with pain—whether because of previous pain experience, cultural background, or individual quirks of personality—will have greater social capability in the face of pain. But if the special value of health is its effect on capability, we're then forced to say that pain matters less when it affects individuals with high capacity for coping. And yet, just as with wellbeing, this seems like the wrong result. The pain of someone with sickle cell disease is not less important than the pain of someone with an aching back simply because a lifetime of illness has conditioned them to deal with it. Likewise, if effect on capability is how we prioritize healthcare and healthcare resources, we'd end up saying that Asian Americans are less in need of palliative treatment for pain, simply because they tend to cope stoically. But that's absurd. We shouldn't deprioritize someone's needs for pain medication, for treatment, and for care more generally simply because they are stoic in the face of suffering.

---

[21] Fuggle et al. (1996), for example, found that children with sickle cell disease were adept at recognizing and dealing with sickle cell pain, and furthermore did not typically adopt a "sick role" for normal childhood illnesses and injuries (saving that instead for their sickle cell events, which are of course much more intense). And in a study of UK adults with sickle cell, Anie et al. (2002) found high levels of so-called "active coping" with severe, recurrent pain.

[22] For discussion, see, e.g., Im (2007) and Kwok and Bhuvanakrishna (2014).

Subjective reactions to pain—including how we cope with and tolerate pain—aren't the only things that matter in assessing the significance or harm of an experience of pain. There is an important sense in which the significance of pain can be rooted in objective biological reality. The pain of sickle cell disease or cancer matters for reasons that go beyond the subjective distress or limits to capability that it causes.

Likewise, a person's *subjective* reactions to pain can be *objectively* unwarranted. If you say that you cannot work because you stubbed your toe, you are *overreacting* to pain. That isn't to say you're lying or intentionally exaggerating. It may genuinely seem to you that the pain of a stubbed toe is too much to bear. But, barring the presence of some sort of neurological hypersensitivity, it makes sense to say that your subjective reaction doesn't match the objective biological state of your body—claiming that you are incapacitated because of the pain of a stubbed toe is an unwarranted subjective reaction to that level of injury, the subjectivity of health notwithstanding.

And yet we can't simply defer to objective measures of health when assessing the severity of a particular health condition or prioritizing particular health needs. Again, so many of the most widespread health conditions in industrialized societies—anxiety and depression, musculoskeletal pain, functional gastrointestinal disorders, headache disorders—either have objective markers that correlate very poorly with experience of illness or have no objective markers at all. And perhaps more importantly, subjective factors are an integral part of the experience of illness itself. How stressed[23] you are is not just a precipitating factor for the flare of a headache or gastrointestinal disorder; it is also, given what we know about pain and about the gut–brain connection, part of that experience of ill health itself. Pain *hurts* more when you're exhausted or stressed. Hypersensitive reactions in your gastrointestinal system are magnified when you're anxious or on edge. And so on.

If you are struggling to cope with back pain because it's hit you at a time when you're already overburdened with childcare, with the social and financial stress of a low-paying job, and with uncertainty in your living situation, then all of that will be part of the illness experience. You will be *less able* to cope with pain, and the pain itself will probably *feel worse*. We cannot simply look at your MRI to determine how bad your experience of pain is. Your experience of pain itself—as well as your ability to cope with pain—is embedded in a complex network of social and psychological factors, some of which are deeply subjective.

---

[23] It's important to disambiguate the complex (and vaguely defined) web of emotions and sensations we call "stress" (or "stress response") from the objective socioeconomic factors sometimes called "stressors" (income or housing insecurity, trauma and abuse, etc.). Both have well-documented impacts on health, but they also—perhaps unsurprisingly—show an imperfect correlation. See, e.g., Christensen, et al. (2019).

5. The Social Components of Health

To summarize the complex interplay between subjectivity and objectivity in understanding health, let's consider a particular example: the health disparities experienced by African American women. Black women in the United States face significant health inequalities relative to other groups. And the way these inequalities play out provides a compelling illustration of the subjective and objective aspects of health—both the ways in which they are interrelated and the ways in which they are each important.

We can't grasp the extent of the health disparities between Black and white women—or the extent of the harm that such disparities cause—without evaluating their objective dimensions. It's not simply that Black women are less likely to feel good. It's not simply that they experience more health-related distress or can accomplish fewer of their valued activities of daily living. All of that might well be true, but it understates the extent of the problem. Black women die earlier than their white peers.[24] They have higher rates of serious illnesses like diabetes, heart disease, autoimmune disorders, and even some cancers.[25] They have a shockingly high maternal mortality rate compared to white women—a difference that doesn't seem to be fully explained by a difference in socioeconomic status.[26] We can't fully appreciate or quantify these harms without talking about objective facts about the body—including states of the body that are more or less likely to cause tissue damage, to cause inflammation, to cause death.

The health disparities faced by Black women also present us with a compelling case in which cultural norms can shape how we experience and interpret the subjective dimensions of health. A growing body of research, for example, suggests that black women's assessment of their own need for care and treatment—most especially mental health treatment—is substantially influenced by the degree to which they endorse the "Strong Black Woman" stereotype.[27] There is substantial cultural pressure on Black women to be strong in the face of adversity and to prioritize the care and needs of others over their own suffering. Evidence suggests that endorsing these norms can lead Black women to, for example, minimize the extent of their own psychological distress and avoid seeking mental healthcare (which, in turn, can increase the rate at which they experience the symptoms of depression and anxiety, even if they wouldn't self-describe as having either condition).[28] There is often a difference between the *perceived* significance of one's suffering or limitations and the actual extent to which one's health is impaired. And research suggests that Black women

---

[24] Fuchs (2016).   [25] For discussion, see Palmer et al. (2022).
[26] See, e.g., Fang et al. (2000) and Joseph et al. (2021).
[27] Hall et al. (2021) and Donovan and West (2015).   [28] Abrams et al. (2019).

may be uniquely positioned to subjectively discount the significance of their own distress.

And yet the solution is not simply to rely on objective measures when assessing the health disparities that black women face. To begin with, the intensity of feelings such as depression and anxiety—the very things that Black women might distinctively downplay—can't be objectively quantified, even if there are objective indicators of their presence. Moreover, evidence suggests that the more subjective elements of Black women's health are often overlooked and under-treated. Black women receive substantially less treatment for pain than other groups, for example (owing at least in part to the common but erroneous perception among healthcare workers that black people are more resistant to pain, combined with the wider phenomenon that women are less likely than men to be taken seriously when they report pain).[29] A significant piece of the health disparity that affects Black women comes from the fact that how they *feel*—their subjective experience of pain, of stress, of depression—is simply viewed differently.

But the impact of subjectivity in Black women's health inequalities is far more pervasive than the undertreatment of pain. We've been taught to pay attention to the social *determinants* of health—the ways in which social factors can play a causal role in the development of health disparities and health problems. We can't understand the prevalence of diabetes without understanding the impact of poverty and urban food deserts, for example, just as we can't understand the prevalence of eating disorders without understanding the impact of gender norms. And yet, I suggest, thinking carefully about the role of subjectivity in health can show us that the idea of social *determinants* of health is inadequate for a full explanation of the social influence on health. We also need to understand the social *components* of health.

Black women are more likely than almost any other demographic group to be victims of domestic violence.[30] They are more likely than almost any other group to be single parents.[31] They work similar hours to other women, but are paid less—often substantially less—than almost any other group, and are more likely than any group except Native American women to experience poverty.[32] What this means, collectively, is that black women experience extremely high rates of trauma and stress.

Trauma and stress, of course, can both precipitate and exacerbate illness. Nor is the ongoing stress of racism, of sexual violence, of unsafe living environments, and

[29] See, e.g., Badredlin et al. (2019), Hampton et al. (2015).
[30] For current statistics, see Connecticut Coalition against Domestic Violence (2022).
[31] See Livingston (2018).
[32] An overview of relevant statistics is available at Bleiweis et al. (2020). See also Wilson and Jones (2018).

so on merely a causal factor in the development of disease. Stress, quite simply, makes many illnesses worse—*objectively* worse. And there's some evidence that this effect is magnified for particularly vulnerable groups such as Black women. Studies have found, for example, a higher correlation between stress and the severity of heart disease in Black women than in white women.[33]

But beyond causation and exacerbation, the cumulative effects of ongoing stress and trauma are also, at times, part of *what it is* to experience illness. Lupus, for example, is an autoimmune disorder that is substantially more common in Black women than in most other groups. (Autoimmune diseases are in general more common in women, and many are more common in women of color especially.) Some of the most difficult and debilitating aspects of lupus are chronic pain and fatigue. And *what it is* to experience things like pain and fatigue can be deeply influenced by things like trauma, stress, and social stability.

If you are already pushing yourself to the max, feeling strung out, and emotionally depleted, that will be part of how you experience the physical sensations of fatigue and exhaustion. If you have experienced violence or trauma, if you feel unsafe in your living situation, or if you are under the constant weight of stigma, that will be part of how your body interprets and processes threat responses[34]— including the threat response of pain. It's not just that you will be less able to "cope" with pain in these circumstances. It's that a disease like lupus will, in all likelihood, quite literally *hurt more.*

The inherent subjectivity of our experiences of health—including objectively important health factors such as pain, fatigue, and stress—show that we need to acknowledge more than just the social *determinants* of health. Social factors, including the complex intersection of race, class, and gender, partly *constitute* health. It's not merely that the lived experiences of racism and sexual violence predispose black women to illness, but also that they can be part of the experience of illness itself, and part of the severity of illness.

6. In Practice: Measuring Health

Perhaps nowhere is the vexed relationship between subjective and objective aspects of health more apparent than in attempts to measure medical outcomes. The goal, ostensibly, of medical treatment is to make us healthier—by eliminating pathology, minimizing the harmful effects of pathology, preventing pathology,

---

[33] Lewis et al. (2009).

[34] Analysis of the Metro Atlanta Heart Disease Study, for example, found that African Americans who reported experiencing race-based discrimination in their workplace were substantially more likely to experience hypertension. See Din-Dzietham et al. (2004).

and so on. Yet history yields plenty of examples of medical treatments that were actively harmful or that did nothing at all to actually address the pathology at which they were targeted. To justify a medical intervention—especially an invasive, time-consuming, or costly one—we need some way of assessing whether and how well it works. And in scenarios in which there are multiple different treatment options, we need to be able to assess which works *better* (in which particular circumstances, for which people). To do any of this, we need some way of measuring health outcomes.

Health outcome measures are an especially salient issue, moreover, in contemporary debates about how best to deliver effective healthcare. An unfortunate but unavoidable fact of the current healthcare landscape is scarcity of resources. Providers, researchers, and patients alike are often thus forced to make choices about the value of different healthcare interventions. Which conditions and treatments do we prioritize in medical research? Which procedures are considered effective enough to be funded? How effective does a treatment need to be to justify the resources used to develop, produce, and deliver it?

I am not, to be clear, attempting to tackle any of these thorny issues of resource allocation in this book. I'm not delving into how we should make judgments about prioritizing healthcare, but rather looking at the upstream epistemic issues of how, if at all, we can make comparative judgments about which health outcomes are better, which conditions are more severe, and so on.

There has been a push, in recent years, toward a model called "value-based healthcare."[35] In value-based healthcare, funding bodies (e.g., private insurance companies, Medicare, the NHS, etc.) compensate healthcare providers (e.g., hospitals, physicians) for treatment not on a fee-for-service basis but rather on a "value-for-service" basis.[36] The gist of the idea is that funding bodies want to compensate good health outcomes, not just pay for treatment, especially since there's a growing concern about the harms of *over*treatment.

With this trend, the need to quantify and compare specific health outcomes has taken center stage. And this is where things get especially tricky. In this section I'm going to briefly discuss the pitfalls of over-reliance on either objective (section 6.1) or subjective (section 6.2) measures of health before looking at three example cases—pain as "the fifth vital sign" (section 6.3), the impact of so-called "functional disorders" (section 6.4), and the epidemiological significance of self-rated health (section 6.5)—which highlight the complex push and pull between objective and subjective factors in how we assess health on an everyday basis. Again, in doing this, I'm not trying to make policy recommendations—I'm interested here in the epistemic issues that arise when we're trying to measure and comparatively assess health status.

---

[35] NEJM Catalyst (2017).   [36] See Hurst et al. (2019).

6.1 Pitfalls of Objective Outcome Measures

Historically, health outcome measures have tended to focus on variables that are—at least primarily—epistemically objective. Key variables for these types of assessments are indicators of "mortality and morbidity." Mortality is perhaps the clearest and most straightforward objective outcome measure. Morbidity is somewhat harder to assess, but researchers have tended to focus on indicators such as: hospital readmission rates, rate of adverse treatment-related events, and indicators of disease activity. So, for example, if we want to know whether insulin therapy is an effective treatment for type 2 diabetes, we might look at the effect of insulin therapy on glycemic status and glycemic control, on rates of diabetes-associated complications such as peripheral neuropathy, and on rates of hospital admission.

The problems with exclusive reliance on such measures, however, are multiple—and in some cases surprising. To begin with the most obvious, many common health conditions—especially mental health conditions but also many chronic pain conditions—lack objective indicators for disease activity. And so the concern is not only that objective measures are ill equipped to evaluate such conditions but also that prioritizing the information we get from objective measures might ultimately deprioritize the seriousness of such conditions.

Perhaps the more striking shortcoming of objective outcome measures, however, comes from the surprisingly weak correlation such measures often bear to how people are able to function, feel, and live. Again, take the example of insulin therapy for type 2 diabetes. Research suggests that it is an effective way of achieving glycemic control, especially in refractory cases where other treatments have failed. However, a substantial body of research also suggests that insulin therapy often fails to make people *feel better*—and can often make them feel worse.[37] People with type 2 diabetes treated with self-administered insulin often report that their experience of their condition is no better, or even is worse, after insulin therapy—and that the therapy itself is frustrating and onerous to them.

Similarly, objective outcome measures for inflammatory conditions such as rheumatoid arthritis look at markers of disease activity—for example, indicators of inflammation in blood work, evidence of increased joint erosion on scans. Therapies then target these markers and would be considered effective insofar as they reduce them. And yet there is an imperfect correlation (and some studies suggest quite a poor correlation) between these objective markers of disease activity and people's reports of pain, functional limitation, and quality of life.[38] By far the more significant predictors of people's self-assessment of their functional ability and quality of life are pain and fatigue—both highly subjective

---

[37] See Davis et al. (2001) and "Quality of life" (1999).
[38] See Madsen et al. (2016) and Sarzi-Puttini et al. (2002).

variables. And, again perhaps surprisingly, reducing disease activity often fails to reduce pain in the ways that might be expected.[39]

In a similar vein, objective outcome measures for cancer treatment have been notorious for examining only mortality rates and "adverse event" rates without assessing the crucial question of how they make people *feel*.[40] Treatment for cancer can be brutal, debilitating, and isolating. Objective measures have often failed to incorporate the experiential aspects of a treatment—how tiring it is, how much time it takes up—into the overall comparative assessment. And in some cases, it's left people with cancer wondering if the treatment is worse than the disease itself.[41]

6.2 Pitfalls of Subjective Outcome Measures

In response to these concerns, there has been a growing push toward "patient-reported outcome measures" (PROMs). PROMs can take the form of generic questionnaires that seek to assess "health-related quality of life" by asking about ability to engage in activities of daily living, pain, mood, and so on.[42] Or they can be condition-specific. PROMs for rheumatologic conditions will often look at things like pain intensity, sleep, and fatigue, whereas PROMs specifically targeted at back pain will ask about perceived ability to lift or sit for more than 10 minutes.

PROMs have some clear advantages over more traditional objective outcome measures insofar as they prioritize the experience of the person who has the condition and focus on what matters to people most: how they feel and what they can do. And yet these types of subjective measures—precisely because of their subjectivity—have pitfalls of their own.

When we're asking about subjective variables like pain and fatigue, or about a person's overall sense of their health and functional capacity, we're invariably asking about how they perceive their own health condition. PROMs have been shown to be influenced by a wide range of factors, both at the level of individual beliefs and attitudes and at the level of cultural and ethnic background.[43] It is to be expected, of course, that different individuals and different groups interpret questions about health, satisfaction, and function in somewhat different ways. What's perhaps more striking is that particular individuals will often interpret the same group of questions differently across time. Again, a major factor here is the phenomenon of *response shift* (see Chapter 2 (section 2.3)). A relatively robust

---

[39] Vergne-Salle et al. (2020) and Zatarain and Strand (2006).   [40] Graupner et al. (2021).
[41] Siddhartha Muhkerjee's (2010) *Emperor of All Maladies* gives a humane and thorough account of much of the relevant history specific to cancer and cancer treatment.
[42] Examples of this kind of measure include the widely used SF-36 and EQ-5D.
[43] See especially Chang et al. (2019) for a major summary and overview of the extant literature on these problems.

finding is that individuals who experience a serious long-term illness will often change both the parameters by which they assess their current health state and their recall of their previous health states.[44] Thus, when they report a change in perceived health status on outcome measures, part of what is in fact being tracked is a change in their health expectations and the standards by which they assess their health. Similarly, people who have recovered from major illnesses like cancer will often rate their health as *better* than they did before getting sick, even if they have major long-term consequences of the illness.[45] As a result of phenomena like this, it can be hard to tell when PROMs are, for example, tracking an improvement in health versus a change in standards. Similarly, it's often unclear whether they're tracking a substantial compromise in health or just a disappointment of very high health expectations.

Another worry for the use of PROMs as a guide to "value-based" care and effective health outcomes is the striking correlation between the amount of medical intervention a person receives and their assessment of their own health outcomes. A growing body of research suggests that people often *perceive* their health to be getting better when they are getting a substantial amount of treatment—including treatment that's been shown to be medically unnecessary, such as advanced imaging for non-specific low back pain.[46] But more treatment is not only costly; it can also be actively harmful. (Medications have side effects, medical procedures can be invasive and damaging, repeatedly taking on the role of the patient can itself be limiting, etc.) Recent research has suggested that people who are most satisfied with their medical treatment outcomes tend to—even when controlling for demographic factors, health status, disease burden, insurance status, and sources of care—take more prescription medication and spend more time in the hospital.[47] And they also, even more alarmingly, have significantly increased mortality.[48]

Findings like this aren't, on reflection, all that surprising. Healthcare is a resource, and in some contexts it's an expensive or status-associated resource. We're conditioned to think that more is better. More stuff, more goods, more attention—we crave it and often perceive things as going better when we get it. But when it comes to healthcare, *more is not always better*. Indeed, over-testing and overtreatment are central problems of modern healthcare. And yet it wouldn't be surprising if we might easily *perceive* a correlation between more care (or expensive care, or fancy care, etc.) and better outcomes, even if that perception isn't ultimately correlated with feeling better and living longer.

With the potential drawbacks of both types of measure in mind, let's now turn to three cases that illustrate the way such tensions can sometimes play out—the

---

[44] Ilie et al. (2019).   [45] Ilie et al. (2019).   [46] Fenton et al. (2012).
[47] Fenton et al. (2012).   [48] Fenton et al. (2012).

role of pain measurement, the assessment of so-called "functional disorders," and the role of self-rated health.

6.3 The Fifth Vital Sign

Perhaps nowhere has the push and pull between objective and subjective health measures been more apparent than in the debate over making pain "the fifth vital sign." Traditionally, in any medical encounter you will have your vital signs measured: heart rate, blood pressure, body temperature, respiratory rate. (Many clinics will also measure blood oxygen levels.) Disturbance in any of these metrics is a key warning sign of an imminent health problem, and monitoring these metrics is central to how healthcare workers assess a person's biomedical stability. The measuring of these variables, moreover, is epistemically objective. Subjective factors can, of course, influence them—stress can increase your blood pressure and excitement can make your pulse race. But if your heart rate is 80 bpm, that is an objectively measurable fact about your body, and it is accessible to all. Likewise, if two people have heart rates of 80 bpm, we can meaningfully say that they have the same heart rate, and if treatment reduces your heart rate from 200 bpm to 80 bpm, we can meaningfully say that it objectively reduced your heart rate, and thus improved your tachycardia.

Reliance on these metrics, however, tells us only part of the story about a person's health. Most especially, it doesn't tell us anything about how much pain or distress they are in. And this led many advocacy groups to point out that, in medical settings, pain (especially cancer pain) was often neglected or undertreated. So much focus was put on maintaining biological function and homeostasis, the argument went, that we forgot to address whether people were suffering.[49]

And so there was a push to introduce measures of pain as "the fifth vital sign."[50] It is now quite likely that if you visit a doctor, you'll be asked—while your vital signs are being taken—to rate your pain on a scale of 1–10. And your answer will be recorded, alongside your pulse and blood pressure, as a key indicator of your health. This shift has led to more attention on pain management, including supportive and palliative care. But it's far from obvious that this shift has been an effective one or led to overall better pain management.[51]

The problem is that pain is *not* a vital sign in the way that blood pressure and heart rate are. Different people interpret the 1–10 scale differently, and the same individual will interpret the scale differently across time. Studies suggest, for example, that major predictors of self-assessed pain outcomes and pain severity

---

[49] For a brief overview of the history, see Baker (2017).   [50] Walid et al. (2008).
[51] Scher et al. (2018).

are a person's beliefs about pain and pain coping.[52] Moreover, in many cases the major predictors of self-assessed pain severity aren't obviously things that are well addressed just by giving people pain medication. Studies have found, for example, that some of the biggest predictors of self-rated pain severity are the emotional variables associated with pain—the depression, anxiety, and fear that can swirl around experiences of illness—and that pain measurements are often more reliably tracking a person's emotional state than anything else.[53]

A substantial body of evidence suggests that moving to a "fifth vital sign" model has not improved pain treatment or health outcomes.[54] A more worrying trail of evidence suggests that it might actually have done substantial harm. In his groundbreaking book *Dreamland*, investigative journalist Sam Quinones makes a compelling argument that the "fifth vital sign" approach to pain assessment can be directly linked to the overprescription of opioids. Clinicians were told that achieving a reduction in the number given on the 1–10 pain scale was an essential marker of a good health outcome, but were told this in an environment in which they had increasingly less time to spend with patients and increasingly fewer options for multi-disciplinary pain care. And they were also told that opioids were especially effective at treating all kinds of persistent pain. It is not surprising, in such a context, that a massive increase in opioid prescribing appears to be quite closely correlated with an increased emphasis on self-assessed pain measures as a key health outcome.

And opioids themselves are another place where we see the complexity of relying on more subjective measures of health outcomes. Opioids, despite their risks, are important drugs that play a key role in the management of acute pain, post-surgical pain, cancer pain, and end-of-life palliative care. The last twenty-five years, however, have seen an astronomical increase in their use for the treatment of non-specific and chronic pain conditions. People with these conditions have advocated tirelessly for their wider availability in the treatment of chronic pain, and people suffering from chronic pain are more likely to be satisfied with their treatment outcomes if they are given these drugs.[55]

And yet, a substantial and growing body of research suggests that *in general* opioids have only a modest effect on chronic pain and are often no more effective than other, non-opioid treatments.[56] That is, for chronic pain conditions, opioids have a very limited impact on how people tend to rate their own pain and ability to function—they don't tend to dramatically lower self-assessed pain scores, they're not very effective in helping people regain function or return to work. And by the lights of what people say about their own pain and ability to function, they don't seem to (again, in general) do any of this more effectively than the best

[52] DeGood and Tait (2001), Jensen et al. (1999), and Sánchez-Rodríguez et al. (2020).
[53] Clark et al. (2002) and Scher et al. (2018).   [54] Levy et al. (2018) and Scher et al. (2018).
[55] Zgierska et al. (2012).   [56] For a good overview, see Chou et al. (2020).

non-opioid treatments. Yet a persistent finding is that people with chronic pain often *perceive* that opioids are essential to their treatment, and will rate their overall health outcomes better if they are given opioids.

And it is hard to deny that this perception—and the popular view of opioids as especially powerful, life-altering medicines for chronic pain—is *at least in part* a matter of *marketing*. As journalist Patrick Raden Keefe (2021) meticulously documents in his book *Empire of Pain*, pharmaceutical companies specifically targeted people with chronic pain conditions (as the biggest and fastest-growing potential customer base for their product) in their marketing campaigns, with the aim of convincing people who were suffering that opioids might hold a uniquely powerful solution to that suffering. In various cases, representatives from pharmaceutical companies surreptitiously joined, and in some cases even secretly founded, chronic pain patient advocacy groups, all with the aim of convincing the people involved that opioids are life-changing medications.

I am not, in discussing this, attempting to take a position on opioid prescribing. I don't doubt that there are individuals with chronic pain for whom opioids genuinely have been life-saving, just as there are people for whom they have been life-ruining. Again, the issue I'm concerned with here is epistemic. The relationship between opioid prescribing and subjective assessment of health outcomes gives us a vivid example of the ways in which people's understanding of, interpretation of, and satisfaction with their own health can be influenced by a wide variety of factors—not all of them clearly related to their actual health state.

6.4 The Curious Case of the Functional Disorder

Yet another place where objective and subjective measures of health routinely clash is in the assessment of so-called "functional disorders." The term "functional disorder" generally refers to a grab-bag of conditions that cause serious distress and limitation in activities of daily living, but for which there is no obvious physiological cause or underlying physiological disease process. And, curiously, many medical specialties have a paradigm functional disorder that is similar in some cardinal symptoms—but not in the underlying pathology—to other chronic disorders (sometimes rather confusingly called "organic disorders") that are treated by that specialty. So, for example, rheumatology treats rheumatoid arthritis and lupus and it also treats the functional disorder fibromyalgia. Gastroenterology treats Crohn's disease and ulcerative colitis and it also treats irritable bowel syndrome. Neurology treats multiple sclerosis and Parkinson's disease and it also treats functional pseudo-seizure and functional movement disorders.

There is a wide-ranging debate about the etiology of functional disorders. Some researchers view them as organic disorders just like any other, but ones for which we don't yet understand the cause, while others view them as "somatic" conditions, caused primarily by a process of central sensitization in response to psychological factors rather than by a physiological process. And it's of course entirely possible that at least some of these diagnostic labels, as currently used, are in fact "wastebasket diagnoses"—how people with distressing symptoms are labeled when no physiological cause for their symptoms is found, but which could ultimately represent a wide spectrum of diverse conditions and different etiologies. I want to be very clear that I am taking no position whatsoever on the etiology of functional disorders.[57]

Two dueling aspects of these conditions, however, do seem fairly well supported, our limited understanding of such conditions notwithstanding. The first is that functional disorders, almost by stipulation, are not associated with major "mortality and morbidity" events in the classic objectivist sense. *What it is* to have a functional disorder, by definition, is to have symptoms in the absence of a clear physiological disease process. And so functional disorders are by their nature not associated with the major disease-related sequelae of their counterpart disorders (and absence of such significant pathology is typically part of their diagnostic criteria). Rheumatoid arthritis can shorten life expectancy, erode joint tissue, and damage internal organs. Fibromyalgia doesn't have a similar physiological impact. Ulcerative colitis can cause intestinal obstruction, require colostomy or other major operations, and can be fatal. Irritable bowel syndrome (IBS) doesn't tend to be associated with these kinds of complications. This is not to say that people *diagnosed* with functional disorders don't sometimes develop such complications—it's just that if/when they do, that's usually sufficient for them to be re-diagnosed with an "organic disorder" instead.[58]

But the second, increasingly well-supported, aspect of such conditions is that they cause immense suffering for the people who have them. An ever-growing body of evidence suggests that people with functional disorders experience high levels of pain and distress and substantially reduced quality of life.[59] Moreover, people with such conditions often rate their own health and functional limitation at similar levels as, or even worse than, those with corresponding "organic" conditions.[60]

---

[57] Although it is important to note that these are not all merely cases in which "unexplained" symptoms are inferred to be psychogenic because we currently do not understand them. In the case of functional neurological disorders especially, it is often the case that patients experience symptoms—such as full body convulsion or apparent paralysis—which directly conflict with known biological patterns and which *cannot* be explained by physiological causes. See especially O'Sullivan (2021).

[58] A substantial percentage of women with autoimmune disorders originally receive a diagnosis of a functional disorder, for example. See especially Maya Dusenberry's (2017) excellent discussion of the use of functional disorders to dismiss women's complaints of physical distress in her book *Doing Harm*. Arguably at least part of the explanation of why functional disorders are poorly understood in general is that they tend to disproportionately affect women.

[59] See the extended discussion on quality-of-life measures in the Appendix to Chapter 4 for more detailed discussion.

[60] See, e.g., Pace et al. (2003), Birtane et al. (2007), and Salaffi et al. (2009).

These two aspects of functional disorders together yield one of the clearest examples of the difficulties of measuring health outcomes, and of relying on either primarily objective or primarily subjective measures. Most patients with functional disorders have had the degrading and dehumanizing experience of being told, in a healthcare setting, that there is "nothing wrong" with them or that they are "perfectly healthy." That is, because objective measures of health assessment don't tend to measure the impact of functional disorders, reliance on such measures easily treats functional disorders as either "not real" or at least "not serious." But the massive amount of suffering—not to mention the inability to work, or the money spent on finding relief—associated with such conditions should surely tell us that they are serious.

And, indeed, if we look simply at subjective measures of health outcomes—such as PROMs—these conditions are just as serious as, if not more than, their inflammatory or autoimmune counterparts. And yet here we run up against a key discrepancy. Someone with IBS and someone with ulcerative colitis might be equally distressed by their respective conditions, experience equal amounts of self-assessed discomfort, and view themselves as equally limited. And yet the person with ulcerative colitis might need to have her colon removed and replaced with a bag, while the person with IBS will not. The person with ulcerative colitis is at a substantially increased risk for colon cancer, but the person with IBS is not. The person with ulcerative colitis might die from her condition, and the person with IBS will not. And so on.

Advocacy groups for those with inflammatory or autoimmune conditions—especially the rarer conditions—sometimes express the worry that growing emphasis on treating perceived outcomes, and thus on treating functional disorders, will deprioritize research into their own conditions, simply because there are so *many more* people with, for example, IBS than with ulcerative colitis.[61] If we measure "disease burden" primarily via people's preferences and self-assessed health status, then IBS carries a far greater disease burden than ulcerative colitis, which would then factor into how research is prioritized. And yet some question whether this is fair—that is, whether it's fair to place the discomfort of something like IBS on a par with the kinds of diseases that can require major medical interventions and involve serious objective health risks.

It is not at all clear how we ought to square this circle. Again, I'm highlighting the difficult epistemic issues in play rather than attempting to recommend any policy. What seems very clear is that it's difficult to do justice to the fact that so much of what we care about when it comes to health is experiential suffering, and

---

[61] Recent research suggests, for example, that roughly 11 percent of the population meets the diagnostic criteria for IBS (although only about 30 percent of those who meet the criteria seek out ongoing medical care). See Canavan et al. (2014).

simultaneously do justice to the more objective biological components of health, including "severity" of health condition.

6.5 Self-Rated Health

Finally, there is perhaps no better illustration of the complex interaction between subjectivity and objectivity in health measurement than overall self-rated health (SRH). An increasingly common question on a wide range of surveys simply asks individuals to rate their health (with no more context added—so "health" means whatever the person takes it to mean) on a five-point Likert scale, with 1 being "very good" and 5 being "very bad."

Research has shown a rather mixed correlation between how people answer this question and what we might call their "objective health indicators"—including the presence of various diseases.[62] Self-rated health isn't unrelated to what we might describe as their "objective health status" but it also, at any given time, isn't a particularly good guide to it. Moreover, there seem to be indications that lots of things other than what we might think of as health-related factors influence people's subjective assessment of their own health. How people rate their own health appears to be influenced, for example, by a wide variety of factors, including personality,[63] gender,[64] and sense of optimism.[65] Broader culture also seems to be a factor; as discussed in Chapter 2, people from former Soviet countries, for example, tend to report much lower self-rated health than their life expectancies and health outcomes would generally suggest.

And more generally, self-rated health appears to be strongly influenced by the beliefs and expectations of the person making the evaluation. Most especially, at least some evidence suggests socioeconomic status has a significant influence on self-rated health. People with more education and higher socioeconomic status seem to have higher expectations for their health; as a result, at least some data suggests that they will often report comparatively larger drops in their health status after adverse health events.[66] That is, they'll rate their health as more compromised, after experiencing a health challenge, than someone with fewer social advantages will after experiencing a similar health challenge, simply because they have higher expectations for their health.

Yet despite wide debate over what they measure—and how it relates to health status at the time at which the question is asked—SRH measures appear to be an unusually powerful independent predictor of mortality.[67] And mortality is

[62] See Murata et al. (2006), Griffith et al. (2011), and Giltay et al. (2012).

[63] See Stephan et al. (2012), Stephan et al. (2020), and Aiken-Morgan et al. (2014).

[64] See Idler (2003).   [65] See Layes et al. (2012) and Grol-Prokopczyk et al. (2011).

[66] See Delpierre et al. (2009); we'll return to this issue in Chapter 6.

[67] See Schnittker and Bacak (2014), Lorem et al. (2020), and DeSalvo et al. (2006).

considered one of the major—if not *the major*—objective health outcomes we can assess. It's not clear why SRH is predictive in this way, or to what extent the predictive effect might be modified by other variables (education, socioeconomic status). But the finding seems, at least given current evidence, to be relatively robust.[68] How you rate your own health at a given time, then, probably gives us limited information about the physiological state of your body or your overall mental functioning at that time. But it tells us a surprising amount—and surprisingly more than many objective measures—about how likely you are to die over an extended period of time.

7. Conclusion

I have argued that dimensions of subjectivity and objectivity represent another kind of Scylla and Charybdis between which our understanding of health must navigate. If we focus too exclusively on objective biological markers of health, we miss out on important subjective components like pain and fatigue. These aspects of health—which are experienced subjectively—matter objectively both to the biological state of our bodies and to our capacity to function. And yet, we can overemphasize the importance of subjectivity in health. We need to be able to say that someone's health is objectively limited even if they are functioning well in spite of it, and we need to be able to say that some people's subjective reactions are unwarranted—even if they are experienced as genuinely distressing or extreme by the person who has them.

But the subjective and objective aspects of health are more than two poles between which we must find a comfortable middle ground. They are also, crucially, intertwined and interdependent. If you are stressed or fatigued, that has an objective biological impact on your body. If your body is damaged, that has a clear and obvious impact on your subjective experiences—even if each person's experience might be somewhat different. And if you experience something like pain or fatigue, part of your experience will be constituted by subjective factors (how you interpret the experience, the meaning you give to it, your expectations), but that experience will itself be objectively important to your body's ability to function. The fascinating—and difficult—thing about health is the way in which these dimensions of subjectivity and objectivity are tangled together, even while they are at times in tension with each other.

---

[68] Bombak (2013).

4

Health and Disability

In examining the inherent tensions in our understanding of health, we've so far looked at the relationship between health and wellbeing (Chapter 2) and the interplay between subjective and objective aspects of health (Chapter 3). In this chapter, we turn attention to the relationship between health and *disability*.

Just as the relationship between health and wellbeing is complex and entangled, so is the relationship between health and physical disability.[1] Just as with wellbeing, it's important not to conflate disability with health—or in this case with *loss of health*. But just as with wellbeing, there's no denying that the two are intimately connected. In this chapter, I'm going to try to tease apart the various threads of this tangle.

In doing this, I'm also going to begin to introduce the shape of the positive account of health I'll present in Chapters 5 and 6. In sections 4 and 5 of this chapter, I offer a model—based on an analogy to David Lewis' contextualist approach to the puzzle of the statue and the clay—that allows us to embrace the inherent shiftiness we encounter in the relationship between health and disability. Although it is in principle separable from the positive account of health I defend in subsequent chapters, this model begins to lay the groundwork for how I think we need to approach the tensions we find ourselves with when trying to theorize health. Rather than trying to clarify away or elide the mess we find ourselves with, I argue, we instead need a framework for understanding why things seem so intractably messy.

1. Disability and Health: It's Complicated

In many philosophical discussions of health, "disability" refers specifically and directly to loss of health, and is often used synonymously or alongside terms like "disease," "illness," and "pathology". The standard assumption in these discussions seems to be that disability *just is* (a type of) reduction in health. Indeed, in his famous analysis of health as normal species function, Christopher Boorse

---

[1] For the purposes here, I will focus primarily on physical disability—although the distinction between physical and psychosocial disability is clearly a blurry one, as has already been highlighted in Chapter 3. Physical disability in particular bears a distinctive relationship to bodily health, and perhaps also to overall wellbeing. In what follows, I won't make any assumptions about whether and to what extent these arguments apply to psychosocial or cognitive disability.

Health Problems: Philosophical Puzzles about the Nature of Health. Elizabeth Barnes, Oxford University Press.
© Elizabeth Barnes 2023. DOI: 10.1093/oso/9780192883476.003.0005

describes a spectrum from full health, through disability, to death, with death being "the most extreme disability."[2]

This way of thinking about disability is highly influential, and not confined to philosophical debate. The DALYs, for example—"disability-adjusted life years"—are one of the most widely used metrics for measuring and assessing health in populations. And a basic assumption of the DALYs seems to be the spectrum model elaborated by Boorse; you can be fully healthy, you can die, or you can live with the reduced health caused by a health condition, which is disability. So to measure the impact of various health conditions, we measure both how many people they kill and how many years "with disability" people live because of them.[3] "Disability" here is used roughly as a synonym for "morbidity." A common assumption in the medical literature is that the impact of a medical condition is a combination of its mortality and morbidity—how likely you are to die from it and how likely you are to suffer significant loss of health from it. In many conversations about health—both philosophical and more broadly—"disability" just refers generically to activity-limiting loss of health that you live with over time.

The standard view, then, sees health as a spectrum—from full health, through disease and disability, to death. Let's call this the *spectrum view*. This way of thinking about disability is sometimes referred to, or strongly associated with, the "Medical Model" of disability. Although there's probably no single view or conception of disability that's singled out by the term "Medical Model"—and it's not a term used by those who defend this kind of view—the basic idea seems be that a "Medical Model" of disability views disability as (primarily or exclusively) a type of health problem or loss of health, contiguous with illness and disease. Likewise, when people say that disability is often *medicalized*, this view of disability is typically the target.

Developed in large part as a reaction to "Medical Model" construals of disability, the broad family of views that fall under the label of the "Social Model" of disability seek to *demedicalize* disability. Again, there is no one view that is "the Social Model" and no single set of commitments that all views broadly classed as a "Social Model" of disability share. But the characteristic commitment is that disability should be understood as a social rather than a biological or medical phenomenon. People have impairments—which are various differences of body or mind that make them statistically atypical. But they become disabled when

---

[2] Boorse (1977), p. 547.

[3] As the World Health Organization explains it: "Mortality does not give a complete picture of the burden of disease borne by individuals in different populations. The overall burden of disease is assessed using the disability-adjusted life year (DALY), a time-based measure that combines years of life lost due to premature mortality (YLLs) and years of life lost due to time lived in states of less than full health, or years of healthy life lost due to disability (YLDs)." www.who.int/data/gho/indicator-metadata-registry/imr-details/158.

societies are designed in ways that fail to accommodate those differences. The limitations and harms of disability are thus a function of social exclusion rather than biological difference. In a more just society, people would still have impairments, but they would no longer be disabled by those impairments.

It's important to note that the claim here is not that impairments don't play a causal role in disabled people's disadvantage. Rather, the claim is that people become disabled because of a mismatch between their bodies/minds and their social environment, and the primary explanation for why this mismatch occurs is that their social environment has (unjustly) been structured in a way that excludes and disadvantages them. The disability/impairment distinction is modeled on the sex/gender distinction. And, clearly, biological sex differences play a causal role in ways in which women are disadvantaged. The different burden that the human reproductive process places on the female body, for example, has historically disadvantaged women in the workplace. But the solution is not to change the female body; it's to offer policies like paid maternity leave and flexible working hours. So the analogous thought is that, when people are disadvantaged by impairments, we ought to modify our social arrangements in a way that would alleviate that disadvantage rather than focusing on the ways in which their bodies/minds might be altered to conform to existing social arrangements.

On this way of understanding disability, disability is not a matter of health at all. Disability, on Social Model views, occurs because of unjust social arrangements, and so should be understood as a social and political issue, not a health issue. And this is often the primary goal for those who want to *demedicalize* disability—the thought is that disability should be seen as an issue of social justice rather than an issue of health inequality or medical need.

To some extent, the debate here may be terminological—someone employing the medicalized use of "disability" needn't deny, for example, that many of the disadvantages encountered by disabled people are matters of injustice, and should be addressed in political rather than medical contexts, and likewise it's within the remit of the Social Model usage that someone could think *impairments* (though not disabilities) are contiguous with illness and disease along a spectrum of reduced health. But the bigger issue—which is not terminological—is how we ought to understand *what it is* to be disabled.

There is a distinctive social category of people we classify as disabled. We assume that the people in this category have something substantial in common with each other. And it's how to understand that commonality—how to understand the explanatory significance of disability as a way of classifying people—that's primarily at stake in these debates. But I'm going to argue that neither the understanding of disability offered in the standard view nor the rival interpretation put forward by the Social Model give us the resources for adequately understanding the complex relationship between disability and health.

1.1 Against the Spectrum View

There's a rich and thorough literature arguing that disability can't be fully understood in biomedical terms.[4] So much of the significance of disability is social—related to how you navigate spaces, how you interact with others, how you can or can't access work, communication, and community, and so on—rather than medical or biological. I won't rehearse these arguments here, however, because they are consistent with the basics of a spectrum view, according to which disability, alongside illness, disease, and injury, is just a type of reduction in health. Disability might just be a particularly socially salient or socially significant way of experiencing reduced health. If that were the case, then talking about a reduction in health wouldn't be sufficient to understand disability—disability wouldn't *merely* be a reduction in health—but disability could still be part of a spectrum, alongside illness and injury, from health to death.

Thus, to criticize the spectrum view, we need to address its specific commitments rather than simply focusing on what it leaves out. If you adopt a spectrum view of health, you think that there is a roughly linear ordering from full health, through diseases and disabilities of increasing severity, to death. The further away from full health you get, the closer you get to health, and to have a disease or disability is to be somewhere in the "not fully healthy but not (yet) dead" middle ground of reduced health. The basics of the spectrum view are consistent with thinking, in line with normal function theorists like Boorse, that full health is just the absence of substantial health conditions. But it's also consistent with the idea that there's a positive dimension to the spectrum as well—the person with no significant health conditions but few positive health behaviors, for example, is not unhealthy, on this conception of the spectrum, but they're also not *as healthy as* the kale-eating, fitness-loving mindfulness enthusiast.

It's important, at this point, to distinguish between the following claims:

(1) Disability always involves a loss of a health.
(2) For any disabled person, x, and person without any substantial health conditions, y, x is less healthy than y.

I think (1) is true—but partly because (as I'll discuss in Chapter 6) whether something constitutes a loss of health is contextually variable and depends on what aspects of health we're assessing in a particular context. The truth of (1) doesn't by itself support the truth of (2), which is a stronger, comparative claim. Disability might always involve the loss of health *along some dimension* or *with*

---

[4] An excellent introductory overview can be found in Watson and Vehmas (2019).

respect to some parameters without it being the case that disabled people are always less healthy, in general, than people without significant health conditions.

I take the comparative claim of (2) to be a characteristic commitment of the spectrum view. Of course, (2) doesn't exhaust the commitments of the spectrum view—and how the details of a spectrum view are filled in will depend in part on what specific view of health it's paired with. But (2) is the characteristic commitment I want to focus on.

As Boorse understands it, for example, full health is normal function (for detailed discussion, see Chapter 1). You function normally, in very basic terms, if there's nothing wrong with you. Illness, disability, disease, pathology—these are all terms we have for negative departures from normal function, and they are all unified by the fact that they reduce species-typical functioning. The more substantial the departure from normal function, the more severe the loss of health.

This model is replicated, and embedded, in widely used assessments of health such as the Health Utilities Index (HUI). In the HUI questionnaires, various descriptions of functional abilities related to different domains (ambulation, dexterity, sight, etc.) are described, starting with a base-level approximation of normal function for that domain and descending in scale of severity of departure from normal function. Here's the HUI classification for ambulation, for example:

1. Able to walk around the neighborhood without difficulty, and without walking equipment.
2. Able to walk around the neighborhood with difficulty, but does not require walking equipment or the help of another person.
3. Able to walk around the neighborhood with walking equipment, but without the help of another person.
4. Able to walk only short distances with walking equipment, and requires a wheelchair to get around the neighborhood.
5. Unable to walk alone, even with walking equipment. Able to walk short distances with the help of another person, and requires a wheelchair to get around the neighborhood.
6. Cannot walk at all.[5]

These classifications are taken to describe "health states," to which respondents are then asked to give utility weightings. Those utility weightings are then used to assess the "health-related quality of life" for the various health states, on a scale from 0–1, where 0 is dead and 1 is perfect health. The weighting of the normal function option (level 1) is always fixed at 1.

---

[5] See www.healthutilities.com/hui3.htm.

What the spectrum view leaves no room for, by setting the parameters this way, is the *healthy disabled person*. If you have a disability, then on the spectrum view you are always in strictly worse health than someone who has no significant health problems. It's worth noting that this is simply an entailment of a view of health like Boorse's. Disability, even if it can't be explained by the idea of departure from statistically typical species function, arguably does always *involve* departure from statistically typical species function.[6] If health is simply normal function, then a person with a disability will always be less healthy than a person without any significant health issues. And given the oversimplifications of the spectrum view, this is perhaps yet another reason to think that normal function views (at least as analyses of health rather than merely as analyses of pathology) fall short.

Marko Cheseto is a successful Kenyan-American marathon runner. He recently ran the Boston Marathon in a time of 2:42:24, placing 483rd out of a field of 26,632 of the best distance runners in the world.[7] Cheseto is also a double lower-limb amputee.

A spectrum view of health—like the standard normal function view or the one embedded in the ambulation scale offered by the Health Utilities Index—construes Cheseto as *less healthy* than any arbitrary person who simply lacks significant functional abnormality. Your average sedentary office worker, even if they're not particularly fit or health conscious, would be healthier, on this view, than someone like Cheseto simply in virtue of having no significant biomedical abnormalities.

Take the specific example of the HUI classification for ambulation. We fix as at "full health" (for the domain of ambulation) anyone who can walk around the block without difficulty and without "walking equipment." Cheseto can't do this—he requires assistive devices to walk and cannot walk at all without them. The HUI will classify him, as a result, as being in a reduced health state (and having lower "health-related quality of life") than someone who can walk normally. Never mind that, with his prosthetics, Cheseto can run blisteringly fast marathons. He's still viewed, on this model, as less healthy (and in this case, less healthy *with respect to ambulation*) than someone who meets the bare minimum of normal function.

---

[6] The potential counterexample to this is disability in old age. Normal function views—as discussed in Chapter 1 (section 2)—have a notoriously difficult time specifying the comparison class that determines statistically typical functioning. It arguably needs to involve an index to age—since, e.g., a normally functioning thirty-year-old female will menstruate but a normally functioning seventy-year-old female will not, and absence of a menstrual cycle is pathological for a thirty-year-old but not for a seventy-year-old. But once we've indexed to age, the problem of common health conditions looms large. The vast majority of people over the age of seventy-five have some significant and activity-limiting health condition, for example, so it becomes harder to call such health conditions (especially ubiquitous ones like arthritis or vision loss) departures from normal function, even if they are properly considered health conditions. Whether the vast majority of people over seventy-five should be considered disabled, or whether being disabled partly involves contravening expectations of what is "normal," is a subject of controversy.

[7] www.runnersworld.com/news/a27183626/marko-cheseto-world-best-boston-marathon.

The spectrum view leaves no room for someone who functions abnormally because of biomedical pathology, but nevertheless functions extremely well.[8] It makes sense, I suggest, to say that the absence of Cheseto's lower limbs is a health condition, and represents a reduction in health (at least along some dimension—see Chapter 6). Injuries are typically thought of as health conditions, as are things that are regularly monitored by doctors and physical therapists, as are bodily differences that require regular use of assistive devices. The condition of Cheseto's limbs is all these things. The problem is not with saying that being a double amputee involves a loss of health or is legitimately understood as a health condition. The problem, rather, is with the assumption that a double amputee is automatically less healthy than someone who lacks any substantial health conditions.

Consider the major things we care about when we care about health—longevity, resistance to disease, fitness, energy, ability to go about our daily lives. Then compare someone like Cheseto to someone who is normally functioning but who doesn't really engage in any proactive health behaviors—someone who can walk without assistive devices, for example, but who is very inactive, doesn't get much sleep, and so on. Someone like Cheseto is likely to have—and have more stably over time—the things we care about most when we care about health. And this is why a claim like (2)—central to the spectrum view—is far too strong. We can allow that Cheseto's disability involves a loss of health, but the comparative claim that he's *less healthy* than someone who simply doesn't have any significant health conditions shouldn't follow.

The ardent defender of the spectrum view might object, at this point, that someone like Cheseto, although very fit, is still less healthy than your average normally functioning person, simply because he relies on assistive devices. His fitness is, thus, in some sense less counterfactually stable. Were the devices not available to him, he would be very limited in his functioning—and thus his mobility (and thereby his health) is more limited, even if he's capable of impressive athleticism.

I discuss these kinds of counterfactual assessments of health in detail in elsewhere (see Chapter 1 (sections 2.1.2 and 2.1.3)), but the main point to make here is simply that reliance on technology for the maintenance of health isn't a way we can differentiate Cheseto's health status from that of people without significant biomedical pathology. In industrialized contexts, we all—every single one of us—rely on technology for the maintenance of our health. We go to the

---

[8] A normal function view like Boorse's does, of course, allow that not all atypical functioning is pathology—but missing your lower limbs clearly *is* pathology in Boorse's sense, and yet it shouldn't follow that someone like Cheseto is thereby less healthy than anyone who lacks significant pathology. For a much more detailed discussion of normal function accounts of pathology, see Chapter 1. For criticism of the tendency to want to normalize disability people in an attempt to make them "healthier," see especially Silvers (1994) and (2017).

grocery store (often in our cars or on buses or subways or bikes) to get food that was harvested by machinery and shipped to our non-farming locations via motorized transport. We drink sanitized water from our indoor plumbing. We have central heating. In the absence of any of these technologies, our health would plummet in a hurry. The fact that Cheseto's protheses are considered medical devices and indoor plumbing is not is sociologically interesting, but it doesn't mean that the health of the "normal functioning" person is not dependent on technology. It's also far from obvious that the health of an extremely fit and capable—but non-normally functioning—person like Cheseto is any *more* dependent on technology than that of your average middle-class office worker.

1.2 Against a Purely Social View

It's in the context of the spectrum view—and its dominance—that various versions of the Social Model have been developed as a contrast.[9] To be clear, defenders of the various formulations of the Social Model of disability don't dispute that impairments often involve loss of health or require medical treatment. But they argue for a distinction between impairment and disability. Disability, according to the Social Model, is the social disadvantage incurred by living with an impairment in a society that unjustly excludes and marginalizes people with impairments. As such, disability should be understood as a social and political phenomenon rather than a biomedical phenomenon. Likewise, the disadvantages associated with disability should be understood as social and political rather than medical. That's not to say that loss of health doesn't create impairment, which can then—in our society—be disabling. But it is to say that the negative effects of disability should be understood as—and addressed as—social and political problems, not health problems.

Again, the differences between the spectrum view and the purely social view might be, at least in part, terminological. The spectrum view uses the term "disability" to refer roughly synonymously to what medical practitioners and epidemiologists call "morbidity"—long-term reductions in health that limit function. The purely social view, in contrast, often simply stipulates that they will use "disability" to refer to the disadvantage created by for those living with impairments in an inaccessible environment. Insofar as this is what the relative sides mean by "disability," they are simply talking past each other.

---

[9] There is no single view that is "the Social Model of disability." Rather there is a family of views, united by broad commitment to two general claims: (i) disability and impairment are distinct; (ii) the bulk of the disadvantage encountered by disabled people is due to lack of access and accommodation for impairments (that is, the creation of disability because of unjust social arrangements) rather than the inherent limitations impairments themselves. For overview and discussion, see especially Shakespeare (2006) and Oliver (2013).

And yet both sides engage in meaningful debate, and both sides appear to take themselves to be referring to the same set of shared paradigm cases—people who are blind, who are paralyzed, who live with long-term illnesses like multiple sclerosis (MS), and so on. So while anyone can use the word "disability" as a technical term of art, I'm going to assume in what follows that there's an attempt here at shared meaning. And so, in making objections to the purely social view, I'm going to assume that they aren't simply changing the subject by using the term "disability" in a highly specific way.

I've argued against the disability/impairment distinction elsewhere,[10] but here I want to focus on its use as a means to demedicalize disability. Impairment, on the purely social view, involves the biological condition of one's body, which might include injuries or illnesses. And impairments can be rightly understood as involving loss of health. But disability, because it is a function of unjust social arrangements imposed on people with impairments, only involves reduced health when it does so via social limitations.

As a hypothetical example, consider Cary, a wheelchair user. Cary's impairment might be rightly understood as a health issue or something that it makes sense to seek medical attention to manage. But Cary is *disabled* only and because it's harder for him to function in society as a wheelchair user. That might have implications for his health—it might be harder for him to travel to see his doctor, or it might be harder for him to get and afford adequate health insurance. But his disability is a moral and political issue, not a health issue. If his health is reduced because of his disability, he needs to write to his congressperson and ask for better healthcare legislation rather than calling his doctor and asking for better treatments.

But to view Cary in this way, we have to neatly separate the issues and limitations he faces because of the biological condition of his body (his impairment) and the issues and limitations he faces because of the way in which society is organized. And in many cases of disability, this distinction breaks down. We're social animals. Our bodies exist in a social environment. For so many aspects of disability, the answer to the question "is this social or biological?" is "both."

Suppose that Cary uses a wheelchair because he has MS. Let's call the demyelinated nerves within his body his impairment. Some of the limitations he faces are obviously a function of his impairment—he might struggle to grip small items or have difficulty with fine motor control and balance, for example. And yet they're also social. We live in a world designed around the assumption that everyone has a certain level of fine motor control, and that makes Cary's life a lot harder. Cary adapts to this with a mixture of strategies—he goes to regular physical therapy, he takes disease-modifying drugs, he uses adaptive equipment and assistive technology (gripping aids, voice-to-text software, smart home devices, easy-grip home

---

[10] Barnes (2018). I also discuss the Social Model in more detail specifically as an account of the nature of disability in my (2016).

items, etc.), and he changes some of his daily life patterns to accommodate the differences in his coordination. He also asks his family to chill out a little. They learn to accept that he sometimes drops things, and they learn to plan extra time in the mornings so he isn't rushed getting dressed.

Which of the limitations that Cary is facing are his disability, and which are his impairment? This is, I suggest, not a good question. Cary is dealing with the effects of physical difference in a social environment. The medical, social, and political are all mixed together, and for many ways of adapting to disability there's a complex mix of strategies that incorporate them all.

Or consider Dara. Dara also has MS, and because she has difficulty swallowing she gets her nutrition from a feeding tube. Obviously, this means that some significant aspects of her condition are medicalized (and rightly so)—the tube has to be surgically placed and monitored for infection, she has to keep careful track of her nutritional intake, she uses the tube for her medications. Likewise, some of the frustrations and limitations Dara encounters are primarily due to the biological limitations of her body. Maybe she simply misses the sensory pleasures of her favorite meals, for example, or maybe the experience of swallowing is painful. But some are due to the combination of the state of her body and her social environment. Food has deep social significance. Being unable to eat ordinary food means being unable to participate (at least in typical ways) in meals with friends and family, unable to share food or drinks with a loved one. All of this has far more significance given that we (contingently) organize so much of our socializing around food than it would if we all simply sucked down liquid ration packs alone in our bedrooms.

And yet, although the limitations Dara encounters are social, it doesn't follow that they are due to prejudice or should be altered. There are some ways that society could change that could make things easier for Dara—we could be more understanding about the reality of feeding tubes, have it be more socially acceptable to talk about them, set them up in a restaurant alongside friends' dinners instead of waiting for the privacy of home. But it's not reasonable to expect that other people stop enjoying sharing food together—and finding deep significance in sharing food together—simply because Dara cannot. And it's obviously reasonable for Dara to feel left out and limited because of this. And so, to a significant extent, Dara's limitations are socially mediated—they're due to the mismatch between her impairment and her social environment—and yet many of the most significant are *not* due to injustice.

Defenders of a Social Model might object at this point by saying that, were society more accommodating and disability-positive, Dara would not experience her inability to share food with others as isolating or limiting. I say this depends a lot on Dara—on what she wants, what she values, and how she prefers to live her life. She might feel excluded just because there's a social expectation that we share food together. But she might feel loss and isolation because food-sharing was a source of genuine joy and meaning for her—a source of joy and meaning that the biological condition of her body has made unavailable to her. To suggest that her feelings of loss are only due to ableism would be to force a very specific narrative on her experience, regardless of what she says about it. And that's something that defenders of the Social Model rightly object to in other contexts.

The basic point I am making here is that, for many disabled people, there is not a neat separation between the physical (the impairment) and the social (the disability). Many parts of experiencing disability—including experiencing limitations and barriers—are a complex mix of the dynamic interplay between—and interconnectedness of—the physical body and its social environment. Indeed, part of the basic biological reality of the human organism is that we are social primates; that is as natural a part of our experience as any other. Moreover, a person can encounter barriers that are at least partly social without those barriers obviously being a form of injustice, or having political solutions, just as a person can encounter barriers that have a biological cause but that are best addressed by social change. Dara's sense of alienation due to her feeding tube, for example, is at least partly social—it's hard for her party because of the social importance of food and sharing meals. But it might be that the best way of addressing that for her is a better medical option (something that allows her to swallow and eat food, for example) rather than a political option (such as a wholesale restructuring of the social significance of food).

While the spectrum view *over*-medicalizes disability—by viewing it as just a medical problem and viewing disabled people as strictly less healthy than non-disabled people—the purely social view risks *under*-medicalizing it. Part of the complex reality of experiencing disability, for many disabled people, is dealing with long-term reduced health, which often includes long-term interaction with and reliance on the institutions of medicine. That doesn't mean such disabilities are fully explicable in biomedical terms, but it does mean that a view that defines disability as purely a matter of social disadvantage—and as having nothing to do with reduced health—risks obscuring a significant part of many people's experience of disability.

1.3 Disability and Health: The Close Connection

Many social categories are associated with reduced health. Black Americans, for example, have reduced health relative to white Americans.[11] But this connection between being Black and reduced health is, quite plausibly, a deeply contingent one.[12] And more strongly, it's one that's contingent on the unjust social arrangements in the United States. While genetic factors explain some health differences

---

[11] For a particularly stark overview, see Hunt and Whitman (2015).

[12] For discussion, see Kaplan (2010).

between the groups of Americans we racialize as black and white—higher incidence of sickle cell disease in people of African ancestry, for example, and higher incidence of MS in people of northern European ancestry—the broad health disparities between Black Americans and other groups are almost certainly a function of things like income inequality, education inequality, urban food deserts, lack of access to affordable and fair housing, and so on.[13]

In a fairer and more just world, there might still be some general health *differences* between the group of people racialized as Black in America and other racial groups—Black Americans might still be less likely than Asian Americans to be lactose intolerant and less likely than white Americans to get skin cancer, but more likely than either group to have a vitamin D deficiency, for example. But the massive health disparities between racial groups, it's generally thought, wouldn't persist if the major social determinants of health were more equitable.

Disabled people, as a group, have reduced health compared to non-disabled people.[14] And some of this disparity is, almost certainly, due to unjust social arrangements. Disabled people are far more likely to be poor, to have limited access to healthcare, for example. But unlike social categorizations such as race or sexual orientation, the major health disparities between disabled people and non-disabled aren't entirely or even primarily explained by contingent factors. Because of the nature of the conditions we group together as disabilities, the class of people we identify as disabled are going to have, on average, a lower level of health than the class of people we identify as non-disabled, even in a more equitable society.

This doesn't, of course, mean that all disabled people are less healthy than all non-disabled people. But many of the conditions that we identify as disabilities directly and non-contingently involve a substantial loss of health. Even in a fairer world, the group of people with such conditions will be, on average, less healthy than an arbitrary group of people without them.

This is why, for example, the right to affordable healthcare is at the center of the disability rights platform. When people are denied access to healthcare, disabled people are more likely to suffer and die. And they're more likely to suffer and die precisely because they have, as a group, reduced health compared to non-disabled people.

2. Disability, Health, and Wellbeing

Let's assume, then, that disability has a close—though non-linear—connection to loss of health. As we've already discussed, there's also a close—though non-linear—connection between loss of health and loss of wellbeing. So, of course,

---

[13] For overview and discussion, see especially Bassett and Galea (2020).
[14] For discussion, see Krahn et al. (2015).

we might naturally conclude that disability has a close—though non-linear—connection to reduction in wellbeing.

And yet it's increasingly common, both within disability rights communities and within academic discussions of disability, to focus on disability as an important—perhaps even a valued—aspect of human diversity, and to *deny* that there is a close connection between disability and reduced quality of life. I have called such views of disability *mere-difference* views of disability.[15] I've argued that mere-difference views of disability are compatible with the idea that disability is a harm with respect to some features or aspects of an individual's life, and compatible with the idea that disability is bad for some particular individuals (depending on their projects, desires, personality, and so on).[16] It's also, of course, compatible with the idea that disability can often reduce wellbeing because of contingent social arrangements, such as discrimination, stigma, and lack of accessibility. But the central contention of such views is that there is not a robust, counterfactually stable correlation between being disabled and having lower quality of life than one would otherwise have, and that disabled people would not automatically be better off if they were non-disabled. This is often coupled with the claim that we should view disability as a social kind—and more strongly as the type of social kind that we seek to eliminate discrimination against but don't seek to eliminate membership in.

Everyone can agree, for example, that it's hard to be gay, and evidence suggests that being gay is *in fact* correlated with reduced wellbeing, at least in many contexts. But we also think that the solution to this problem is to treat gay people more justly, not to seek to eliminate the biological and social variables that determine same-sex and same-gender attraction. We value the range of human diversity, while seeking to eliminate the ways in which being different from the norm can disadvantage you.

Mere-difference views can, in essence, be seen as advocating this type of position for disability. Even while we acknowledge that disability can be associated with hardship and harm—some due to social disadvantage, some due to the basic biological reality of disability itself—the central claim is that being disabled does not by itself make a person's life go worse for them, and that we should treat disability as a way of being different from the norm without automatically being worse off or unfortunate. Thus ADAPT, one of the largest and most successful grassroots disability rights organizations in America, writes in their mission statement:

Many might think it is a tragedy to acquire or be diagnosed with a disability; it isn't. The tragedy is when we are denied our civil rights, our dignity and our right to choice. This is what fuels our righteous efforts. In ADAPT, we strive to

---

[15] See Barnes (2014) and Barnes (2016), chapter 2 especially.   [16] Barnes (2016), chapter 3.

promote this energy and this love...In ADAPT, we strive to advance beyond "mere tolerance," where tolerance might be defined as a mere acknowledgement of someone's right to exist but may not require more effort than that. Indeed, we dismiss tolerance—it is lacking, it is empty, it is not enough. Instead we argue for embrace. We advocate for embracing disability.[17]

This idea of *embracing* disability and rejecting the assumption that it is somehow an inherent detraction from a good life—a tragedy—is what I take to be the heart of mere-difference views.

It's worth noting that there's substantial empirical evidence to prop up at least some aspects of the mere-difference position—although findings vary *significantly* depending on how widely we construe "disability", by the types of disabling conditions studied, as well as on multiple social factors.[18] In the Appendix to this chapter, I do a deep dive into some of the current empirical research—often confusing and conflicting—on the relationship between disability and self-reported quality of life. The main upshot is that it depends dramatically on what kind of conditions we're including under the broad heading "disability" and that in the abstract "what is the relationship between disability and quality of life?" is probably not a helpful question. Non-specific chronic pain conditions (such as chronic back pain, headache disorders, or fibromyalgia) and mental health conditions (such as generalized anxiety or depression), in particular, tend to be substantially correlated with reduced self-reported quality of life. And such conditions are, increasingly, the most common reasons that people in Western countries do things like seek disability benefits or ask for workplace accommodations. Nevertheless, there's substantial evidence that many of the physical conditions we think of as "paradigm cases" of disability (such as spinal cord injury, achondroplasia, or blindness) are compatible with high levels of self-reported quality of life, and often don't have much correlation with reduced self-reported quality of life. Moreover, the best predictors of self-reported quality of life, in the context of many such conditions, tend to be factors like social support and inclusion rather than the physical "severity" of the condition.

Such research is limited in scope, of course. It doesn't tell us that disabled people actually have good lives; it only tells us that they perceive themselves (or even more cautiously, tell others that they perceive themselves) to have good lives. And it also doesn't tell us the role that disability plays in quality of life. It's possible for people to be wrong about their own quality of life, and it's possible for people to have high quality of life in the face of harms and setbacks. Disabled people might just be wrong in their self-assessment—maybe due to status quo bias

---

[17] https://adapt.org/our-journey.

[18] In previous work—especially my (2016)—I didn't fully engage with this complexity, and probably over-stated the empirical support for the view I was defending. Sorry about that.

or adaptive preference. Or disabled people might be telling us that they have high quality of life in spite of the badness of disability, or through overcoming the badness of disability. I've argued against both such interpretations—and in favor of mere-difference views—elsewhere, and I won't rehearse those arguments here.[19] The main takeaway, for the purposes here, is that any defender of a mere-difference view of disability denies that there's a close, robust connection between disability per se and reduction in wellbeing, and there's both philosophical and empirical support for such a view, at least for many significant and substantial forms of disability.

A mere-difference view of *health*, however, looks like a disaster. Loss of health may not have a perfect or linear correlation to reduction in wellbeing, but—as discussed in Chapter 2—there's substantial reason to think that there's a close connection. Health permeates so many aspects of our lives—what we can do, how we feel, our plans for the future. While loss of health doesn't automatically or directly correlate to loss of wellbeing, it seems obvious that there's a connection between the two. And, as discussed, there's also a robust—though imperfect—empirical correlation between health and life satisfaction.

But a mere-difference view of health is not merely implausible, it's also morally and politically objectionable. Health inequalities are one of the biggest negative consequences of socioeconomic disadvantage. Loss of health is a major reason why lack of environmental safety regulations and increased pollution are so harmful. Negative consequences to health are one of the main things we can point to in arguing for the importance of a fair and just healthcare system. And so on. Health *matters* to us, morally and politically, and loss of health is often one of the central negative consequences of social injustice.

We can perhaps too readily imagine someone who embraces a mere-difference theory of health going to Flint, Michigan, for example, and telling residents there—poisoned by their own drinking water due to lack of governmental oversight and concern—that while of course adjustment to a loss of health can be a difficult process, they have not been made intrinsically worse off, and there's nothing automatically worse about a population being less healthy. Embrace the loss of health! Unhealthy pride!

I'm going to go out on a limb and suggest we shouldn't do this. There can at times be a fine line between disability acceptance and Panglossian optimism. A mere-difference view of health seems dramatically on the wrong side of that line.

We need to be able to say, for example, that one of the worst aspects of racial inequalities in America are health inequalities. Black Americans get type 2 diabetes at much higher rates than white Americans.[20] And when they have diabetes,

---

[19] See my (2016), chapter 4.
[20] www.cdc.gov/diabetes/pdfs/data/statistics/national-diabetes-statistics-report.pdf.

they are two to three times more likely than white Americans to have a limb amputated. Of course, a mere-difference view can grant that life transitions—especially forced ones—are hard. Even the most ardent defender of a mere-difference view of disability doesn't deny that it's difficult to become disabled, or that someone wrongs you if they cause you to become disabled without your consent.[21] But there are limits to what a mere-difference view can say is wrong in a situation like this.

There's nothing intrinsically better or worse, I maintain controversially, about living in Boston or living in Philadelphia. You can have an equally good life in either city, which is of course consistent with some people preferring one to the other, and with living in one being better for some individuals than living in the other—depending on your career, your habits, your sandwich and beer preferences. And it's also consistent with Boston being better along some dimensions than Philadelphia, and Philadelphia being better along some dimensions than Boston. Compared side by side, though, a Boston-based life and a Philly-based life are equally good ways to live.

But if you'd built your life around Philly—your friends were there, your job was there, you supported all the local sports teams, you really identified as being a Philly person—and you were suddenly forced, without any say in the matter, to relocate to Boston, where you don't know anyone and you don't know your way around and you don't have a job, this would be a major, life-altering hardship for you. And that's true even though there's nothing intrinsically better about a Philly-based life than a Boston-based life, and it's true even though you might ultimately adapt to your new city, and come to be just as happy living a Boston-based life and doing Boston-based things as you were living a Philly-based life and doing Philly-based things. If you were a Philly person, and you're somehow forced to uproot your life and move to Boston, that's bad for you, and that'll obviously be hard for you—but not because there's anything intrinsically worse about living in Boston.

This is, very roughly, how a mere-difference view of disability conceptualizes becoming disabled. There's nothing intrinsically worse about living a life with a disability. But obviously if you've built your life around being non-disabled, and you strongly prefer to be non-disabled, then a forced change to disability will be incredibly difficult and painful. And we can think that's bad for you, while maintaining the state you end up in (being disabled) is *by itself* a worse state, and even while allowing that you might adapt and end up perfectly happy with your life as a disabled person.

But this, I submit, is an inadequate way to view loss of health. Losing health is not like a forced move from one perfectly good city (that you prefer, that your life

---

[21] See Barnes (2014).

is built around) to another. People in worse health—groups and populations in worse health—aren't merely in a different state. They have been made *worse off* in some especially salient way.

Here, then, is the basic puzzle that any defender of a mere-difference view of disability finds themselves with. Mere-difference approaches, when applied specifically to loss of health, look awful. And yet disability is strongly counterfactually correlated with loss of health.

3. Some Unsuccessful Solutions

The easiest way out of this puzzle is to give up a mere-difference view of disability. We can grant, for example, that disabled people can have high quality of life but maintain that it's the high quality of life that people can often have in spite of (and by "overcoming") negative things in their lives.[22] For myself, I don't find these accounts of wellbeing in the context of disability satisfactory, but I'm not going to discuss their shortcomings, or positive arguments in favor of a mere-difference approach, here. Instead, I want to focus specifically on what options there are for someone who wants to maintain that disability is mere difference but deny that loss of health is mere difference.

3.1 Deny the Correlation

One option is to resolutely dig in one's heels and deny a close connection between disability and loss of health. There are two main ways of doing this. The first is to, via something like a Social Model, maintain a robust distinction between *disability* and *impairment*. If disability is entirely a social phenomenon—a way people are treated or socially positioned—rather than anything to do with the physical body, then we can say that impairment is strongly correlated with loss of health, but disability is not. And that's because disability is not about the state of the body at all. It's about how you're treated, how you're disadvantaged, how you're socially positioned, and so on in response to norms and assumptions about how bodies *should be*. But none of that itself involves loss of health, except for contingent reasons (such as lack of access to healthcare or medical stigma).

---

[22] Disability, in this sense, might be like the range of hardships—divorce, bullying, poverty, etc.—that you can experience in your life and that can, in the right context, be instrumentally related to things you value, and consistent with high levels of wellbeing overall, because you gained something good in a way that's causally related to something negative or bad. For reasons why I find such accounts of disability unsatisfactory, and inadequate for what many disabled people say about their own experiences, see Barnes (2016), chapter 3.

As I discussed earlier, though,[23] I don't find the disability/impairment distinction plausible. Especially when it comes to embodied issues of health, I don't think we can make a neat separation between the social and physiological/biological—and I think trying to make that separation obscures a lot of the ways in which the biological and the social are interwoven in the everyday experiences of many disabled people. Being disabled, at least in many contexts, is *partly* a matter of the biological condition of your body: the pain you might be experiencing that's causing you distress, the jerking limb that's distracting you while you're trying to write, the stream of medical appointments that are taking up *so much time*. And so—absent a revisionary story about health—it seems hard to deny that disability is strongly correlated with reduced health.

The other main way of denying the correlation between disability and reduced health, then, is to give a revisionary account of what health is. If we adopt a strongly subjectivist theory of health, for example, as discussed in Chapter 1 (section 4), then we can perhaps say that provided a disabled person feels "at home" in her body, as long as she generally *feels good* and has the ability to do the things she wants to do—as so many disabled people do—then that's just what it is for her to be healthy. And so, if we adopt a view of health like this, we can deny the strong correlation between disability and reduced health. Whether a disabled person is in a state of reduced health because of their disability will depend on how they feel, and on how they respond and adapt to their disability—all things that are subjective and that depend on a particular individual's context and pattern of responses.

We've already discussed the significant problems with a purely subjective theory of health in Chapters 1 and 3. But for the purposes here, the main problem with this response is simple: if we change the meaning of "health" to refer to a subjective state of feeling well, such that many disabled people are perfectly healthy, then we can no longer explain why it's so morally and politically important to ensure that disabled people have good access to healthcare resources. The right to healthcare, and the need for better, more expansive, and more equally distributed healthcare resources, is at the center of the disability rights platform because disabled people tend to have higher healthcare needs than the average population, and tend to be more vulnerable to restrictions or inequalities in healthcare access. To put it another way, the strongly subjective approaches to health tend to focus primarily on the subjective wellbeing—including the subjective positive experiences of embodiment—of disabled people. And it's of course right to emphasize that disabled people are often happy, content with their bodies, out in the world living fit, active lives, and so on. But if we insist that health *just is* these subjective aspects of wellbeing, then we either have to deny that disabled people have unique health needs and vulnerabilities—and that seems clearly false.

---

[23] As well as in substantially more detail in Barnes (2018).

3.2 Healthy versus Unhealthy Disabled

Rather than denying the correlation across the board, Susan Wendell (2001) has put forward the intriguing suggestion that, in discussing the relationship between disability and wellbeing, we need to separate out two different types of disability: healthy and unhealthy.[24] Healthy disabilities, as Wendell construes them, are those that are relatively static in their physical presentation—that is, they don't require regular ongoing medical intervention to manage, they aren't degenerative, they don't have flare/remission cycles, or similar. She also takes a hallmark of healthy disabilities to be a lack of "constitutional" manifestations such as fever, fatigue, and chronic pain. In contrast, unhealthy disabilities are those that require ongoing medical intervention or are degenerative and those that have fluctuating or constitutional manifestations. And while Wendell readily admits that this distinction will be vague, she argues that most disabilities are classifiable on either side of the divide. She takes the paradigm instances of healthy disabilities to include things like being deaf or blind, being an amputee, or having a spinal cord injury. Paradigm unhealthy disabilities, in contrast, are those that arise from conditions we might typically class as *illnesses*—MS, muscular dystrophy, lupus, rheumatoid arthritis, or similar.

Capturing the distinction between healthy and unhealthy disabilities, Wendell argues, is crucial for capturing differences in disabled people's relationships to their bodies, and to the nature of disability more generally. The healthy disabled may be right when they say that their disabilities are a natural part of human diversity that don't automatically make their lives worse and that shouldn't obviously be "cured." They might also be right in objecting to the idea that their disabilities are medical problems and in maintaining that the bulk of the bad effects of their disabilities are caused by discrimination rather than the biological state of their bodies. But the same might not hold for the unhealthy disabled. Their experience of their bodies might be more painful and complicated. It might be that their disabilities really do detract—independently of discrimination or social barriers—from their wellbeing, or at least contribute to it instrumentally only via the hard lessons of suffering. And perhaps a cure for such disabilities might be desirable. Moreover, such disabilities do have distinctively medical aspects, even while granting that they incur many social barriers as well. In short, Wendell argues that a mere-difference view of disability is plausible for some disabilities (the healthy ones), but she suggests that the picture might be more complicated

---

[24] Wendell seems to want to draw this distinction between disabled *people* rather than between *disabilities*—she describes it as the distinction between the "healthy disabled" and the "unhealthy disabled." But this can't quite be right, given that someone can be an unhealthy disabled person for reasons that have nothing to do with disability—e.g., you can be a deaf person with acute lymphoma. So in what follows I'm going to characterize the distinction as applying to disabilities rather than to disabled people.

for others (the unhealthy ones), even while we allow the rich experiences people can have with and around them.

There are two main worries for this approach. Firstly, I am unconvinced that the "healthy versus unhealthy" distinction between disabilities, if it could be drawn in the way Wendell suggests, would tell us anything deep about the complicated relationship between disability and wellbeing. Many people who fall clearly on the "unhealthy disabled" side of Wendell's proposed divide have been among the most eloquent and vocal proponents of the idea that disability is a valuable part of human diversity that can create rich and unique experiences (experiences that go well beyond the instrumental value of what we learn through suffering).[25] Many people on the "unhealthy" side of the divide have also been among the most politically active critics of the idea of cures as an unequivocal good or primary goal for disabled people. If we think the testimony of disabled people is one of the major reasons to support something like a mere-difference view of disability—allowing that disabled peoples experiences of disability can and do vary significantly—then it's not clear why we should think that a mere-difference view of disability is substantially more plausible for the "healthy" disabilities than for the "unhealthy" ones.

But my more substantial worry—at least for the purposes here—is whether the healthy/unhealthy disability distinction can be drawn in the way Wendell suggests. Wendell grants that the distinction is vague, and allows that some disabled people might cross the boundary in the course of their lifetime (e.g., someone who was previously "healthy disabled" in the aftermath of polio might become "unhealthy disabled" due to post-polio syndrome). But she nevertheless thinks it is substantial enough to do important theoretical work. Arguably, though, in characterizing the distinction Wendell overstates the "static" and "non-constitutional" nature of many of the conditions she considers paradigm healthy disabilities. A person who uses a prosthetic due to a limb amputation, for example, will often need ongoing physical therapy and will regularly need their prosthetic refitted due to changes in their affected limb. Many amputees also experience chronic pain—this can include the famous "phantom limb pain," as well as pain that results from the of the musculoskeletal effects of adapting to a prosthetic limb, as well as soreness and skin irritation at the site where their organic limb attaches to their prosthetic one.

Similar points can be made for many other disabilities that fall on the "healthy" side of Wendell's proposed distinction. Many people with spinal cord injuries require complex ongoing medical care; many people with achondroplasia

---

[25] The works of Harriet McBryde Johnson (2005) and Nancy Eiesland (1994) are two prime examples. Eiesland defended the most resolutely mere-difference stance that I have found to date: she was a Christian who believed in bodily resurrection, and she believed that her resurrected body would be disabled in heaven (or at least would have the physical differences that marked her out, in our context, as disabled).

experience chronic back and leg pain; many people with cerebral palsy need regular medical and physical therapy and most have chronic pain; and so on. Indeed, many disabilities that Wendell places on the "healthy" side of the divide because of what she describes as their static nature are associated with some of the most significant aspects of reduced health, such as shortened lifespan. People with spinal cord injuries, for example, have a substantially lowered life expectancy relative to population norms.[26] In contrast, some of the conditions she places firmly on the "unhealthy" divide—although they uncontroversially involve distress and suffering—actually involve relatively few major objective markers of reduced health. Many chronic pain conditions, for example, as well as chronic fatigue syndrome—her own experience, which Wendell discusses evocatively, as one way of being "unhealthy disabled"—involve a large amount of pain and fatigue but don't cause reduced life expectancy, major organ damage, or similar.[27]

Wendell is surely right that many people with what she describes as "unhealthy disabilities" have complicated relationships with their bodies—they deal with pain, they have to regularly see doctors, their bodies change in very unpredictable ways. And none of that would go away in a world without ableism. But I think Wendell underestimates the extent to which this is true of the everyday reality of *most* disabilities. There is doubtless a spectrum, with some disabilities more associated with medical or constitutional aspects than others. Perhaps congenital deafness is the paradigm example of a disability with little further impact on a person's overall health, and MS a paradigm example of a disability with profound and degenerative impact on a person's health. But most disabilities involve reduced health along *at least some dimension*, and most disabilities involve specific forms of medical care and treatment. I'm thus skeptical that a "healthy versus unhealthy" distinction between disabilities can be drawn successfully, or at least drawn in a way that will illuminate the thorny relationship between disability, health, and wellbeing.

4. A Metaphysical Interlude: The Statue and the Clay

Part of the problem we're running into here is that in talking about disabilities, we're talking about something that starts to evoke a kind of duck/rabbit phenomenon: how we perceive it changes depending on how we're looking at it. Many of the physical conditions that we classify as disabilities can legitimately be viewed as medical problems. And from that perspective, it makes sense to say that we should seek to treat them as best we can, to minimize the harm they cause,

---

[26] See Middleton et al. (2012).

[27] The distressing exception to this, in the case of chronic fatigue syndrome, is that people with the condition seem to be at somewhat higher risk of death by suicide. See Roberts et al. (2016).

reduce or eliminate them when we can to promote better health, and so on. But when we consider these conditions as part of people's overall experience of their bodies, they can also be viewed as rich and complicated social phenomena, from which perspective it makes sense—at least if you're an advocate of a mere-difference view—to say that they don't detract from a rich a valuable life, that they're an important part of human diversity, that they should be embraced as part of the spectrum of weird and wonderful ways that human bodies can be, and so on.

These perspectives on the same bodily states both make sense. And it's important to recognize both. As many disability rights activists have pointed out, for example, progress on disability rights stalls when we respond to any claim of disadvantage from a disabled person with "the quest for a cure" for their physical condition or by seeking better treatment for their physical condition—rather than by prioritizing things like building ramps, making our doorways accessible, or providing more flexible working environments. At the same time, though, a disabled person could be rightly annoyed if, instead of therapeutic exercises to help improve her balance and coordination, her physical therapist handed her a pamphlet on Disability Pride.

The trouble is that these two ways of conceptualizing the same physical state of a human body seem directly incompatible with each other. Viewed as a richly embedded social phenomenon, we want to say that such conditions should be embraced and destigmatized. Viewed as biomedical pathology, we want to say that such conditions should be treated, minimized, and prevented where possible. And we can't truly say both of those things at the same time. You can't embrace and normalize something you're trying to reduce and prevent.

This kind of dual-perspective puzzle has a famous, though much more abstract, analogue within metaphysics: the puzzle of the statue and the clay. And, perhaps just because of my philosophical background, I have found thinking about this puzzle illuminating in trying to tease out some of the confusions we find ourselves in when trying to reconcile the relationship between health and disability.

In brief, the puzzle goes like this. Imagine a beautiful statue made from a lump of clay. Intuitively, there's a single object we're considering—a statue made from clay. The problem is that what is true of the object when we consider it as a statue seems to conflict with what is true of the object when we consider it as a lump of clay. The statue can't survive being squashed down into a ball—if you squashed it, you'd destroy the artwork. But the lump of clay can, quite easily, survive being squashed down into a ball—it would be the same lump of clay, just formed into a different shape. The form and shape seem essential to the statue, but not to the lump of clay. Similarly, the statue could have been made with slightly different material, whereas the lump of clay could not. Had the artist grabbed a slightly different handful of clay from her stash, she could still have made the very same statue, but she would not have used the very same lump of clay in the process. The material constitution of the lump of clay is essential to it, but not to the statue. And the artist *created* the statue when she molded the clay, but she didn't create the lump of clay. The artist didn't somehow find or set free a statue that had been hiding in the lump of clay the whole time—she brought the statue into being by molding the lump of clay in a particular way. But the lump of clay, in contrast, *has* been there the whole time (and would still have been there even if the artist had decided not to create the statue, or to create a different statue).

So what is going on? We have what looks like a single object that both can and can't survive being squashed, that could and couldn't have been made of different parts, that does and doesn't predate an act of artistic creation, and so on. That's not a way philosophers like objects to be. And so there's a small cottage industry of attempted solutions to the puzzle of the statue and the clay. And several of the most prominent of these solutions are, I think, helpfully analogous to ways that people have tried to understand the relationship between disability and loss of health.

4.1 Two-Thing Solutions

One prominent way of getting out of the statue/clay puzzle is to insist that, in fact, there are really two things where there are appears to be only one. On one prominent articulation of this response, the statue is *constituted* by the clay but it is not identical to it.[28] The statue is the thing with a certain form and shape, individuated by its aesthetic properties and causal history. The lump of clay isn't individuated by any of these things—it could have (and has had) different aesthetic properties, different shape, and needn't have any particular causal connection to an artist. The lump of clay is the material that constitutes the statue, but they're two different things—they have different properties, and while the statue is in fact constituted by this particular lump of clay, it could've been constituted by others.

We can think of a two-thing solution like this as—roughly—analogous to a disability/impairment distinction. There are conflicting things we want to say about particular embodied states—the ones we recognize and classify as disabilities. The disability/impairment distinction tries to ease this conflict by maintaining that, when we are discussing a particular embodied state of this kind, there are in fact two things here, not one. There's the impairment, which is just the physical/biological state of the human body. And then there's the disability, which is the social entity that's created by that particular physical state being socially positioned in specific ways.

---

[28] See, *inter alia*, Thompson (1998) and Baker (1997).

4.2 Dominant Sortal Solutions

Another option for dealing with the statue and the clay is to maintain that a single object—the lump of clay shaped into a statue—falls under the sortal terms "statue" and "lump of clay," but that "statue" is the *dominant sortal*.[29] Some pieces of clay are most saliently statues, and some are not. When the artist takes a random lump of clay and molds it into a beautiful statue, the lump of clay that was goes out of existence and a new thing, the statue, comes into existence. Now, of course, the statue is made of clay, and it's made of the very same clay that was the lump. And in some sense, the statue is still a lump of clay—it's just a lump of clay that is a very particular, statue-like way. This new object simply couldn't survive being squashed—squashing it would destroy the statue and create a (mere) lump of clay.

There might, though, be other cases in which the distinction is less clear, or for which "lump of clay" is the dominant sortal for a thing that is also a statue. Suppose that an artist, making a modernist commentary, grabs a lump of clay, leaves it shapeless, and calls it "Statue." It's then in some sense a statue, but it's perhaps more saliently just a lump of clay. And it's harder to say, in a case like this, that the artist, just in the act of baptizing the lump "Statue," has destroyed the lump and created a new object. Could this modernist statue survive being squashed? Who knows—the ontology of modern art is a subject for a different day—but it's obviously a trickier question than the case in which the artist takes the lump of clay and sculpts it into a bust of Socrates.

The dominant sortal view can be seen as—again, roughly—analogous to Wendell's healthy/unhealthy disability distinction. There are some physical states that are rightly considered biomedical pathology in isolation, but once you embed them in a life—and in the complex social context of how we view, think about, and treat people because of their bodies—they become *disabilities*. Disability is the dominant sortal. And sure, disabilities are in some sense grounded in or constructed from biomedical pathologies, but once they're disabilities—once they take on that full social meaning and inherit all those complex social properties—they become something new and different.

And yet, there are some things that we might correctly call disabilities for which this transition is less clear, and for which it's less obvious that "disability" really is the dominant sortal. Just as "Statue," the modernist commentary, feels a lot more like a lump of clay than the bust of Socrates, these conditions feel a lot more like illnesses, like diseases, like things it's appropriate to think about in biomedical terms, than do conditions like deafness or achondroplasia.

---

[29] See especially Burke (1992) and (1994).

4.3 Lewis on Context and Counterparts

Hopefully these analogies will now better allow me to situate my own preferred response to the puzzle. The relationship between disability and reduction in health, I suggest, can be seen analogous to David Lewis' approach to the relationship between the statue and the clay.[30] To be clear, I don't intend to endorse a Lewisian account of the relationship between health and disability—whatever that might be—or to take on board all of the technical details of Lewis' preferred solution. Rather, the *overall basic gist* of Lewis' response was illuminating for me as I thought through this issue, and I've found the structural analogy helpful in framing my own view. (If you like esoteric puzzles, join me in some exploration; if you don't, and this is confusing rather than helpful, don't worry—we'll zoom back out shortly.)

Lewis' response to the statue/clay puzzle is metaphysically quite minimal. According to Lewis, there is a single object: a statue made from clay. The explanation of how we can say apparently contradictory things about one and the same object is then, for Lewis, primarily a semantic and not a metaphysical story.

For Lewis, there is just one thing, the statue/clay. But, depending on the context, we can refer to the object in ways that make different things about the object salient. We can refer to the object qua statue, and in doing this we invoke a particular *similarity relation*. For the fans of esoterica: on Lewis' view, x is possibly F just in case x has a counterpart that is F. But which objects are x's counterparts is determined by the context, and more specifically by which similarity relation we invoke in a context. Could the statue/clay survive being squashed? Well, when we refer to it qua statue we invoke a similarity relation that makes salient its aesthetic properties. It's counterparts, in this context, are things like Rodin's *Thinker* and Michelangelo's *David*. None of these things are squashed, amorphous lumps. And so, in this context, no—the statue can't survive being squashed.

But we could also refer to the statue/clay qua lump of clay. In that context, we invoke a similarity relation based on the object's material properties. The statue/clay, according to this similarity relation, has counterparts that are statues, counterparts that are pottery bowls and mugs, counterparts that are amorphous lumps, counterparts that are pretty much anything you can do with a lump of clay. What's salient, when we're speaking about the statue/clay qua lump of clay, is the fact that the object is made of clay, not that it's been molded into a statue. And so, in this context, yes—the statue could survive being squashed, or made into some entirely different shape altogether.

---

[30] This approach is developed throughout Lewis' systematic philosophy, but see especially Lewis (1971), (1983), and (1986) (especially chapter 4).

The key insight of Lewis' solution to the statue/clay puzzle is that the very same thing can have an apparently different nature—a different essential nature,[31] a different modal profile—depending on the context in which we're viewing it. Crucially, for Lewis, this is not just because extrinsic features of the object change as the context changes. Familiarly, a 6-ft man might be tall at a philosophy conference, but not tall at a basketball game. That he's tall in one scenario but not in the other is no great mystery—whether you're tall depends on who you're being compared to, and if we change who you're being compared to we might change whether you have the property "is tall." But for Lewis, the change in the object's character arises from how we view and speak about the object, not (merely) from changes in the external environment. I can view the statue/clay qua statue, or I can view it qua clay. And in doing so, I change the context, and change the salient metric of similarity. The resulting picture is not simply one in which an object can have different things truly predicated of it in different contexts, but one in which the very same object can have different *natures* in different contexts.

When we view the statue/clay qua statue, we create a context in which what's most salient about the object—what's essential to it, what makes it the thing it is— are its form, its aesthetic properties, its causal relationship to an artist, and so on. When we view the statue/clay qua clay, we create a context in which what's most salient about it—what's essential to it, what makes it the thing it is—are its material constitution, its physical makeup, and so on. There's only one object. But what is true about the nature of that object depends on the context in which we're viewing it.

There is also, for Lewis, a way of viewing the object according to which we can say what the object is like simpliciter—there is a maximally expansive context that considers all the potential similarity relations and all the ways the object could be viewed. From this maximal context, we can say that it has such and such aesthetic properties and such and such material properties, and so on. And we can say that it has these counterparts according to this similarity relation and these other counterparts according to this other similarity relation. But this isn't the same as viewing the object from a context in which it is every way it can be. It isn't, in this context, essentially statue-shaped, or essentially material, even though it has those natures in more restricted contexts. There is simply no answer when describing what the object is like simpliciter (that is, from the maximally expansive context) to the question "can the object survive being squashed?". To answer a question like that, we have to narrow our focus. It can't be squashed if we consider

---

[31] For the nerds: on Lewis' view, changing the context doesn't change what properties an object has, but it does change what we can truly say the object is essentially like. Changing the context also changes whether, for example, a particular property an object has is picked out by the predicate "squashable."

it qua statue; it can if we consider it qua lump of clay. And there's no further, more fundamental way of answering that question.

5. Disability, Health, and Context

5.1 A Lewisian Take on Disability and Loss of Health

You might be scratching your head slightly, at this point, over what counterpart relations of similarity for statues have to do with disability, but bear with me. What I'm proposing is that analogy to the Lewisian-style solution to the statue/clay puzzle can help us think about the relationship between disability and loss of health. Let's consider a particular disabled person. That person is in a specific bodily state such that they're disabled (in this particular context) in virtue of being in that state—on my view, because that state falls under a specific cluster concept of norms, social positions, and judgments.[32] Now consider that particular bodily state (it might be the overall effects of an incomplete L3 spinal cord injury, or it might be the overall effects of a genetic difference like achondroplasia, or it might be the overall effects of ongoing autoimmune process like lupus, or etc.). Just as we start with what looks like a single object—the statue made from clay—but seem to want to say different things about the statue than we do about the lump of clay, here we start with a single state—the physical state of a body—but seem to want to say different things about the disability and the loss of health.

For the purposes here, I'll refer to the specific loss of health involved in the physical states we classify as disabilities "biomedical pathology." Biomedical pathology is something we typically want to minimize, treat, and prevent—or prevent the progression of—where possible. And what I'm arguing is that biomedical pathology bears a statue/clay-type relationship to disability, and that a broadly Lewis-style approach is the right way to conceptualize that relationship.[33]

---

[32] Physical disabilities are, on my view, states of the body. But what unifies the various, disparate things we classify as "disability" is their social significance: they're physical states that are unusual, that we tend to view as especially "defective," that single you out as "different" specifically with respect to how your body functions, that affect the way you use your body in everyday life to navigate social environments, that make assistive technology beneficial, etc. There might also be physical things these traits have in common—they might, in fact, all be statistically atypical or all be departures from normal species functioning, but the explanation for what, if anything, unifies the relatively odd group of physical traits we call "disability" together, on this view, is social rather than biological. See my (2016), chapter 1.

[33] Again, what I'm arguing is that a *Lewis-style* approach is helpful here, not that we should endorse the details of counterpart theory. And more broadly, much of what I want to argue is that we need resources that allow us to embrace, rather than try to explain away, the kinds of duck/rabbit tensions we find with puzzles like this in talking about health. In Chapter 6, using the work of Delia Graff Fara, I attempt to show how a certain kind of contextual shiftiness can help us in talking about health more generally.

There's a single thing here—the particular physical state of a human body. But we can view that thing qua disability or we can view it qua biomedical pathology. When we view it qua disability, we create a context in which its social- and identity-related features and its embedded social context are most salient. Its salient counterparts, in this context, are other ways that bodies can be that are stigmatized, that we're trying to promote inclusion and justice for, that provide unique and rich social experiences that depart from the norm, that can shape a person's sense of who they are—trans bodies, femme male bodies, brown-skinned bodies (in white-dominated contexts). In this context—with these counterparts— it's true to say that disability should be embraced, destigmatized, and treated as a part of the spectrum of human diversity. It's true that we should focus on changing our norms and our social arrangements rather than changing people's bodies. And so on.

But we can view the very same thing—the very same bodily state—qua bio-medical pathology rather than qua socially significant axis of difference. And viewed qua biomedical pathology, we create a context in which its most salient features are the (present and future) harm it can cause to the ongoing function and survival of the human organism. In this context, its counterparts are other physical states that can cause similar types of harm—cancer, infectious disease. And so in this context—when viewed qua biomedical pathology—it's true to say of bodily states like these that we ought to work to minimize and treat them where we can, and offer prevention and cures where we can. It's true to say that they're harmful.

The very same thing—the very same physical state of a body—can, on this view, have different natures depending on the context in which we're considering it. When we're thinking about someone's disability as a complex social phenomenon, we create a particular kind of context. When we're thinking about the very same bodily state in terms of reduced health and biomedical pathology, we create a very different kind of context. And what is true of the state when viewed qua disability is different than what is true of the state when viewed qua biomedical pathology.

A note of clarification before moving on. I'm framing this, for convenience, about a contrast between a biomedical context and a socially embedded context, but I'm not in any way assuming that social norms aren't part of medical contexts or part of how we determine harms in medical contexts. Medical contexts are themselves thoroughly social. I'm also simplifying in assuming there's a single biomedical perspective and single socially embedded perspective, when obviously there are many. And there are, of course, many other perspectives at stake as well, which might view the same physical condition somewhat differently, such as the perspective of public health policy, of governments considering how to distribute benefits and resources, and so on. We'll discuss these contrasting perspectives more a bit later, but hopefully this simplification will make sense as a way of illustrating the main idea. I'm also not, in presenting this view, suggesting that ordinary speakers use the term "disability" in this way, or make these kinds of perspectival distinctions. What I'm arguing is that there are distinctions to be drawn here, and that it would be helpful to draw them. Nothing about this, though, hinges on how we do or should employ the ordinary-language term "disability."

5.2 An Analogy to Aging

If all this seems a little too convoluted, or a little too much like semantic trickery, then let me suggest that this is also a helpful (and somewhat more familiar) way to think about a closely related issue: aging. Aging is a physical process, but we can view that same physical process from a context in which we highlight its social features, or from a context in which we highlight its biomedical features. So consider, again, a particular physical state of a person's body—say, of your average sixty-year-old—that we'd classify as an aging body.

We can—legitimately—view this bodily state as a type of biomedical pathology. The physical process of aging involves, quite literally, the slow decline and decay of the human body, ending inevitably in death. The average sixty-year-old (or even a very healthy sixty-year-old) will almost certainly have some health issues, which will likely be linked to aging. And viewed in this light, it makes sense to try to stall and minimize the effects of aging as much as possible, in order to help people live healthier lives for longer. And it makes sense to recognize the harm that aging can routinely cause—the pain, the functional limitation, the increased fragility.

But we can also—legitimately—view aging from a social perspective. And viewed in that context, we want to emphasize that our society overvalues youth and wrongly stigmatizes age. We likewise want to emphasize that there are a lot of good things about getting older, and that many people are happier and more content with their lives as they age, find a lot of value in aging, and wouldn't want to turn back the clock to their younger years. That is, from a social perspective, we want to embrace and destigmatize aging.

And so, again, we have the same type of picture. Viewed qua biomedical pathology, aging is something we should combat, treat, and minimize. Viewed qua biomedical pathology, aging is harmful. But viewed qua social phenomenon, aging is a valuable and natural part of the human experience, and we need to work to make society more inclusive of older people (rather than trying to make older people more like younger people). And while we can of course acknowledge that aging is associated with loss of health, viewed qua social phenomenon (where what's salient is people's overall wellbeing or quality of life), it's true that aging isn't by itself or intrinsically harmful, and that it can often be a rich and valuable experience.

When it comes to aging, both these perspectives are legitimate. Aging has the same kind of duck/rabbit shift that disability does—where it can seem like the nature of the thing we're looking at shifts depending on how we view it. A Lewisian-style "shifting perspective" approach allows us to explain this. And, more strongly, it allows us to explain how the set of things we want to say about aging in a social context and the set of things we want to say about aging in a biomedical context, although they are incompatible with each other, are *both true*. Just like we can say that, in one sense, the statue can't survive being squashed, but in another sense it can, we can also say that, for example, in one sense aging is harmful and in another sense it's not. And we can maintain that these are *both* reasonable ways to view aging, and neither is somehow more true or more legitimate than the other.

5.3 Contexts and Justice

A common criticism of the Lewisian solution to the statue/clay puzzle is that, for Lewis, contexts are cheap.[34] We can gerrymander almost any context—and any corresponding similarity relation—we like. So, sure, it's *true*, once we fix the context appropriately, that the statue can't survive being squashed. But it's not true in any deep or substantial way. Nothing about that context is better or more privileged than a context in which the statue could be squashed (or, indeed, a context in which the statue could've been a secret robot—possibility is wide open, for Lewis, and it's how we choose the context, rather than any deep truths about the nature of the statue itself, that constrains what it could or couldn't have been).

This is a place where a broadly Lewis-inspired approach will arguably do *better* applied to a normatively weighty issue than applied to an esoteric puzzle like the statue and the clay. We can easily shift and gerrymander contexts in whatever way we like, in the state/clay case, because neither metaphysical nor semantic facts determine which context (or corresponding similarity relation) is best. Refer to the statue qua clay, refer to it qua statue, refer to it qua thing in a museum, refer to it qua thing that metaphysicians spend too much time thinking about—refer to it however you like.[35] It doesn't really matter, and no context is better than any other.

The way we view things like disability and aging, in contrast, is constrained by normative and political considerations in a way that discussions of statues are not. There are, in at least most cases, no deeply important moral constraints on how we

---

[34] See especially Fara (2009).
[35] This is consistent, though, with Lewis maintaining that there are broader contexts in which it would be misleading to refer to the statue/clay qua lump of clay, or in which we'd communicate better by referring to it qua lump of clay. If you're at a fine art auction, it makes sense to refer to the statue/clay qua statue, and would be misleading and unhelpful to refer to it qua clay, for example.

speak about statues/lumps of clay. But of course there *are* deeply important moral constraints on how we speak about disability or about aging. And these moral constraints can give us guidance for which context is the right context, the best context, the appropriate context.

Again, let's consider aging. Take, as a specific example, someone with the ordinary sorts of joint and ligament changes you'd expect to see in a sixty-year-old—some stiffness, osteoarthritis in a hip, and so on. There are contexts where it's important to view this bodily state qua biomedical pathology. There are real and serious physical harms and physical discomforts associated with aging, and there are plenty of times when we need to focus on ways to alleviate them. If you were to go to your doctor complaining that the pain and stiffness in your hip was limiting your activities, you'd be rightly frustrated if she responded, "You know, we really need to stop prioritizing youth and treating aging as a bad thing—a lot of people find as they age that a slower pace of life really suits them, and that there's a lot to be gained from that perspective! Did you know that life satisfaction often increases past middle age?" All that might be *true*. But your hip hurts, and you would really like it to hurt less, and it's really frustrating to you that the pain and stiffness in your hip is limiting the hiking you can do. This is a scenario in which the morally best context seems to be one in which the bodily state in question is viewed qua biomedical pathology—a context in which aging is bad for you and we need to do what we can (pain relievers, physical therapy) to help mitigate its effects.

In other scenarios, though, the social context might be the much better context—again, based on normative rather than semantic considerations. Suppose that a younger hiking enthusiast derisively says that she wants to get all the "good hikes" in while she can before she gets too "old and frail" and can only do "grandma trails." You might immediately point out that, in fact, a lot of people end up really valuing the slower pace that aging imposes—they become less competitive with themselves, things like hiking become more about the journey and less about the conquest, they learn about spotting wildflowers or the best light for macro photography or all the subtle details that passed them by before. Life satisfaction, you point out—including satisfaction with recreational activities—often increases past middle age. Here, you're trying to combat stigma and highlight the ways in which an aging body, when embedded in the rich fabric of someone's life, is fully compatible with, and sometimes conducive to, high quality of life. Here, it makes sense—focused on overall quality of life and wider social norms—to insist that aging needn't be bad and needn't be harmful. You'd be callously ignoring someone's suffering if you insisted on this context for someone seeking medical care, but you'd be reinforcing harmful stigmas if you insisted on the biomedical context when talking about aging across the board.

Note that this shift goes beyond the kind of reconceptualization sometimes described as "conceptual engineering."[36] A defender of conceptual engineering might, for example, argue that we need to shift our understanding of concepts like "bad" or "harmful," depending on our purposes. There is a sense of "harmful" in which the aging knee is harmful, but another sense in which it is not harmful. What I'm advocating is something stronger. It is *how we view the physical state in question* that is shifting, just as—for Lewis—it is how we view the single statue/lump entity that is shifting. We are considering a single thing—an aging body. And we are holding fixed many of our basic norms, values, and concepts (just as when we consider the statue/clay, we're holding fixed what it is to be squashable).[37] But what we can say about that single thing depends on the (appropriate) context in which we view it.

Is the aging body harmful to the person in question? There's no God's-eye, perspective-free way to answer this question. Viewed from the perspective of the person's overall wellbeing, it's not. Viewed from the perspective of biomedical pathology, it is. Perhaps we employ different senses of "harm" in different contexts—aging might be harmful to you in a more restrictive, medically salient sense, but not harmful to you when harm means "negatively affecting your overall quality of life." "Harm" is a loose enough concept that this kind of shifting isn't implausible. But what I'm arguing is that sometimes, as the analogy to the Lewisian approach would maintain, our basic sense of "harm" can remain the same, but that the reference changes depending on the context. We don't change our idea of what it is to be squashable in different contexts, for Lewis; it's just that in different contexts that same basic idea refers to different properties. Maybe "harm" generally just means something like "bad for the individual," but in the medical context what's especially salient are the goods associated with health, and so things that reduce health count as bad for the individual, and thus as harmful. But in a broader context where we care about a person's overall wellbeing and the course of their life, what's most salient is the overall character of their experiences; and in that context, reduction in health won't automatically count as bad for the individual, and thus won't automatically count as a harm.

Here, we can imagine an ethicist stamping her foot and saying, "yes, but I want to know what's *bad*." What I'm arguing is that often it's a little more complicated than this. For some bodily states associated with aging, the social context will nearly always be the right context. Take, for example, aging skin. We certainly can view aging skin as biomedical pathology—as the accumulation of UV damage,

---

[36] See especially Cappelen, Plunkett, and Burgess (2019). We'll return to the issue of conceptual engineering in section 5.4, and again in Chapter 5.
[37] Again, for Lewis we're holding fixed the basic idea but changing which property is picked out— that is, holding fixed the sense but shifting the reference.

collagen degeneration, and reduced cell turnover that occur in the human epidermis as it degrades over time. And we can treat it as such, with creams, peels, and injections. But most of time, the best approach to normal skin aging is acceptance and destigmatization. Viewed qua biomedical pathology, we can truly say, "Wrinkles are bad—they're a basic sign of your skin degrading." But more often than not, the morally and politically better thing to do is to view skin aging qua embedded social phenomenon, from which perspective we can truly say, "Wrinkles aren't bad! Wrinkles are beautiful! Embrace the lines around your eyes—they're the marks of all the times you've smiled." That is, more often than not, the morally and politically better thing to do is to adopt a context in which we can destigmatize and embrace rather than a context in which we can pathologize and treat.

On the other hand, consider something like the neurodegeneration that underlies Alzheimer's disease. We can gerrymander social contexts in which we seek to destigmatize Alzheimer's (although even in such contexts, "embrace" is probably a step too far). But more often than not, the most morally and politically helpful response to Alzheimer's is arguably to adopt the biomedical context—to look for a cure and to seek better treatments. That context is, of course, fully compatible with discussing the injustices of long-term care homes, or the awful ways we often treat older people with dementia. But it treats as essential to the neurological changes associated with Alzheimer's that that are biomedical pathology—something that's gone wrong and something that's harmful and something that we should seek to eliminate.

For many bodily states associated with aging, though, the morally best way to view them will be a complex mixture of different contexts. Take the physical states associated with menopause, for example. Somewhat perplexingly, menopause is both over- and under-medicalized at the same time; we often pathologize it far too much, but we also don't understand how to treat the very real and distressing symptoms that many women experience nearly as well as we should. There are contexts where what's most helpful is to normalize menopause, and to emphasize the way in which many women find that the physical changes involved ultimately don't decrease (and sometimes even increase) their quality of life. There are also, conversely, contexts in which what's most helpful is to emphasize the physical suffering that many women endure in menopause—the hot flashes, the headaches, the low energy—so that we can better understand their biological basis and find better ways of minimizing them. Navigating between these contexts—between the places where it's best to embrace the normal changes of an aging female body and combat the stigma that women are in decline once they can no longer bear children versus the places in which women's health and women's suffering is too often ignored by contemporary medicine—is a delicate task. And there's no God's-eye view—no ultimate "master context"—from which we can decide which context is the best to employ in any given scenario.

Why can't we just zoom out and consider the most expansive context? Why isn't the context that considers both the biomedical perspective and the socially embedded perspective thereby the correct or best context? The problem is that, especially in the normative case, a more expansive context isn't automatically a better or more informative context. Recall that in the statue/clay case, for Lewis, if we consider the maximal context—every way the object could be—we lose the ability to say that the object has any particular nature essentially (the way we can if we view in qua statue or qua clay). Again, this might seem like semantic trickery in the Lewisian case. But in the normative case, I think there's much more to the idea that we can sometimes lose important information by trying to zoom out and trying to "see all sides."

In the doctor's office, age-related changes are harmful pathology that should be minimized. In your everyday life—with your family, at your dinner table, in your hobbies—they may well be valuable (if surprising) parts of your overall wellbeing. Which way of viewing your body is correct? Well, each is important. Can we then learn the truth by just combining them? Not obviously. It might be that to live your best life in an aging body, you need to—for the most part—tune out the medical perspective, and consider yourself to be mellowing like a fine wine. You will then, occasionally, need to put your patient hat on and go to the doctor about that pesky hip. More often than not, though, it's a perspective you can (and perhaps should) actively ignore. But at the same time, you need your doctor to actively pursue that perspective. Your doctor needs to rage against the dying of the light. And you would rightly be frustrated if your doctor pointed to your x-ray and said "it could make sense to think of these changes as part of mellowing like a fine wine." The point here is something akin to *role morality*. Just because there are multiple, contrasting perspectives on normatively significant matters doesn't mean we can arrive at what is best or most informative by combining them into a single perspective. Often specificity is illuminating, and morally important. (We'll return to this issue in Chapter 6.)

The same is true, I'm arguing, of many of the physical states that we classify as disabilities. For some bodily states that would mark a person out as disabled, the morally and politically best way to view them might almost always be qua biomedical pathology. Consider, for example, the neurodegeneration associated with amyotrophic lateral sclerosis (ALS). Someone living with ALS will be classed as a disabled person, and there might be times when the most helpful thing to do is focus on accessibility, destigmatize the use of wheelchairs and assistive devices, and so on. But the vast majority of the time, I contend, the best way to view ALS is qua (devastating) disease—a something that's harmful and something we want to cure.

For many—perhaps most—of the physical states we class as disabilities, though, it's much more of a mixed bag. When someone with cerebral palsy goes to a doctor or physical therapist, it makes sense to view her condition as pathological—to say that her muscle spasticity and functional limitation is harmful to her, the result of an injury to her brain, and to focus on ways to minimize it. But when the same person attends a Disability Pride rally, it makes sense for her to celebrate the ways in which her body is different, and to emphasize that the state of her body—far from making her miserable—is fully compatible with a rich life, and deeply entwined with some of the experiences she values the most. Her functional differences are bad and harmful when viewed by the physical therapist, but not bad and not harmful when viewed by the Disability Pride activists. And what I'm arguing is that *both claims are true*, and true of the very same physical state.

Note that this is not—by any means—to say that all the claims made about disability in medical contexts are true, or to deny that that many assumptions about disability in medical contexts are ableist. It's a much more minimal claim: it's true in these contexts that the physical states that we classify as disabilities are pathological, are harmful, should be treated, should be prevented where possible. But this is, of course, compatible with there being many false, and ableist, assumptions about the nature of disability, or the quality of life had by disabled people, at play in medical contexts. The person in the medical context can be saying something true when she says that the physical state is pathological or harmful, without thereby being right in everything else she says.

Just as we can give a complete description of what the statue/clay is like simpliciter, we can also give a complete description of what the physical state is like simpliciter. The particular state is, for example, directly causally related to reduced health, impaired physical functioning, and physical discomfort. It is also compatible with high overall levels of wellbeing, is something that can be a valuable part of a person's sense of self, and is something that can give rise to rich and unique social experiences. But often, in order to address questions like "is this state harmful?" or "should this state be minimized or prevented?", we have to get more specific. Just as we can't answer "can this object survive being squashed?" without narrowing our focus, we can't answer questions like "should this state be prevented?" without narrowing our focus. There's no context-neutral way of answering a question like that. But, in this case, which context we employ will make a significant normative difference.

Which context we employ, then, will be a matter of which context does the most good—which perspective it's most useful to adopt in accomplishing what Sally Haslanger (2012a) would call "our legitimate political and social goals." It won't be a matter of the deep truth about disability. Both sets of claims, on the view I'm defending, are equally true of disability (depending on the context), and neither is a more accurate description.

5.4 Contexts and Conflict

Again, in the picture I'm describing, the contextual shift is not over what "bad" or "harmful" means. Let's go back to the statue/clay. By shifting the context to view the statue/clay qua statue or qua clay, we aren't shifting what we mean[38] by terms like "squashable" or descriptions like "existed before it had the form it currently does."[39] Rather, we're shifting how we view the object—what things are most salient about it, and thus (in Lewisian terms) what its counterparts are. When we view the object qua statue, its aesthetic properties are most salient, and so its salient counterparts are other statues. None of these counterparts are formless lumps, and so in this context we say the statue isn't squashable—it couldn't survive being squashed down into a lump. Likewise, all these counterparts are created by artistic endeavor, so viewed in this light the statue's causal history is essential to it—it didn't exist (as a formless lump, waiting for its statue parts to be discovered) before the artist made it.

In contrast, when we view the same object qua lump of clay, what's most salient are its material properties, and so its counterparts are other pieces of clay, in all manner of shapes and arrangements. Plenty of these counterparts are lumpy, and so it's true in this context that the statue/clay is squashable. Likewise, it's true that it existed—as a lump—before the artist molded into a statue shape. But in making this contrast, we're holding fixed our concept of what it is to be "squashable." What's changing is how the single object—the statue/clay—is viewed.

This makes the proposal I'm presenting somewhat different from some popular extant models of contextual shift and normative disagreement. There's been a lot of discussion, in recent literature, on theories of *conceptual engineering* and *metalinguistic negotiation*.[40] The rough idea behind such projects is that we can often diagnose places of normative disagreement as places where we're engaged in a metalinguistic dispute about which concepts to deploy. If you say "Formula 1 is a sport" and I say "no, Formula 1 is not a sport," we could be having a factual disagreement about the nature of Formula 1 auto racing. But more likely than not, I'm not directly contradicting your assertion. Rather, we're having a disagreement over what concept of "sport" is the best concept to deploy. This (very roughly) is metalinguistic negotiation, and what I'm arguing is that disputes about the nature of disability go deeper than this.

---

[38] "Meaning" as in "sense" or "content"; we are, for Lewis, changing the reference (or the character). The basic sense of "squashable" remains "has squashed counterparts," but which property this picks out changes depending on the context.

[39] This is in contrast to standard "contextualist" approaches. The reason that the 6-ft man is tall at the philosophy conference but not tall at the basketball game is simply that we're shifting what we mean (as in the sense or content) by "tall."

[40] As in, e.g., Plunkett and Sundell (2013), Plunkett (2015), and Thomasson (2017).

Consider a case in which a doctor and a disability rights activist are disagreeing over whether a particular disability is harmful. Everything I've said is fully compatible with there being *some* conceptual disagreement in a case like this. But on the model I'm presenting, they can fully agree about what concept of "harmful" should be deployed. Let's stipulate for the sake of argument that they both agree it means something as straightforward as: x is harmful to y iff y would be better off without x. The doctor and the activist can disagree—without either saying something false—about whether the physical state in question is harmful but mean the same thing by "harmful." If the activist views the physical state qua socially embedded entity, then what's most salient is its complex role in the person's overall social experience and wellbeing. If the doctor views the physical state qua biomedical pathology, then what's most salient is the discomfort and pain, the disruption to function, and the potential threat to long-term survival. Viewed qua social entity, we can truly say that the condition isn't harmful—that the person wouldn't be better off without it or better off the more it was minimized. Viewed qua biomedical pathology, we can truly say that the condition is harmful—that the person would be better off without it or better off the more it was minimized.

Sometimes, apparently conflicting statements like this can be a matter of so-called *faultless disagreement*. When the fine arts dealer looks at the statue/clay and says it can't be squashed, she's correct. When the clay aficionado looks at the same object and says it can be squashed, she's also correct. Neither is making a mistake, both are saying something true, both have the same concept in mind when they say "squashable," and both are viewing the object appropriately. Likewise, in some cases the activist is correct to say, of a particular bodily state, that it's not harmful, the doctor is correct to say that it is harmful, and neither is making a mistake.[41]

But in other cases, disputes arise precisely because there's a question about which way of viewing the particular bodily state is best or most appropriate in the given circumstances. Again, this is a normative issue and not a semantic one. On this view, we can say true things if we refer to a physical state qua disability, and true things if we refer to it qua biomedical pathology. But the truth of natural language utterances is relatively easy to come by; communication is harder. If someone refers to a physical state qua biomedical pathology during a conversation about civil rights, we might rightly object that they are *over-medicalizing* disability. Perhaps they've gerrymandered the context in a way that allows them to say something that, strictly speaking, is true. Nevertheless, what they say might be unhelpful—it might lead us to focus on individual biological functioning rather

---

[41] Again, this is not, importantly, to say that everything a doctor might say about the condition is thereby correct. Plenty of doctors have lots of ableist beliefs that are false in any context. The claim is simply that some of the things the doctor wants to say about the disability will be true when it is viewed qua biomedical pathology, even though they are false when it is viewed qua socially embedded identity category.

than mechanisms of social inclusion, for example. And, although it might be true, it could also, in such a conversation, have plenty of implicatures that are false. Referring to disability qua biomedical pathology in such a conversation might, for example, imply lots of negative stereotypes about disabled people—that they are less happy, that the problems they encounter are the result of natural bad luck rather than social inequality, and so on.

None of this need be implied, though, if a doctor refers to the same physical state qua biomedical pathology in the course of conducting an exam or planning a treatment strategy. Part of what matters is whether a the context we employ is the right one for the particular purposes and interests at hand. And so, sometimes, there will be disputes about which context is the appropriate context to employ.

Take, for example, a hypothetical case of deaf parents of a deaf child, in a dispute with their pediatrician about whether to give their very young child a cochlear implant. The parents think that deafness, qua disability, is a rich and valuable part of their lives, and that there's no need to minimize it. The pediatrician thinks that the same physical state, qua biomedical pathology, is harmful, and that when treatments exist to minimize its long-term impact on children they should be utilized. Their surface-level dispute is over whether the physical condition is harmful to the child. But the deeper dispute is over which context is the right context to deploy—whether it's best, in making this decision, to view the physical condition qua socially embedded entity or qua biomedical pathology.

Disputes like this can be especially difficult because, I'm arguing, we can't solve them simply by zooming out to see the "big picture." We have to act, and acting requires us to take a view on what the right way to view this particular trait (deafness) in this particular context (the child's early linguistic development) is. It's important to appreciate that there might be multiple, legitimate perspectives on a question like this—that both the doctor and the parents could be making good judgments, even while disagreeing.[42] But in viewing a conflict like this, we're often asking the much murkier question of which context is normatively best—most just, most fair, most helpful. And sometimes the answer to that question might be unknowable. It might, for example, require a kind of God's-eye, impartial, and fully informed perspective on the overall situation that we just can't inhabit. More strongly, sometimes there might simply be no fact of the matter. That is, sometimes it might simply be indeterminate what the right thing to do is, at least in part because it's indeterminate what the best way to view the state in question is (and no further hidden information would settle the matter).[43] Perhaps partly because of the inchoate nature of health itself (more on this in Chapter 5), there's just no fact of the matter, in some cases, about the normatively best way to

---

[42] For an argument that context is always important in assessing moral reasons when thinking about disability, see Francis and Silvers (2013).

[43] As in, e.g., Schoenfield (2016).

view the type of physical state that can irremediably harm health but also be deeply and richly embedded in a person's identity and community. Such indeterminacy in the normative appropriateness of contexts can, I suggest, give rise to genuine and unresolvable moral dilemmas.[44]

Similarly, consider particular disputes over whether discussion of a case of disability is "over-medicalized." One upshot of the view I am pushing here is that, even if you are embedded in your own perspective, it can be wrong to ignore the fact that other perspectives exist and are legitimate. Some disability activists (especially the Very Online) will at times object to any biomedical framing of a disability, or any discussion of disabling condition as harmful, treatable, or something we seek to prevent. And this seems to simply over-apply their own perspective on the issue. But there will be plenty of cases of media representation where things get much murkier. Consider a story about the growing number of people relying on assistive devices. To what extent do we embrace and normalize their use, and to what extent do we focus on addressing and minimizing the causes that make them increasingly necessary? It's hard—it's often impossible—to do both at the same time. But they may both be good things to do, and different groups may have good reason to aim at each goal.

Things only get more complicated when we consider the wide range of perspectives we can take on the physical conditions we call disabilities. For the sake of simplicity, I've been contrasting two—a socially embedded one and a biomedical one. But there will also be the perspective of public health and public policy, the legal perspective, the caregiver perspective, and so on. Consider, for example, the difference between viewing a disability qua social kind and viewing a disability qua public health matter. The latter perspective makes salient, among other things, long-term use of resources in the context of scarcity in a way the former does not. It might be true from the perspective of the flourishing disabled person that her disability shouldn't be prevented, for example, because what's most salient from her perspective is the impact that disability has on her individual wellbeing and quality of life. But it might be equally true from a public health perspective that such disabilities should be prevented where possible, at least in part because ensuring a good life for the people that have them often requires a level of medical resources and accommodation resources which would be unsustainable if it was required by a wider portion of the population.

Sometimes, such differences in how disability is viewed will produce faultless disagreements. But sometimes they'll produce deep conflicts over the best way to view disability in a particular circumstance. It might be objected that the epistemic picture this leaves us with is overly pessimistic—with deeply entrenched conflicts that it's difficult to adjudicate. But I submit that this is a feature, not a bug, of the

---

[44] Which isn't to say we couldn't reason or act in the face of such indeterminacy—see especially Williams (2014) and (2016).

account in question. The practical reality of reasoning about disability and health often does involve difficult, seemingly intractable disputes, where's its simply very hard (perhaps sometimes impossible) to know what the best way forward is.[45] This account attempts to give a diagnosis, but is skeptical about the possibility of a cure.

6. Grief, Loss, and the Good Life

Context-shifting and clashing of perspectives aside, however, we are still left with an odd tension. The physical conditions that typically mark someone out as disabled almost always involved reduced health, at least along some dimension. It isn't right to think of disability as just a matter of reduced health, or to think of disabled people as automatically being less healthy than non-disabled people.[46] But being disabled very often involves substantial loss of health, and even in contexts where we're focused on the overall wellbeing of the person embedded in their social context, that's something that's not only true but that needs to be acknowledged. Even at the Disability Pride rally, we have to acknowledge the close relationship between disability and health, if only to argue more aggressively for changes to healthcare policy.

Can something like a mere-difference view of disability accommodate this, without sliding unacceptably—at least within a given context—into a mere-difference view of health? To explain why I think it can, let me return to the analogy I made to grief in Chapter 2. The central claim made by mere-difference views should, I argue, be interpreted as a claim about overall wellbeing: there is not a robust, counterfactually stable, or intrinsic relationship between disability and reduced wellbeing. This is fully compatible with disability often reducing wellbeing in social situations that are unjust or exclusionary for disabled people. And it's also, crucially, compatible with disability being the kind of thing that would reduce wellbeing even in ideal circumstances *for some people*—depending on their goals, their aims, their preferences, their personality, and so. Similar things are true about even the most rich, rewarding, and treasured things involved in our lives—something can be really good or valuable without being good for everyone.

Even when disability does not reduce overall quality of life, however, I've argued that it can be considered a harm with respect to *some specific aspects* of a person's wellbeing.[47] Saying that being disabled is, overall, mere difference doesn't

---

[45] Again, thinking that there's indeterminacy doesn't preclude decision-making—as in Williams (2014) and (2016). For discussion of how we might reason about and adjudicate between clashing normative perspectives, see, e.g., Lindauer (2020).

[46] We'll discuss the complexity of comparative health judgments in more detail in Chapter 6.

[47] See my (2016), chapter 3.

require saying that every aspect of being disabled is mere difference any more than, for example, saying that being a philosopher increases my quality of life means that department meetings increase my quality of life.

We make a mistake if we ignore the close relationship between disability and health, but we also make a mistake if we overly medicalize the experience of being disabled—if we assume that *what it is* to be disabled is just to experience a collection of "symptoms," or to have a specific biomedical pathology, and so on, rather than the complex and richly textured experience of navigating the world in a body that falls outside the norms for how we expect bodies to function. Reduced health will be part of almost any experience of being disabled, but it won't be the whole thing.

In speaking about wellbeing in the context of disability—and in making the case for a mere-difference view—I think that here it's important to draw a distinction between *absence* and *loss*. Some ways in which disabled people experience the world differently from non-disabled people involve simple absences. If a deaf person cannot experience auditory enjoyment of music, that's an absence. Furthermore, that's an absence of something valuable. And it's an absence of something valuable that might be all-things-considered bad for some people. (Acquired deafness, while fully compatible with a flourishing life for many people, was devastating for Beethoven.) But we don't, I suggest, need to explain what *compensates* a deaf person for this absence in order to understand how she might have a full, rich life (a life that is not worse than the life of a non-deaf person).

Think of the components of wellbeing as ingredients we're putting in a shopping basket (and wellbeing as the meal we're making with those ingredients). The simple fact that an ingredient is good and many people enjoy it (or even find it essential) doesn't mean that someone else's basket is obviously lacking if they don't include it. You need enough ingredients to make a rich and nutritious meal, and there's only so much you can fit in a single basket or make into a single meal. I might always reach for hot sauce—I might be baffled that other people don't reach for hot sauce—but we don't need to explain what it is in my friend's basket that *compensates* for lack of hot sauce, as long as she's got what she needs to make a great meal.

In contrast to mere absence, I suggest that loss is something stronger. To belabor the metaphor, if absence is simply picking different ingredients for your shopping basket, then loss is closer to having something go wrong in the making of the meal. Maybe the onions burned when they should've sautéed, or maybe your oven will just never get quite hot enough. But something in the process goes off the rails. This doesn't mean the meal itself will automatically turn out worse. Sometimes you can compensate—if you added too much vinegar, just add a pinch of sugar and let everything simmer a little longer. And sometimes you can even work out a better alternative—your banana bread comes out mushy in the middle, so you slice it thinly and pan fry it. (If you gain nothing else from this book, try this.) But you have to work around a problem, and integrate it into the meal you're making.

To unpack this, let's return to talking about grief. The extent (and the badness) of grief, I argued in Chapter 2, isn't well assessed by looking at overall wellbeing. Grief, in all its complexity, is compatible with a very good life (maybe even necessary for a good life), but in ways that don't diminish the harm of loss.

Consider, again, having dogs. When you get a dog, you do so with the expectation that you will outlive it. And so, at least if you are a dog person, you do so with the expectation that you will experience grief and loss. Now here are two very stupid things someone could say to you when your dog eventually dies. Either the loss of the dog is not, in fact, bad for you, because overall you are better off for having gotten the dog. Or, in fact, you really never should have gotten the dog at all, because grieving a dog is bad for you. Either way, we can chastise you. How can you expect sympathy for losing the dog since you obviously knew it was going to die? How can you say it's so hard to lose a dog when you also say that you love having dogs and invariably having dogs involves mourning them?

These options, of course, seem to be missing something fundamental about the nature of grief. The loss of a dog doesn't cease to be a harm because *on the whole* having a dog enriched your life. Nor can we quantify the extent to which you have experienced loss simply by the overall reduction in your wellbeing. What we can say, though, is that loss, in this sense, is bad for you. And in order to understand why this doesn't make having a dog bad for you, we have to understand how the overall experience of having a dog compensates for that experience of loss. It's not enough to say that sure, your dog died, but then you save money on dog food and vet bills so ultimately things are fine. Instead, to understand how such (predictable, inevitable) experiences of grief and loss are compatible with the good life, we have to explain how dogs are, in some sense, "worth it"—how the inevitable pain you'll experience when a dog dies is integrated with (and in many ways inextricably bound up with) the experience of loving and living with a dog. And those little shooting stars are absolutely worth it, says this dog person.

We need similar conceptual structure, I suggest, in talking about the loss of health associated with disability.[48] Yes, disability can be and often is compatible with living a rich, flourishing life, such that it makes sense to think about disability (at least in many contexts) as mere difference. Yes, disability is constitutively connected to loss of health. No, loss of health is not mere difference. More strongly, loss of health is bad for the people who experience it, but (as argued in Chapter 2) it's a mistake to think that we can capture this badness in terms of its effect on overall wellbeing. To say that the loss of health associated with disability

---

[48] By "loss" here I mean reduction in health from expected norms. Someone with a congenital disability will still have loss of health, on this view, even if they have never *experienced* a marked reduction in their health.

is compatible with flourishing, however, we need to be able to explain the ways in which other things about being disabled are integrated with—and perhaps constitutively connected with—the harms of loss of health, such that they can in some sense *compensate* for this loss when it comes to overall wellbeing (at least for some people).

Consider Juan, the wheelchair user with cerebral palsy. It's not enough, I suggest, to simply explain how it is that Juan is happy, despite the fact that the loss of health involved in his physical condition entails, among other things, persistent pain and reduced life expectancy. A mere-difference view of disability needs something stronger—it needs something like the claim that Juan wouldn't automatically be better off if we could keep other factors in his life the same, but remove his disability.

For some things about Juan's experience of wellbeing, we can simply explain how, although they constitute absences, his life is filled with so many other things that those absences don't in any way reduce his wellbeing. Sure, Juan can't run marathons or hike the Appalachian trail. But he's really into electronic music and really enjoys adaptive dancing programs. His life is full. He's not harmed by those absences, even though they're the absence of valuable things that are central to other people's lives.

When it comes to his life expectancy or his pain, however, we need to say something more specific. Even if his *overall* wellbeing is not reduced—because overall wellbeing is a complicated blend of so many things—there's at least some sense in which these things harm him and are bad for him. They aren't just absence; they're loss. And so to say that Juan's being disabled is mere difference, we need to explain how these harms—which are constitutively connected to his experience of disability—are integrated into his broader experience of being disabled, such that they are in some way compensated for. That is, we need to say how it is that there are other things about Juan's experience of being disabled—maybe the different way of viewing and navigating the world, maybe the distinctive joys of inhabiting an unruly body—which in a sense "make up for" (in terms of his overall quality of life) the harms of loss of health.[49] If loss of health is harmful, then we need to be able to say how it is integrated with (and perhaps at times even bound together with) positive aspects of disability in order to support a mere-difference view.

We also, of course, need to be able to say how what we owe to Juan, in virtue of his reduced health, is in no way diminished by his high wellbeing. All too often, the case for healthcare rights for disabled people is made based on the idea that

---

[49] And, crucially, in saying this we're explaining things about Juan's individual wellbeing. Another person with the same or similar physical condition might experience their disability as an overall harm (and might experience no sense in which the badness of loss of health was compensated for) simply because they had different goals, aims, preferences, etc.

good health is essential to "minimal wellbeing," and that reduction in health always involves reduction in wellbeing. As we discussed at length in Chapter 2, this conception of the relationship between health and wellbeing (and of the value of health to our lives) is far too simplistic. For someone like Juan, health and wellbeing probably come apart more than they do for the average person. If Juan was denied adequate healthcare, his health would decline. But his wellbeing might not—or might not decline as much as you'd expect. He's coped with reduced health all his life, and he's adapted his life accordingly. But his health still matters. He is still harmed if he loses health, and we still owe him the best health outcomes we can reasonably achieve for him. And this is true even if his overall experience of being disabled—connected as it is to physiological conditions that inevitably reduce his health—is something that doesn't diminish his quality of life.

7. Conclusion

The conditions we typically label "disability" can be deeply embedded in people's identities, communities, and overall way of being in the world. They can also be things that directly, and sometimes dramatically, involve loss of health. And giving due weight to both these aspects of disability is yet another Scylla and Charybdis that our understanding of health must navigate between. It's doubtless true that both our "common-sense" understanding and the institutions of medicine often over-medicalize disability—assuming that you can fully explain or understand disability as nothing more than a collection of "symptoms," or as a medical problem. It's probably also true that activists sometimes under-medicalize it in response—minimizing its embodied aspects (some of which are inevitably medical) and relying on an artificial distinction between the social and biological.

To adequately talk about disability—and the complex relationship between disability and health—we need to instead have some way of doing justice to the kind of duck/rabbit phenomenon we often encounter. It seems right, in some cases, to treat these conditions as pathology, as harm, as loss. It seems right, in other cases, to embrace the experience of living with them as a form of diversity and an axis of social justice. We ignore either at our peril, but we can't seem to capture both at the same time.

What I've tried to sketch, in this chapter, is a model for theorizing this kind of shiftiness—and perhaps, more generally, for theorizing the instability and ambivalence that many disabled people feel about their relationship to medicine and medical treatment.[50] The model I offer here is roughly inspired by—and roughly

---

[50] See especially Eli Clare's (2017) *Brilliant Imperfection: Grappling with Cure*.

analogous to—David Lewis' solution to the famous puzzle of the statue and the clay. But the more general point, the details of that model aside, is that very often—and perhaps especially often when it comes to our health—we have to take on this kind of shifting perspective reasoning. We want to say both that aging is a process of physical decline and decay and that it is fully compatible with high quality of life, and something we should embrace. We want to say that scars that disfigure someone are a harm and a terrible consequence of their injury but also that they can beautiful. And so on. How we view states of the body, especially states of the body tied both directly to our health and directly to our overall way of being in the world, can be filled with deep tensions, and require shifting perspectives. And there's arguably no one master perspective from which we can say everything we want or need to say, or know everything we might want to know. We just have to live with the tensions that result.

Appendix: Empirical Research on Disability and Subjective Wellbeing

The empirical literature on the relationship between disability and self-reported wellbeing is complex and often contradictory. And it is complex and contradictory enough that nearly everyone can find a survey or study to support their preferred stance (a type of cherry-picking which I've fallen victim to myself). In what follows, I give an overview of some of the central complexities, as well as potential diagnoses for some of the contradictions. But I am philosopher, not a psychologist, economist, or rehabilitation specialist. The reader is thus invited to take everything I'm about to say with their preferred amount of salt. What philosophy can provide, I think—and what is well within the remit of philosophy to explore—is insight into places in which the concepts we employ, the definitions we use, and the ideological frameworks we presume can affect how we collect and interpret empirical data. And, of course, philosophy can also provide skepticism. And skepticism is probably the biggest upshot of this overview. The answer to the question "what is the relationship between disability and average self-reported quality of life?" is almost certainly "we don't know," followed closely by "and that's probably not a very good question."

To begin, we'll look at variation in what is meant by "quality of life" (section 1). Part of the reason that empirical studies can show quite different associations between disability and quality of life is that they often mean different things by "quality of life," and some measures (especially for "health-related quality of life") are such that many disabilities automatically reduce the thing that is being measured. The next—and arguably most significant—variable is what counts as "disability" (section 2). As we'll see, there are many different groups of people that might be examined when we're considering the relationship between disability and quality of life, and many different things we might mean by "disability." And then (sections 3 and 4), we'll look briefly at differences in what is meant by "severity" (for studies that look at correlation between "severity" of disability and quality of life) and variation in which co-factors (especially socioeconomic variables) are controlled for.

After discussing these axes of variation, we can begin to bring some of the major discrepancies in findings into clearer view (section 5). It's probably not surprising that what connection we find between disability and self-reported quality of life varies radically based on which conditions we're looking at under the broad heading of "disability." It probably *is* surprising, at least to many people, that reduction in quality of life doesn't seem to be correlated with what we might generally think of as the medical severity of a condition (although it's more correlated with a person's own assessment of the severity of their condition). The famous studies citing little to no negative effect on quality of life from disability, for example, tend to focus on single conditions, or specific types of conditions, and they tend to be what we associate as "paradigm disabilities"—spinal cord injury, absence of a sensory modality, and so on. Studies that look at the population more generally—where by far the most common sources of self-identified disability or receipt of disability benefits are mental health disorders and non-specific chronic pain disorders—tend to find a much stronger correlation between disability and reduced quality of life. Overall, the picture that emerges seems to be one in which the disabling conditions that take the greatest *self-reported* toll on happiness, life satisfaction, and self-reported quality of life are things like depression and chronic pain—conditions that are increasingly common and that we're less likely to think of as "disabilities" in the traditional sense.

From there, we'll look at research on hedonic adaptation (section 6). Again, it seems that what kind of disabilities we're considering plays a significant role in what findings we get. But it also seems that both personal and—perhaps more significantly—socioeconomic variables play a lasting and substantial role. The net result of looking at all this together, I suggest, is that we need to be wary of any *general conclusions* about the relationship between "disability," broadly understood, and self-reported quality of life.

1. What Is Being Measured: Quality of Life

Empirical studies on the relationship between disability and quality of life tend to rely on individuals' self-assessment. But what people are asked to assess can vary greatly. Within the medical and rehabilitation literature, studies often look specifically at "health-related quality of life," for example. Respondents are given questionnaires such as the SF-36, which ask them to rate their health and their mood (including whether they feel "full of pep"), as well as their ability to undertake various "ordinary" activities. It's important to note that many physical disabilities—especially ones that limit mobility—will automatically lower the score on a measure like the SF-36, simply because the respondent can't climb stairs, walk, kneel, and so on. Whether such limitations should be directly understood as detracting from quality of life, or even from *health-related* quality of life, is of course an ideologically loaded question. The energetic, kale-smoothie-chugging wheelchair basketball enthusiast might view themselves as having not only high quality of life but relatively high health-related quality of life as well. And yet their score on a measure like the SF-36 will be substantially lower than it otherwise would be because they can't do a lot of typical activities. That being said, questionnaires like the SF-36 provide tangible and meaningful ways to assess how limited a person feels because of their health. Because the measures like the SF-36 also include domains for life satisfaction and mood, they can thus be a good tool to measure whether functional limitation is correlated with lower mood or reduced happiness. They can also be an especially useful tool for tracking changes across time. For this reason, they're commonly used when assessing things like the outcomes of specific medical interventions. When considering whether to recommend hip replacement or non-surgical treatment for someone with arthritis, for example, something like the SF-36 allows for the comparison of outcomes specifically related to individuals' sense of things like functional limitation and pain. If we look at the baseline scores on an SF-36 for people with hip arthritis, and then compare the scores of people six months after hip replacement and six months after conservative treatment, and the hip replacement cohort have substantially more improvement in what activities they can do, how limited they say they are, or how they rate their own health, then that's some reason to think that a hip replacement is the preferred treatment option. But calling any of that a proxy for quality of life simpliciter is obviously quite a theoretically weighty move, and it's striking that in medical literature "quality of life" and "health-related quality of life" (along specific measures) are often treated as synonymous.[1]

In contrast, studies on the self-assessed quality of life of disabled people from within economics or psychology tend to focus entirely on self-assessed *subjective wellbeing*. Measures of subjective wellbeing can vary, though, in what specifically is being assessed.[2] Some surveys ask respondents to rate their quality of life. Some surveys focus instead on *life satisfaction*, asking respondents how satisfied they are with their lives (sometimes broken down into life satisfaction along various dimensions plus overall or general life satisfaction, sometimes focusing exclusively on general life satisfaction). And some focus more directly on positive mood or affect, asking respondents to rate themselves as "very happy," "somewhat happy," "somewhat unhappy," and so on.

An important caveat, of course, is that while such studies are attempting to measure subjective wellbeing, what they're in fact measuring is what respondents will tell researchers about their own subjective wellbeing. And while there's reason to think these two things are plausibly correlated, there's also reason to think that

[1] Interestingly, empirical research also suggests that individuals view overall quality of life and health status/health-related quality of life as distinct phenomena, and weight overall quality of life (including mental health and emotions) as more important. See Smith et al. (1999). For a representative example of using health-related quality of life measures to assess and compare overall quality of life in the context of disability, see Boakye et al. (2012).

[2] Some researchers also defend the idea that things like life-satisfaction and self-reported happiness are separate but related domains of subjective wellbeing, and that subjective wellbeing can be assessed by combining these various domains. I won't wade into any of those details, and the nuances don't particularly impact anything I'm saying here.

they come apart. We know that self-deception and self-delusion can play a substantial role in what people are willing to say about themselves—if most people want to be happy and want to be satisfied with their lives, there might be pressure to say that you are somewhat happier than you really are or somewhat more satisfied than you really are, simply because you don't want to admit the truth (to yourself or to others). It's also hard to assess how much connection there is between, for example, how happy individuals say they are and how much of a particular positive psychological state they are in fact experiencing. Expectations, habituation, previous experience—these can all influence how we assess our own happiness. In the television show *Crazy Ex-Girlfriend*, the central character Rebecca will often repeat to herself, when she gets something that she thinks that she wants (a promotion at work, a proposal from her boyfriend), "This is what happiness feels like, this is what happiness feels like, this is what happiness feels like." Rebecca, in these circumstances, will typically say that she is happy. But what the viewer knows—that Rebecca often does not—is that she is *not* happy. Instead, she's someone who's previously experienced profound depression, and now tries to convince herself that the absence of severe distress combined with getting what she thinks she wants is what happiness really is. And again, what the viewer knows is that Rebecca can convince herself of this fairly well because Rebecca has never really been happy. Rebecca's statements that she is happy seem to express something quite different from other characters' statements that they are happy, and yet they'd of course be marked as the same result on a survey.

So there is some reason to be skeptical that personal ratings of subjective wellbeing correlate all that well with, or give us good intrapersonal comparative data for, subjective wellbeing itself (similar to the reasons we have to be skeptical of interpersonal comparisons of pain measurements, as discussed in Chapter 3). And that's without considering what the relationship is between subjective wellbeing and wellbeing itself. It's well documented that people in all sorts of difficult circumstances will say they are happy. Whether this means they are in fact comparably happy to those in easier circumstances who also say they are happy is a tricky question. And whether their happiness means that they *in fact* have high quality of life is an even trickier one—one that no empirical study can answer for us.

2. What Is Being Measured: Disability

Things get yet more complicated by the fact that researchers often focus on very different markers of disability when studying the relationship between disability and self-reported wellbeing. In medical literature, "disability" typically refers to degree of self-reported functional limitation. So, for example, the Oswestry Disability Index—a survey initially designed to measure functional limitation in low back pain but now used widely across a range of conditions involving chronic pain—asks respondents to assess the extent to which pain influences and limits a wide range of activities of daily living. Scores are then assessed to give a percentage rating and corresponding severity index—0–20 percent disabled is "minimal disability," 21–40 percent disabled is "moderate disability," and so on. Assessment of disability will also sometimes be condition-specific and attempt to measure functional limitation accordingly. Measures like the Harris Hip Score or the Hip Disability and Osteoarthritis Outcome Score, for example, are often used to assess "degree of disability" specifically for conditions involving the hip (and to assess outcomes after procedures like a hip replacement). Respondents are asked to assess their limitations with a range of everyday activities that hip pathology is likely to impact (walking, climbing stairs, putting on socks and shoes, etc.) and these assessments are combined with objective measures of range of motion in the hip joint to tabulate a total score. Similarly, widely used health measures such as the Health Assessment Questionnaire (HAQ) often include sub-scales specifically to measure disability (for the HAQ it is the Disability Index) and again these tend to be a measure of perceived functional limitations on activities of daily living (dressing, hygiene, climbing stairs, etc.). In general, medical usage of the term "disability" tends to refer to functional limitation (including functional differences, such as using assistive devices) on typical everyday activities. Disability, as understood in this context, is thus something that comes in degrees, and can be domain-specific (that is, you can have significant hand-related disability or shoulder-related without having significant disability *overall*).

The literature from psychology and economics is, unsurprisingly, far less focused on the specific functional status of individuals. But research within these fields offers little uniformity over what is being studied under the broad label "disability." There are four main strands, each of which will yield somewhat different populations.

The first is people who have recognized legal status as disabled, or who receive disability benefits, or similar. These are people who count as disabled under the law, but who might not self-identify as disabled. The second is people who self-report functional limitation or significant ongoing health problems. The British Household Panel Survey (BHPS), for example, uses responses to the SF-36 to assess disability. Similarly, the American Community Survey (ACS) asks a range of questions about limitations in everyday activities due to an ongoing health condition, and then uses responses to assess a "percent disability." Again, people classed as disabled on such measures might not self-identify as disabled, and they also might not meet the legal standard of being disabled (or they might meet it, but simply haven't sought it out). Some surveys don't ask questions about specific functional limitations, but instead look at responses to self-identification questions such as "do you consider yourself to have a disability?" or "do you consider yourself disabled?". And finally, some surveys directly target people with a specific condition (or group of conditions) that is generally recognized as a disability.

It's important to note that each of these methods of classifying a specific group of people as disabled—to subsequently survey about their happiness or life satisfaction—will yield interestingly different results. Someone who is HIV positive but largely asymptomatic, for example, is protected from discrimination under the Americans with Disabilities Act (ADA), but wouldn't be classified as disabled on a survey measuring everyday functional limitations. Similarly, a student with ADHD would count as having a disability under the ADA, and might be used to saying of themselves that they have a disability (since they might regularly apply for classroom accommodations, for example) and yet they might not be classified as having a disability on health-related quality of life measures like the SF-36. Many older people, in contrast, do count as having a disability according to measures of functional limitation, but would never self-identify as such, especially given the extent to which such functional limitation is often seen as a part of normal aging. Similarly, many people who report high levels of functional limitation don't self-identify as disabled, at least in part because their everyday limitations are caused by conditions we don't typically think of as "disabilities"—back pain, migraines, and so on. Indeed, research shows that only a small percentage of the people who population surveys like the ACS or BHPS classify as disabled self-identify as disabled.[3] On the flip side, many people who have conditions we typically think of as disabilities, and who self-identify as disabled, don't have formal or legal recognition of that status, simply because such certifications can often be time-consuming and they might never have felt the need for it.

3. What Is Being Measured: Severity

Studies also sometimes address whether and to what extent the *severity* of disability is correlated with quality of life. And again, different measures of severity are used, yielding highly different results. Sometimes—almost always within studies that focus on a particular condition—differences in objective disease state or impairment are measured. So, for example, studies of people with vision loss can look at objective differences in visual acuity, studies of people with lupus can look at objective indicators of disease activity, studies of traumatic brain injury or stroke can look at objective measures of neurological impairment, and so on. With such measures, the idea is generally that greater objective impairment means a more severe condition—the less visual acuity the more severe vision loss, and total blindness is the most severe form of vision loss, for example.

---

[3] See especially Bogart et al. (2017) for discussion of contextual and personal variables that influence whether people self-identify as disabled.

In contrast, many measures of severity look instead at self-reported functional limitation. Here, the idea is that the more limited a person reports themselves to be, the more severe the disability. It's worth noting, though, that there seems to be poor correlation between things like objective disease activity or objective neurological impairment and perceived severity of functional limitation.[4] As discussed in Chapter 3, however, functional ability can be highly subjective, and can also be influenced by many things—social support, living situation, income, personality—other than a particular impairment. At the same time, many of the most common causes of health-related impairment—especially chronic pain and mental health conditions—don't have any objective measures or tests, and so we don't have anything other than first-person reports of how bad or limiting the condition is to go by.

And finally, some studies look at an official government classification—such as Germany's disability severity classification—to assess severity. Such classifications are typically a mixture of an individual's reported functional limitation and the assessment of medical professionals, which may or may not include objective testing (blood work, imaging, etc.). Two people with the same level of biomedical impairment don't automatically count as having the same level of disability on such an assessment—it depends on their situation, what they say about their limitations, and so on. And by far the most common causes of disability certification—musculoskeletal pain and mood disorders—rely to a great extent on first-person reports to assess their severity. (For more extended discussion of the role of subjectivity in cases such as these, see Chapter 3.)

Again, it's important to bear in mind that these different interpretations of "severity" can yield highly different results for who is (most) severely disabled. Within specific conditions, self-reported functional limitation often shows weak correlation with objective measures of impairment or disease activity.[5] And there is likewise little correlation between what are typically viewed as more "severe" conditions or illnesses—especially in medical contexts—and which conditions tend to be reported as most functionally limiting by the people that have them. Lupus, for example, is generally considered a more severe disease than fibromyalgia, especially because of its potential for major organ involvement and life-threatening complications. Likewise, lupus often progresses over time, and has objective measures of disease activity, such as inflammation and degenerative tissue changes (whereas fibromyalgia does not). And yet people with fibromyalgia typically report themselves as significantly more functionally limited.[6]

---

[4] See, for example, Mailhan et al. (2005).

[5] Even for conditions that can be progressive—such as rheumatoid arthritis—research has shown that it's a mix of social and biological factors that determine functional limitation, with social and psychosocial factors (education, social support, working environment, locus of control, etc.) often being shown to be the bigger predictor (Yelin et al. (1980)).

[6] See Wolfe et al. (2010).

It's also important to note that very little research exists investigating the subjective wellbeing and life satisfaction in cognitive disabilities. This is likely due—at least in large part—to the fact that the methodology of giving surveys to individuals in order to elicit self-reports about their own sense of quality of life and life satisfaction is not well suited to understanding the experiences of at least *many* forms of cognitive disability.

4. What Is Being Measured: Co-Variables

Any study of the relationship between disability and quality of life is complicated by the fact that disability is often linked to other factors that themselves have strong associations with quality of life and subjective wellbeing. Disability is highly correlated with unemployment, with lower levels of education, with lower socioeconomic status, and so on.[7] And depending on what is meant by "disability," this correlation can sometimes be a bidirectional causal connection.

Such a bidirectional relationship often isn't present (or present to the same degree) when we look at individual conditions that are traditionally considered disabilities, especially congenital ones. But it's a major factor when we look at population studies for who meets the legal standard of disability or who self-reports functional limitation. For example, people with achondroplasia are more likely to be unemployed, to have lower levels of education, and to have lower socioeconomic status than people without achondroplasia,[8] but there's no reason to think that these factors *cause* achondroplasia (which is a genetically inherited condition). In contrast, by far the most common causes of self-reported functional limitation or application for disability benefits in Western societies are chronic musculoskeletal pain conditions and mood disorders.[9] And for these conditions, a large body of evidence suggests that social determinants of health such as socioeconomic status, employment, race, and education are major drivers in their development.[10] There's also a significant body of evidence that suggests that such variables are independently correlated with life satisfaction and subjective wellbeing.[11] (Unsurprisingly, the same types of economic and social hardship which can cause disability are also linked to lower levels of subjective wellbeing.)

For studies that track life satisfaction or happiness over time—especially those looking at whether these measures adapt after disability is acquired in

---

[7] See, for example, Australian Institute of Health and Welfare (2009).

[8] See Gollust et al. (2003).   [9] See, e.g., Social Security Administration (2017).

[10] The social correlations are often quite vivid. Racial differences, including differences in the racial makeup and socioeconomic status of a person's neighborhood, are a significant predictor of chronic pain outcomes, for example (Green and Hart-Johnson (2012)). See Bonathan et al. (2013) for a helpful overview of the relationship between chronic pain and socioeconomic factors. And see Chen et al. (2019) for a good overview of the association between socioeconomic factors and mood disorders.

[11] For an excellent overview and discussion, see Helgeson (2003).

adulthood—it's also worth emphasizing that disability acquired in adulthood can often be a precipitating factor for things like unemployment and loss of income. And such factors are themselves independently associated with subjective wellbeing.[12]

When examining the subjective wellbeing of disabled people, studies vary strikingly in whether, and to what extent, they examine and control for these variables. Especially for variables that appear to be strongly independently linked with subjective wellbeing and self-reported quality of life, this variation can of course influence what's being studied.

5. What Current Research Suggests: Subjective Wellbeing

It's fair to say that headline findings about the relationship between disability and wellbeing appear inconsistent. Contradictory findings aren't all that surprising, though, when we consider the heterogeneity of which populations and which range of conditions are being studied. Reported results range from disabled people having subjective quality of life that is not statistically different from non-disabled people[13] to the presence of a disability being one of the most significant predictors of reduced subjective quality of life.[14]

A huge amount of the variance here, though, may arise from variance in how "disability" is being construed and which populations are being studied. Studies that show relatively high quality of life almost always focus on quality of life in *physical* disability. And, more specifically, such studies often select participants either based on the presence of a specific physical condition (spinal cord injury, blindness, etc.) or by recruiting from physical rehabilitation programs (where participants all tend to have the kinds of physical condition we often think of as paradigm examples of disability—spinal cord injuries, limb loss, etc.). And so, for example, studies have shown subjective wellbeing that doesn't significantly differ from non-disabled comparison groups in individuals with as diverse a range of conditions as spinal cord injury,[15] osteogenesis imperfecta,[16] spina bifida,[17] cerebral palsy,[18] hemophilia,[19] myelomeningocele,[20] muscular dystrophy,[21] and limb amputation.[22]

---

[12] See, e.g., Knabe and Rätzel (2011) for discussion and literature review.
[13] For a good summary and overview, see Bronsteen et al. (2008).   [14] See Lucas (2007).
[15] See Benony et al. (2002).   [16] See Widmann et al. (2002).   [17] See Barf et al. (2007).
[18] See Dickinson et al. (2007).   [19] See Poon et al. (2012).   [20] See Cope et al. (2013).
[21] Researchers compared quality-of-life assessments for Duchenne muscular dystrophy, a progressive degenerative form of muscular dystrophy. Self-reported quality of life was not correlated to the severity of disability and although self-reported physical function was severely reduced, "vitality, role-emotional, social function, and mental health scores were nearly normal (67–98), and did not differ between groups" (Kohler et al. (2005), p. 1032).
[22] See Tyc (1992).

Perhaps a somewhat more consistent finding, however, is that self-reported wellbeing is moderately lower[23] in the presence of physical conditions we typically think of as disabilities but more strongly linked to social variables (social support and inclusion,[24] access to work,[25] access to community activities,[26] etc.) than it is to physical factors associated with the disability itself (disease activity, degree of physical impairment, etc.).[27] An emerging body of evidence also suggests that positive disability identity,[28] disability acceptance,[29] and a sense of social acceptance are major predictors of quality of life in the context of many of the conditions we traditionally consider physical disabilities. The objective "severity" of the condition is often a poor predictor of self-reported wellbeing and life satisfaction.[30]

The story appears to be somewhat more complicated, however, for physical disabilities associated with an ongoing (often degenerative or progressive) disease process, such as multiple sclerosis, Parkinson's, lupus, and rheumatoid arthritis. Again, a substantial body of evidence suggests that the major predictors of self-reported quality of life in the context of such conditions are things like social support and social inclusion, as well as disability acceptance,[31] although the data on the link between severity and self-reported quality of life is somewhat more mixed.[32] But the extant literature also suggests that it's not uncommon—though by no means universal—for such conditions to take a heavy toll on emotional wellbeing and life satisfaction.[33]

---

[23] So, for example, in an overview of the literature on subjective wellbeing in spinal cord injury (SCI) Dijkers (2005) concludes that, on average, people with SCI have lower subjective quality of life when compared to non-disabled controls, but that the measured differences are typically small and there is wide variation within both groups, such that "stating that most people with SCI have lower [subjective wellbeing] than those with whom they are compared would be wrong." The review also concludes the degree of objective impairment has little to no correlation with subjective wellbeing, while social factors are strongly correlated.

[24] See Fuhrer et al. (1992) and Kim et al. (2015).

[25] See Kim et al. (2018) and Corrigan et al. (2001).

[26] Kinney and Coyle (1992), for example, found that the largest predictor of life satisfaction in adults with physical disabilities was integration in leisure activities (which was itself influenced by other social variables).

[27] See, e.g., Pierce and Hanks (2006) and Chen and Crewe (2009).

[28] See Forber-Pratt et al. (2017).   [29] See Li and Moore (1998).

[30] Abrantes-Pais et al. (2007), for example, in a comparative study of high spinal cord injury, low spinal cord injury, and able-bodied controls showed no difference in life satisfaction between the three groups, and people with high spinal cord injury reported better mental health than able-bodied controls; both spinal cord injury groups reported lower levels of physical functioning and physical health.

[31] See, e.g., Motl et al. (2009), Krokavcova et al. (2008), Schwartz and Frohner (2005), Koelmel et al. (2007), and Chen and Crewe (2009).

[32] Several studies of people with lupus, for example, have shown little to no correlation between disease severity or disease progression and self-reported quality of life (Kiani and Petri (2010) and Kuriya et al. (2008)). But studies of Parkinson's disease often show a mixture of biological and social factors as the main contributors to life satisfaction, as in Rosqvist et al. (2017).

[33] For an excellent overview of the complexities in self-reported quality of life in multiple sclerosis, see Isaksson et al. (2005). See also Livneh and Antonak (1997).

If we widen the lens yet further to include a broader range of physical conditions, the story begins to look very different. By far the most common physical conditions associated with self-reported functional limitation, unemployment, application for disability benefits, and so on in the US and Europe are chronic pain conditions—including chronic low back pain, osteoarthritis, fibromyalgia, and migraine.[34] For conditions like these, self-reported quality of life tends to be substantially lowered.[35] And people with these conditions often report much lower quality of life than people with other physical impairments, including impairments we'd tend to think of as more medically "severe." People with fibromyalgia, for example, have been shown to rate their quality of life and functioning worse than people with cancer,[36] people with permanent ostomies,[37] and people with rheumatoid arthritis.[38] Similarly, one study found that people with arthritis rate their quality of life as lower than people on hemodialysis.[39]

The relationship between severity of the condition and subjective wellbeing is also more complex in chronic pain conditions. Severity for such conditions is typically measured by self-reported pain intensity and functional limitation. There are often no objective tests for the severity of such conditions (as in headache conditions, fibromyalgia, etc.), and in cases where there is an associated measurable physical impairment (degree of joint degeneration in arthritis, degree of disc pathology in spinal conditions), there is surprisingly little correlations between the findings on objective imaging and self-reported severity of pain and functional limitation.[40] The measured severity of chronic pain conditions is thus, quite simply, a matter of how badly the person says it hurts, and how limited they report themselves to be.

As discussed in more detail in Chapter 3, what such self-reports of severity are measuring is complicated by the fact that people often don't distinguish between the questions "how much pain are you currently experiencing?" and "how badly is your pain distressing you?". It's also not always clear that these two questions are fully separable, since pain is *both* a sensory and an emotional process. So it's perhaps unsurprising if there's a close connection between self-reports of pain severity and self-reports of emotional wellbeing or life satisfaction. That being said, while there are some outlier findings, the overwhelming body of evidence suggests that the more severe a person rates their own pain, the lower their life satisfaction and subjective wellbeing.[41]

---

[34] See Osterwise et al. (1987) and Dahlhamer et al. (2018).
[35] Boonstra et al. (2013), McNamee and Mendolia (2014), Becker et al. (1997), and Kwi-Ok and Nan-Young (2008).
[36] Kaplan et al. (2000).   [37] Burckhardt et al. (1993).
[38] See van Ittersum et al. (2009).   [39] Laborde and Powers (1980).
[40] Finan et al. (2013), Cubukcu et al. (2012), Maus (2010), and Vagaska et al. (2019).
[41] Stålnacke (2011), White et al. (2002), Baker et al. (2011), and Moore et al. (2010).

Interestingly, though, the finding that ongoing pain leads to severe reduction in subjective quality of life doesn't seem to generalize to all forms of chronic pain.[42] Many of the physical conditions in which people often report high quality of life also frequently involve chronic pain, including spinal cord injury, amputation, achondroplasia, cerebral palsy, just to name a few. What the literature seems to suggest, though, is that there is a marked difference in subjective wellbeing between conditions that have ongoing pain associated with them and what we might call *primary pain conditions*—conditions where chronic pain is the primary (or even the only) salient aspect of the condition.

There is also a complex chicken-and-egg situation when it comes to primary pain conditions and subjective wellbeing. Evidence suggests that such conditions negatively impact subjective wellbeing but also that factors that reduce subjective wellbeing—stress, trauma, depression—can be major triggers in the development of such conditions.[43] Their presence is also closely linked (although the cause/effect relationship is unclear) to depression and anxiety,[44] which are of course themselves closely linked to social factors that have a substantial impact on quality of life.[45]

This then brings us to the next set of conditions we can consider when evaluating what current research says on the relationship between disability and subjective wellbeing: mental health conditions. Overwhelmingly, data suggests that the most common mental health conditions—depression,[46] anxiety,[47] PTSD,[48] bipolar disorder[49]—have a substantial and long-term negative effect on self-reported subjective wellbeing. Indeed, mental health appears to be one of the biggest predictors of self-reported wellbeing and life satisfaction, even when other important variables (such as socioeconomic status) are controlled.[50]

---

[42] It's worth noting that "chronic pain" is sometimes used to refer to pain that lasts beyond the period that would be expected by an initial injury, or pain that persists in the absence of an obvious biological cause. In that case, pain associated with conditions like osteogenesis imperfecta or cerebral palsy, although chronic in nature (that is, enduring over time), wouldn't be "chronic pain." "Chronic pain" as used in the more restricted sense might well be a distinct entity and have a distinct relationship to subjective wellbeing.

[43] Pain conditions appear to be both more common and more severe/more distressing in areas where there are fewer socioeconomic resources, for example (Brekke et al. (2002)).

[44] Fishbain et al. (1997).

[45] Moreover, there's a more vexed relationship between factors like disability acceptance and disability identity—which seem to bolster quality of life in things like spinal cord injury or cerebral palsy—for conditions such as these. Perhaps because of the complex relationship (and potential looping affects) of expectation and interpretation on certain forms of chronic pain, at least some research suggests that integrating a chronic pain condition into one's identity can be actively harmful. These issues are, of course, incredibly fraught, and some of the research seems to actively stigmatize people with chronic pain conditions. But for a thoughtful and nuanced discussion, based largely around her own experience navigating chronic pain, see Isobel Whitcomb's article "When Chronic Pain Becomes Who You Are" (2022).

[46] Indeed, the connection between measures of depression and reduced life satisfaction is close to linear (and may also be close to analytic) (Koivumaa-Honkanen et al. (2004); Headey et al. (1993)).

[47] Mendlowicz and Stein (2000) and Rapaport et al. (2005).

[48] Olatunji et al. (2007).   [49] Vojta et al. (2001) and Michalak et al. (2005).

[50] Lombardo et al. (2018).

Interpreting this data is especially difficult, though, since in many cases part of the symptomatology of mental health conditions includes discounting future pleasures, diminished interest in current pleasures, and so on. That is, in many cases part of *what it is* to have a mental health condition will directly involve and affect responses to the kinds of questions being asked on subjective wellbeing surveys. So the issue of whether responses to such surveys are accurate measures of subjective wellbeing becomes especially complex in the context of mental health conditions. And what any of this says about the relationship between mental health and overall quality of life is of course a further theoretical question. But what the extant data seems to support quite strongly is that people with mental health conditions tend to rate their own life satisfaction, quality of life, and happiness substantially lower than people without such conditions. And, of course, the development of mental health conditions is closely linked to other factors that have a strong association with subjective wellbeing, such as stress and trauma.[51]

Mental health conditions are second only to chronic musculoskeletal pain as the most common causes of self-reported functional limitation, as well as of government-recognized disability status in most Western countries. Their incidence (or at least the incidence of their formal diagnosis and recognition) has also been growing at a remarkable rate over the last thirty years, to the extent that they are offered as a partial explanation for the so-called "healthy population–high disability paradox"—the observation than in wealthy countries where objective indicators of physical health are improving, rates of disability in the working-age population are also increasing.[52]

Putting this all together, we start to get some picture of why radically different answers might be arrived at from asking the question "what is the impact of disability on subjective wellbeing and self-reported quality of life?". When disability is construed as a functional limitation because of a mental or physical condition, or as a condition that qualifies a person for disability benefits, then by far the most common forms of disability in Western countries are chronic pain conditions and mental disorders.[53] And so when we measure these populations broadly, we unsurprisingly find that people with disabilities have substantially lower-than-average subjective wellbeing. Likewise, we find that (self-reported) severity of disability is closely linked with lower levels of subjective wellbeing.

---

[51] Reiss (2013).   [52] Ferrie et al. (2014).

[53] The Institute for Health Metrics and Evaluation reports that, for example, the most common forms of disability in Germany are "low back pain, major depressive disorder, falls, neck pain, and other musculoskeletal disorders" (2010, p. 2). In Germany, mental health disorders are among the most common causes of application for disability benefits (Wilken and Breucker (2000)).

In the UK, the most common cause for receiving disability benefits is mental disorders (Viola and Moncrieff (2016)). The most common reason for receiving the more specific Disability Living Allowance is arthritis, followed by low back pain and other musculoskeletal pain conditions, followed by learning difficulties and psychosis (Datablog (2012)).

In the US, the most common causes for receipt of social security disability benefits are musculoskeletal pain conditions and mood disorders (Social Security Administration (2017)).

If instead we look specifically at the relationship between physical conditions we have typically tended to think of as "disabilities," then the relationship to subjective wellbeing is much more complex. It's not at all unusual to find little or no difference between people with such conditions and people without them. And subjective wellbeing in the context of such conditions seems more likely to track social factors than the objective severity of the condition.

An interesting "middle ground" appears to be chronic degenerative diseases, such as multiple sclerosis and rheumatoid arthritis. People with such conditions often report better quality of life than people with primary pain conditions, and the relationship between quality of life and disease severity is at the very least quite complicated. Likewise, there are strong associations between social support and improved quality of life. But there does seem to be a robust tendency for people with such conditions to report, on average, report *moderately* reduced quality of life.

6. What Research Suggests: Hedonic Adaptation

Researchers have been interested not just in what people with disabilities say about their quality of life but also how what they say changes over time. Disability, when acquired in adulthood, is often thought to be an especially salient case of *hedonic adaptation* (or lack thereof).[54] When disability is acquired, happiness and life satisfaction tend to be substantially reduced in the immediate aftermath. So a significant body of research has focused on how much of that subjective wellbeing people recover over time—with results ranging from "almost none," to "some," to "pretty much all."

Again, a major determining factor here seems to be what kind of disability we're talking about. Studies examining large population cohorts—where, again, the most common forms of disability will be chronic pain and mental health conditions—have often shown poor hedonic adaptation. The much-cited Lucas (2007) study, for example, which looked at responses over time from the German Socio-Economic Panel Study (GSOEP) and the BHPS, found that respondents rated their happiness as substantially lower after the onset of disability and recovered very little happiness over time. The most severely disabled were the least happy, where "severity" is self-reported functional limitation.[55]

---

[54] It's worth noting that literature on disability and hedonic adaptation comes from two distinct points of entry—those primarily interested in understanding and improving the experiences of disabled people (mostly rehabilitation psychology and counseling) and those primarily interested in the broader phenomenon of hedonic adaptation, who see disability as an interesting test case in whether happiness returns to a stable "set point" after major life events.

[55] Severity is self-reported in the BHPS and based on self-reports of government certification in the GSOEP. (Although it's objective whether a person has such a certification in Germany, such certifications are applied based at least in part on a person's self-reported limitation—e.g., whether pain is too severe to carry out everyday activities, etc.) Somewhat perplexingly, despite objective evidence that

Studying similar changes in population surveys over time, Oswald and Powdthavee (2008) found evidence of moderate, though incomplete, adaptation. Individuals who reported a new disability reported lower levels of subjective well-being than they previously had, and over time tended to recover some—around 30–50 percent—but by no means all of their self-assessed happiness. Individuals who self-reported the most severe disabilities recovered the least amount of happiness, where "severity" again means self-reported functional limitation.

And specifically studying people with chronic musculoskeletal pain, McNamee and Mendolia (2014) found a severe reduction in self-reported wellbeing after the onset of disability, with limited adaptation over a three year period. Similarly to Oswald and Powdthavee, they found some evidence of hedonic recovery over time, but the effects were quite small (although, interestingly, women adapted somewhat more than men). And again, those with the most self-reported pain seem to fare the worst.[56]

In contrast, many studies of specific acquired conditions have shown total or near-total adaptation. Sample sizes in condition-specific studies are often relatively small, but they include as diverse a range of acquired conditions as spinal cord injury,[57] limb loss,[58] colostomy,[59] and hemodialysis.[60] A more consistent finding of condition-specific studies, though, has been that social factors— especially community integration, family support, and disability acceptance[61]—are the biggest predictors of positive adaptation over time.[62] In contrast, severity—

people with mental health disorders make up a large percentage of people receiving disability benefits in the UK, Lucas assumes they are a relatively small percentage based on self-reported scores of anxiousness and depressed mood (p. 725).

[56] McNamee and Mendolia (2014).

[57] Dorsett and Geraghty (2004), for example, found significant increase in depression shortly after spinal cord injury, followed by a trend of significant decrease in depression scores and high levels of self-rated adjustment. See also the discussion in Kennedy and Rogers (2000). Research has tended to show higher levels of depression and anxiety in people with spinal cord injury in the acute post-injury phase, with levels gradually improving over time and eventually returning to general-population levels.

[58] Horgan and Maclachlan (2004), in a systematic review of the extant literature, conclude that better data is needed but that in general research suggests that psychological status in amputees returns to normal after a two-year period (although depression and anxiety increase immediately after the disabling event), and that major predictors of life satisfaction are social support and quality assistive devices.

[59] Researchers found hedonic adaptation to at or near previous levels of self-reported life satisfaction a year after colostomy (with social support a major variable in the adaptation process) (Ito et al. (2012)).

[60] See Riis et al. (2005).

[61] See Li and Moore (1998) for discussion of the importance of disability acceptance and the ways in which acceptance is itself linked to family support and community integration.

Disability acceptance can itself be influenced and mediated by a wide and sometimes surprising range of social factors. One study, for example, found that the onset of Parkinson's disease led to a substantial decline in life satisfaction in men, but was not associated with decline in life satisfaction in women (Buczak-Stec et al. (2018)).

[62] See especially Hammell (2004) for a comprehensive literature review of the relationship between social factors and quality of life in high spinal cord injury, as well as a nuanced discussion of the complexities of evaluating quality of life in the context of disability.

where "severity" typically means the objective severity of the injury or impairment—has tended to be poorly correlated[63] with adaptation.[64]

Indeed, although results are somewhat mixed, a growing body of evidence suggests that acquiring a physical disability creates a distinctive type of social precarity. Factors such as income,[65] education,[66] employment,[67] and community and family support[68] then become even more important than they might otherwise be to subjective wellbeing. And so while people can and do adapt to disability, recovering happiness is often contingent on a safety net of social factors that aren't equally distributed.[69]

In a similar vein, there is a growing strand of thought within discussions of hedonic adaptation that in looking at average levels of subjective wellbeing we miss crucial information about how happiness changes over time. A much-discussed finding from the hedonic adaptation literature is that marriage does not have a long-term impact on happiness—people enjoy a brief "honeymoon" phase and then happiness levels return to baseline.[70] However, a growing trend of criticism for such findings is that in focusing on the average level of happiness for married people over time, we obscure importantly distinct "happiness trajectories". Yes, some people who marry display the typical hedonic treadmill pattern of adaptation. For a significant group of married people, though, happiness doesn't revert—that is, marriage provides a lasting boost to happiness. This is counterbalanced by a significant portion of people for whom happiness, after marriage, declines substantially. To summarize this entirely unshocking result: for

---

[63] Although, once again, the story seems to be significantly more complicated, and the objective parameters of the condition somewhat more closely related to life satisfaction, in progressive or degenerative conditions like multiple sclerosis or Parkinson's. Here again, research suggests that many of the biggest predictors of subjective wellbeing are social (Chen and Crewe (2009)).

But research also suggests a somewhat closer link to disease status and disease progression. One hypothesis is that such conditions require a continued and ongoing process of adaptation to loss, as the condition worsens over time. And even very positive adaptation requires the continued negotiation of a worsening health condition—or, as Michael J. Fox has described his Parkinson's, learning to live with "the gift that keeps on taking." See especially Fafchamps and Kebede (2012) and Rosengren et al. (2016).

[64] Interestingly, Ville et al. (2001) found that in individuals with spinal cord injury, the clinical severity of the injury was not correlated with subjective wellbeing, but the person's own self-assessment of the severity was.

[65] Smith et al. (2005) found that wealthier people lost substantially less subjective wellbeing after the onset of a disability.

[66] Smith et al. (2005) found that socioeconomic resources—most especially education and income—were the biggest predictors of a "resilient" recovery trajectory (that is, a trajectory in which subjective wellbeing returns to baseline) after the onset of a disability.

[67] A Norwegian study of upper-limb amputees found that self-assessed quality of life was substantially lower than in non-disabled controls, but that the effect was mediated by employment status (with many amputees having lost their jobs after their accident) (Østlie et al. (2011)).

[68] See Li and Moore (1998).

[69] It's thus also crucial to note that adaptation to disability will likely look very different in societies with less healthcare and welfare infrastructure, more reliance on physical labor for work, etc. (Fafchamps and Kebede (2012)).

[70] See Easterlin (2003).

some people marriage is fine, for some people marriage is great, and for some people marriage is awful. And so on average happiness reverts to baseline, but the story behind those averages is substantially more complicated than simple hedonic reversion to baseline. Likewise, some research suggests that factors like gender and whether there are young children involved have a significant impact on happiness trajectories in marriage (and, again, unsurprisingly, that the presence of young children heightens gender differences in happiness).[71]

Similarly, a growing body of research suggests that there are multiple distinct "adaptation trajectories" following the onset of a physical disability.[72] Some people return to their previous levels of happiness, or even report that their lives are richer and changed for the better.[73] Others recover some of their happiness slowly over an extended period, but never feel that they are as satisfied with their lives as they used to be (and their sense of life satisfaction can often fluctuate—there's little reason to think such a recovery process is linear or stable). And another distinct group suffers a substantial drop in subjective wellbeing, and recovers very little, if at all.[74]

Social factors such as employment, education, and family support can be major predictors of which trajectory a person follows. So too can personal factors, such as resilience, hope, and acceptance.[75] Regardless, the explanation for how and to what extent a person's subjective wellbeing adapts to the onset of disability is inevitably complex, and there is no one-size-fits-all story to be told.

7. Conclusion

This is by no means an exhaustive summary of the literature—where an ever-growing body of evidence shows a wide range of findings. It is, rather, my gestalt impression of the major contours of that body of evidence. But perhaps the single biggest takeaway of that gestalt impression is that there is no single, informative answer to the question "what is the relationship between disability and subjective wellbeing?".

For one thing, the type of condition seems to matter a great deal—and not always in the ways you might expect. If *all* you cared about was your subjective

---

[71] For an overview and discussion on happiness trajectories in marriage, see Stutzer and Frey (2006) and Mancini et al. (2011).

[72] See Livneh and Martz (2003) for a good overview and discussion. And see, e.g., van Leeuwen et al. (2011) for an example of specific application of an application trajectory model to spinal cord injury: distinct "adaptation trajectories" for people with spinal cord injury.

[73] That is, adaptation sometimes goes beyond a simple return to levels of happiness had prior to disability. See especially Dunn et al. (2009).

[74] For discussion of different adaptation trajectories, see also Livneh (2001).

[75] See Dunn and Brody (2008) for an overview and discussion of factors that shape the adaptation process emphasizing how both social and personal variables can shape the adaptive process, with distinct adaptive trajectories.

wellbeing, then current research suggests that—at least looking at statistical averages—you have reason to prefer a broken back over non-specific low back pain, for example.[76] And while we often dismiss mental health condition as less "serious" or "severe" than physical diseases, research shows that, at least from the perspective of self-reported wellbeimg, they can take a far more devastating toll.

Socioeconomic factors matter hugely as well. A white, educated, insured, financially well-off woman (I write as I nod to myself in the mirror) might enjoy high levels of subjective wellbeing in the context of substantial disability. The same quality of life, though, might be much harder to come by for someone who has fewer socioeconomic advantages.

And finally, personality and personal variables are a major contributor. Disability is life-shaping and life-altering. People are different. Different people react to major life circumstances in dramatically different ways.

It's important that we don't stigmatize disability by ignoring the high levels of subjective wellbeing and life satisfaction had by many disabled people. And certainly many disabled people have been harmed—in medical decision-making, in policies, in norms—by ignorance of the extent to which people can and do live happy and satisfied lives with substantial disabilities.

But at the same time, repeatedly emphasizing the high subjective wellbeing and hedonic adaptation of many disabled people can itself be stigmatizing for those who are suffering. Many people don't seem to have reduced subjective wellbeing because of disability, but many people do. And too much emphasis on high quality of life in the context of disability can begin to seem insensitive to their suffering— or worse, subtly blaming them for not adapting or coping better.[77]

Often, when there is unclarity or conflicting data, we say that we need more research. In the case of the connection between disability and subjective wellbeing, we certainly need more research. But we also need better questions. There are many different things that can fall under the broad label "disability," depending on the context, and they probably don't share any kind of unified relationship to subjective wellbeing. Asking "what is the relationship between disability and subjective wellbeing?" where "disability" is meant in its broadest sense—and where we aren't also considering other important social, cultural, and economic factors—is probably asking a question that obscures more than it illuminates.

---

[76] This should not be construed as life advice.

[77] Anecdotally, I've had several friends with long-term health conditions—most of them involving chronic pain of some kind—confess that they feel like a "bad disabled person" because of how negatively their condition has affected their quality of life.

5

Ameliorative Skepticism and the Nature of Health

Thus far, I've been arguing that, in attempting to explain health, we're often pulled in different, competing directions. We want to say that health is intimately connected to, yet not the same thing as, wellbeing (Chapter 2). We want to say that our subjective experience is part of what constitutes our health, but also that our health has an objective reality independent of our subjective reactions to it or our experience of it (Chapter 3). We want to say that being disabled is strongly correlated with loss of health, and yet we might want to speak about disability and reduced health in very different ways (Chapter 4). And so on.

Having explored some of these tensions in detail, I'm now in a position to defend my own positive (or, as it happens, not so positive) view on the nature of health. I'm going to argue that we need a skeptical account of health—that the tensions I've been describing are baked into our understanding of what health is and that nothing can resolve them. But in arguing for this, I'm going to suggest that we need a middle ground between traditional forms of metaphysical skepticism—error theory, eliminativism, fictionalism—and various forms of realism. Health, I argue, is real. We are tracking real things when we talk about health, and the things we're tracking are interrelated and interdependent in such a way that we need to talk about them as a unified whole (the way that our idea of health attempts to do) rather than simply theorizing them individually. But there is also no way of giving a fully coherent, extensionally adequate theory of health. The resulting view is what I'm calling *ameliorative skepticism* about health. In this chapter, I outline what that view looks like—with the hope that it can be helpful as a way of understanding the confusions we encounter in grappling with health. And then, in the final chapter (Chapter 6), I explore why health can still be an important thing to focus on and evaluate, even if there's no clear, coherent thing that health is.

1. Health: The Core Problem

In Chapter 1, I argued against what I take to be the major contenders for philosophical theories of health. Rather than offering my own alternative, however, my take on this is that the overlapping projects these theories are engaged in—the project of trying to explain what health really is, to characterize the

Health Problems: Philosophical Puzzles about the Nature of Health. Elizabeth Barnes, Oxford University Press.
© Elizabeth Barnes 2023. DOI: 10.1093/oso/9780192883476.003.0006

difference between health and pathology, to give a definition or necessary and sufficient conditions for health (or pathology, or disease, etc.)—are doomed to failure. And here I want to explain the heart of that skepticism.

Chapters 2 to 4 argue that there are serious puzzles and tensions in how we think about health. In this section I want to try to crystalize those puzzles—to make the case that those tensions are rooted in a core problem. The way we think about health is unstable. What we're trying to capture when we talk about health is something that we can't capture.

The German economy and the Yellowstone ecosystem are both healthy. They are quite obviously, however, not healthy in the same way—health for an economy is not the same thing as health for an ecosystem, and there's no master theory of health that will give us an informative story about health for both economies and ecosystems. And yet, though the German economy and the Yellowstone ecosystem aren't healthy in the same way, there's a roughly similar idea that's being applied to each when we say that they are both healthy. There's no property that they share or specific feature that they have in common when we say they are healthy. But when we say they are both healthy, we communicate roughly something like: they are flourishing, they are functioning well, they are thriving. Of course, it's a different thing for an economy to flourish than for an ecosystem to flourish. And in either case, there's probably no reductive, fully informative account to be had about what its flourishing consists in. But we know what thriving economies are like, and we know what thriving ecosystems are like, and we can understand the analogy between the two, and say that both are healthy. We don't, of course, thereby assume that there's one thing that is health had by both the German economy and the Yellowstone ecosystem. But by saying that both are healthy, we apply the same intuitive gloss to each.

This is, roughly speaking, what is meant when we say that "health" in its broadest usage is a concept "unified by analogy" or a "family resemblance" concept. It would be foolish to try to give a unifying theory of health such that there's a single property or set of specific factors that make it the case that both the German economy and the Yellowstone ecosystem are healthy. But we get the gist.

When we talk about healthy economies or healthy marriages or healthy ecosystems, the idea we're gesturing at is something generally in the neighborhood of flourishing or thriving. A healthy economy is a thriving economy. A healthy ecosystem is a flourishing ecosystem. (I'm using "flourishing" and "thriving" as synonyms, trying to give a similar gloss on the same basic idea.) Health isn't the same thing for an economy as it is for an ecosystem, but by saying that each is healthy we're saying, roughly, that each is a flourishing instance of its kind.[1] And we typically

---

[1] This isn't to say that, e.g., we mean exactly the same thing by "economic health" and "economic flourishing"—maybe the latter is slightly more positive or has a little more normative evaluation packed into it, etc. The claim is just that health and flourishing are more or less the same/highly similar *general idea* when applied to something like an economy.

do think there are more specific, informative things we can say about what it means and what it takes for an ecosystem to be healthy (to be flourishing to be thriving), or for an economy to be healthy (to be flourishing, to be thriving), even if there's not a more general theory of health or flourishing or thriving that applies to both economies and ecosystems.

This same idea flourishing or thriving is also, arguably, what unifies the idea of health for many types of living organism. A healthy oak tree is a thriving oak tree. A healthy penguin is a flourishing penguin, insofar as we understand what it is for penguins to flourish. And this idea of thriving, even for things like oak trees, is not merely a matter of species typicality or species-normal function. If rising global temperatures slowly started to reduce the height and longevity of oak trees, it would become normal for oak trees to be unhealthy, rather than shifting the parameters of oak tree health.

Likewise, we can't simply say that health for living organisms is determined by the natural selection history of a species. There are, for example, some species that are by and large not very healthy. At least some research suggests, for example, that the panda is fairly poorly adapted to long-term survival. It eats only bamboo—which has extreme amounts of insoluble fiber and relatively limited nutritional value—despite having a digestive system more associated with carnivorous diets. As a result, most pandas have substantial nutritional problems and constant diarrhea. It thus seems plausible, given our current understanding, that both species normality and the history of natural selection for pandas have led to an entire species that is fairly unhealthy. (We can, of course, talk about what it means to be healthy for a panda—but healthy for a panda is not all that healthy.)

Similarly, we can't simply say that health for a living organism is its evolutionary robustness or adaptability (its likelihood of surviving long term, either as an individual or as a species). And that's because this often depends on complex factors that are external to how we evaluate an organism's health. Even a very healthy penguin might be relatively less likely to survive—and penguins generally less likely to survive as a species—because of global warming. But global warming doesn't mean that a particularly fit penguin is in fact unhealthy, even if it might threaten that penguin's long-term survival. And it doesn't automatically mean the species currently has reduced health, even if its long-term prospects for survival and reproduction don't look as rosy as they used to. The health difference between penguins and pandas—and the fact that penguins as a species are healthier than pandas—can't be explicated in terms of which group is likelier to survive and reproduce long term.

Even when we're talking about health for oak trees or penguins or pandas, we seem to be appealing to something that's partly normative or evaluative. That is, the central thing we seem to be trying to capture when we say the oak tree is healthy is that it's thriving (for the kind of thing it is)—just as an economy can be thriving or an ecosystem can be thriving. And that idea seems to go beyond a purely naturalistic evaluation of species norms or adaptation.

The idea of health (and the normativity involved) seems to become more complicated, though, when we apply it to more complicated creatures—animals with their own beliefs, projects, desires, and so on. The healthy oak tree just is the thriving oak tree. But the healthy dog isn't quite the same thing as the thriving dog. If you took away my border collie's frisbee, his wild runs, his tricks and games, but gave him basic amounts of exercise, food, and shelter, he would still be healthy—but he wouldn't be thriving. To really thrive, to live his best life, Breccan needs a project—preferably a frisbee-shaped project. He also needs lots of belly rubs, attention, and to be able to take up at least 80 percent of the bed. Perfectly healthy dogs can be bored, lonely, or frustrated. For complex animals like dogs, what we seem to be trying to capture with talk of health is something like "physical thriving." A fit and nourished dog might be thriving physically, even if it is frustrated or bored, for example.

But we can't quite give the same gloss on health when it comes to humans (and perhaps other primates) because of the complex relationship between mental and physical health. When we talk about health for humans, we're talking about more than our physical status. We're also talking about our mental states, our ability to function, our emotional stability.

A person in an uncontrolled manic state due to bipolar disorder, for example, might be physically thriving and yet not be in good health overall. To assess a person's mental health, we need to assess more than their physiological function or their "physical thriving." And this is true regardless of whether we can find a reductive biological explanation for mental illnesses. As Nomy Arpaly (2005) persuasively argues, there's obviously a sense in which the mental states that constitute mental health conditions are "just brain states," in the sense that we are all physical organisms and everything we experience is realized in our bodies. But they might be "just brain states" only in the same way that love, existential dread, or ennui are "just brain states." The distinctiveness of mental health is wrapped up in the distinctiveness of the mental itself—especially its connection to meaning and representation. Part of how we assess mental health is the extent to which our mental states are justified or unjustified, warranted or unwarranted, accurate or inaccurate, in a way that wouldn't make sense for non-mental physiological processes. Mental states have representational content—they are *about* things—in a way that makes mental health itself distinctive and makes the assessment of mental health point beyond simple physiological functioning. (Your blood pressure, for example, can be high or low, it can be a risk factor, it can be causing your lightheadedness—but it can't ever be unwarranted or inaccurate.)

It's tempting, then, to say that overall health for humans is just physical health plus mental health. That is, health for humans is just a combination of *physical thriving* and *mental thriving*. What we seem to be trying to capture includes both physiological and mental function, as well as subjective experiences tied to both (pain, fatigue, stress), and the overall capabilities linked to both. The problem here, though, is that if we say health is physical and mental thriving, the intuitive gloss we're putting on health creeps incredibly close to our overall wellbeing.

Something like flourishing or thriving is also, crucially, the central idea we aim to capture when we talk about wellbeing. Although they diverge widely on the specifics, most theories of wellbeing agree that to have a high level of wellbeing is to be living a life that is going well for you. Whether we construe human flourishing as a subjective mental state like happiness, as a matter of having your preferences satisfied, or as matter of living a life that contains objectively valuable things, the idea of wellbeing centers around the difference between surviving and thriving.[2]

So far, I've made the case that the basic gloss or intuitive idea we're trying to capture in the broad variety of cases where we talk about health is something like thriving or flourishing. For some animals, like dogs, we can arguably say that health is physical thriving. But for humans, we need to be able to say that health is some (complex combination of) physical and mental thriving or flourishing. So if the basic idea we're trying to explain when talking about wellbeing is also thriving or flourishing, it starts to look like we're encountering, yet again, the collapse between health and wellbeing. And yet most people are easily able to recognize that health and wellbeing aren't the same—that we can compromise our health for the sake of our wellbeing, or experience harm to our wellbeing without experiencing a loss of health. Likewise, part of why it's so useful (and important) to talk about health is that it allows us to narrow our focus to a specific aspect of a person's life. The disabled person may be living a flourishing life, for example, but still have reduced health, and that's an important distinction to be able to draw.

When talking about health, we seem to be aiming at something broader than simple physiological function (as in health for plants or penguins) but narrower than the overall flourishing of the person (as in wellbeing). And this, I contend, is the essential instability of our understanding of health. There's no stable ground for us to capture that is broader than basic physical function but narrower than overall wellbeing. There is no clear "medium place" that is more expansive than

---

[2] In what follows, I appeal to an intuitive gloss on wellbeing, and to a distinction between health and wellbeing, but I should note that it's no part of this argument that we assume any particular account of wellbeing or think that we can clearly define what wellbeing is. Indeed, I think wellbeing is a very plausible candidate for ameliorative skepticism—we need to talk about it, and the things we're tracking when we talk about it are real, but any attempt to give a specific account of it goes hopelessly off the rails because we're pulled in so many different directions. Our understanding of wellbeing needs to account for our subjective experiences but be more than just fleeting mood, it needs to allow that individuals might sometimes be mistaken in their own assessment of their wellbeing without being paternalistic or imperialist, etc. The pull between the objective and the subjective, and between the normative and the phenomenological—where all these things are inter-related and inter-dependent—might plausibly land us in a situation where no one understanding of wellbeing can give us everything we want and need. And if that is right, then arguably part of our inherent confusion—both conceptual and metaphysical—over health is inherited from our inherent confusion over wellbeing. We understand health partly in terms of its connection to our broader wellbeing—but we're not quite sure what wellbeing is.

basic physiology but less expansive than quality of life. And yet for health to do the work we need it to do, we both need to be able to understand how health and wellbeing come apart and need to be able to understand how so many of the most important aspects of health—how we feel, our ability to function, our capacity to perform the activities of daily living most important to us—go well beyond (and often don't neatly correlate with) our basic physiological functioning.

In Chapter 1, I suggested an interpretation of Philippa Foot according to which health is the flourishing of the human organism. And while I don't think this succeeds as an explanatory theory of what health *is*, here I want to suggest that it comes fairly close to characterizing the basic idea we're *aiming* at when we're talking about health. That is, I think the general idea we're trying to get at when we talk about health for human beings is something like the flourishing or thriving of the organism, rather than the person. What this means, in effect, is that unifying idea of health for human beings is wellbeing—since flourishing or thriving are the basic gist of our understanding of wellbeing. But it's the wellbeing of the human organism rather than the person.

The wellbeing of the person requires us to evaluate projects, desires, and so on. And someone might reasonably sacrifice some health for the sake of a meaningful project or a strong desire—the familiar cases in which we sacrifice some health for the wellbeing of the person as a whole. Likewise, things might compromise your wellbeing without thereby compromising your health. If you really wanted that promotion, it might harm your wellbeing to lose it without thereby harming your health. You can be sad, frustrated, or disappointed without being unhealthy, physically or mentally. (Indeed, the ability to feel negative emotions in response to setbacks, and then move on, is often a sign of mental health.) Similarly, if you really wanted that promotion, it might increase your wellbeing to get it—because it allows you to pursue meaningful goals, or have a sense of purpose, or be happier—even if the longer hours and increased workload have a negative impact on your health.

What we're aiming to capture by health is, more narrowly, the wellbeing of the organism—the mechanistic thing, the instance of a particular species. And we're aiming to capture the wellbeing (flourishing, thriving) of that thing, as opposed to measuring the overall flourishing of the human life. That is, we're trying to zero in on something like physical and psychological functional capacity—in a way that's more normatively loaded than looking at something like statistical typicality (after all, the majority of a population could develop a disorder) and isn't quite evolutionary adaptation (since some disorders might've been adaptive at one point), but is rather something more strongly normative.

And so, again, I suggest that what we're aiming at here is something like wellbeing of the organism. The idea, then, is that while the wellbeing of the person and the organism are closely interwoven (and the former arguably can't survive without the latter), they aren't the same thing. Something can be good for the person as a whole, but not for the organism. And something can be good for the organism, but not for the person as a whole. When we talk about health, we try to narrow in on a specific type of thriving—thriving qua instance of *Homo sapiens* rather than thriving as a person. This notion of thriving, though, extends beyond simple physiological functioning because it includes psychological function, subjective experience, and so on.

This rough idea of wellbeing (flourishing, thriving) as applied to the human organism, rather than the human person, is also, crucially, at the heart of the skeptical diagnosis I'm offering. I think something like "wellbeing of the human organism" is the basic intuitive gloss we aim to capture when we talk about health. That is, when we're trying to explain what health for human beings is, or trying to give an analysis or theory of health, I think something like the idea of wellbeing (flourishing, thriving) applied to the organism rather than the person is what we're trying to explain. But I also think that "wellbeing of the human organism" doesn't make sense. The basic gloss or intuitive idea behind what we're trying explain when we talk about health for human beings is inherently in tension and unstable.

Focusing on the wellbeing—flourishing, thriving—of the organism rather than the wellbeing of the person relies on the idea that we can separate (at least conceptually) the two. And I simply don't think that we can—at least for the way that "organism" is being used in this context. Let's suppose that we can make a distinction between persons and organisms (I'm skeptical, but let's grant it for the sake of argument). *Wellbeing* for the person as distinct from *wellbeing* for the organism, though, is a much more robust distinction.

We're trying to distinguish the thriving of the organism from the thriving of the person. To do that, we need to separate out something that includes the functioning of both the body and the mind, without treading into the territory of the thriving of the person as a whole. And I don't think there's any stable middle ground—anything broader than our physiology[3] but narrower than our overall wellbeing—on which to mark this separation. Our genetic makeup massively influences our personality. Our neurobiology shapes and alters so much of who we are—from mood to personal characteristics to memory. A growing body of evidence supports the major role that both social factors (socioeconomic status, work, housing, education) and things like mood and stress play in our longevity and susceptibility to disease. Low thyroid levels or a bad breakup can both cause depression, and either way it just feels like depression. When it comes to our ability to thrive or flourish, we simply can't separate—again, even conceptually— the person and the organism.

It is, of course, precisely considerations such as these that push the World Health Organization toward defining health as "a state of complete physical, mental, and social wellbeing." But, as discussed at length in Chapters 1 and 2,

---

[3] Even framing it this way is overly simplistic, since clearly some aspects of our physiology and physiological functioning will be part of the thriving of the person as a whole.

this is to conflate health and overall wellbeing. And we mischaracterize health—and overgeneralize its significance to our lives—if we conflate it with wellbeing simpliciter. So the essence of the worry is this. It seems that to understand health—without erring into the realm of healthism—we need to characterize it as something like the flourishing or thriving of the mechanistic thing we inhabit: the flourishing of the organism rather than the flourishing of the person as a whole. But we can't actually make this kind of robust separation.

This tension lies at the root of our confusion about health. We need health to be both mechanistic—about our biology, about the biomedical workings of the human organism—but also about the more complex experiences of social creatures with emotions, thoughts, and goals. We need health to be objective—something we can quantify, measure, and compare—but also include and capture subjective dimensions of our experience. We need health to be about wellbeing (the flourishing and thriving of creatures like us) and yet, somehow, not the same thing as our wellbeing. And there is nothing that can do all this work for us.

The idea we're trying to capture, when we talk about health, is fundamentally unstable. We zero in on a range of things that matter—biological function, social capability, subjective experience, social norms, and so on—but if we prioritize any one of them at the expense of the others, we lose the distinctiveness of health. Nor can we simply list these features and be done, or say that health is some conjunction of them (as in standard hybrid accounts—see Chapter 1 (section 6)). What we're seeking, when we talk about health, is a way of capturing how these features are combined and entangled. And yet there doesn't seem to be a way of capturing this entanglement that is stable.

2. Ameliorative Skepticism

Many of the puzzles I've presented in Chapters 2, 3, and 4 center around the basic question of how we should understand health. What is its relationship to our wellbeing and quality of life? What is its relationship to our objective physiological functioning, or to our subjective sense of what we can do and how we feel? What is its relationship to disability? In short: what exactly is this thing, health, that we take to be so important?

Sally Haslanger ((2000), (2012a)) has argued that, in asking the philosophical question "what is x?", we should sometimes be asking how we *ought* to understand or theorize x. More specifically, Haslanger argues that we should sometimes be asking how we should understand x in order to best achieve our goals.[4] And so, for example, Haslanger's own definition of gender is as follows. A person, S, is a woman if and only if:

---

[4] See especially Haslanger (2012a) and (2012e).

(1) S is regularly and for the most part observed or imagined to have certain bodily features presumed to be evidence of a female's biological role in reproduction;

(2) that S has these features marks S within the dominant ideology of S's society as someone who ought to occupy certain kinds of social position that are in fact subordinate (and so motivates and justifies S occupying such a position); and

(3) the fact that S satisfies (i) and (ii) plays a role in S's systematic subordination; that is, along some dimension, S's social position is oppressive, and S's satisfying (i) and (ii) plays a role in that dimension of subordination.[5]

Haslanger, crucially, is not arguing that is what most people *in fact* take themselves to mean when they talk about women. Nor is she arguing that this is the only or universally correct way to understand what it is to be a woman.[6] Rather, she's arguing that this hierarchical social structure exists in the world, and that it's something that's closely related to how we think and speak about gender, and something that our understanding of gender actually tracks (even though we might not intend for it to, and even though we might not realize that it does).[7] Moreover, we easily overlook the hierarchical (and non-naturalistic) aspects of this social structure that we are in fact tracking in our understanding of gender, often to pernicious effect.[8] And so, she further argues, it would be best—at least for many of our goals and aims, in many contexts—if we understood gender as referring to this structure. Understanding gender in this way would "unmask" the social reality of our world and give us a better understanding of the social structures that shape the way we think, speak, and organize ourselves.[9]

There has been much subsequent debate about how, exactly, we should understand the idea of ameliorative projects. The explanation above is slanted toward my own interpretation. An ameliorative approach to something like gender, I've argued, should be understood as giving a theory of what the social world is like—aspects of which have perhaps been hard for us to appreciate because of our norms and beliefs about gender. The sense in which theorizing gender in this way helps us accomplish our "legitimate political goals" is then at most an *optimistic bet*. It's trading on the hope that if we understand social reality for what it is, it will help us make social progress.[10]

Stronger, and less realist, interpretations of the ameliorative project instead maintain that in describing how social reality in fact *is*, we should consider how we

---

[5] From Haslanger (2000, p. 42); reprinted as Haslanger (2012b) (author's source material).

[6] Importantly, Haslanger also does not intend to exclude trans women who do not "pass" as cis women from being women—she proposed this definition at a particular time, to make a particular point about the nature of gendered social position, but has since amended the approach. See Haslanger (2020). See also Jenkins (2016).

[7] Haslanger (2012b) and (2016).   [8] See especially Haslanger (2012c) and (2012d).

[9] Haslanger (2012c) and (2012e).   [10] Barnes (2017) and (2020b).

want it to be.[11] So, for example, when we are asking what gender is, we should consider what political and moral goals we want to accomplish. The answer to the question "what is gender?" is then determined by what way of answering the question will help us reach those goals. That is, the truth about what gender is, on this account, is determined in part by which way of thinking about gender can best help us to accomplish our political goals.

I worry about the both the plausibility and the stability of these stronger views, but those debates are orthogonal to the issues at stake here. Regardless of how you interpret the ameliorative project, the chief idea is that when philosophizing about socially embedded kinds like gender, part of what we consider is the normative impact of theorizing. Part of what we're asking, when giving an ameliorative account of x, is what a theory of x is *for*—why theorizing x matters, what role(s) we expect a theory of x to play, and how theorizing x might help us make address some of the problems related to x. And ameliorative analyses, thus far, have focused on trying to find a workable theory of x that best addresses such factors.

But my contention here is that sometimes, when we're adopting this kind of stance, there is no good answer to the question "what is x?". When we're asking what a theory of x is for, the answer might be that it is for inherently confused things, or for contradictory things, or for impossible things. An ameliorative analysis asks the question "what do we want x to be?"—but sometimes theorizing might lead us to the conclusion that there's no account of x that will give us everything we might want or need from a successful theory of x.

In some instances, we reach a philosophical impasse because we're trying to explain something that isn't there to be explained—nothing can give us everything we want from a theory of phlogiston because there's no such thing as phlogiston. Likewise, we might sometimes run into trouble because there is at best a loose "family resemblance"—there's no point in trying to develop a master theory of games, Wittgenstein assures us, that unifies chess, soccer, and Minecraft under a single account.[12] Instead, there's a grab-bag of different things that we can roughly group together as "unified by analogy." (Indeed, "health" in its broadest usage is often given as a paradigm example of a concept unified by analogy—there's no one thing that it is for both an ecosystem and a marriage to be healthy, but there are some vaguely analogous things we're gesturing to when we ascribe health to each.) And we might sometimes run into trouble because we're caught in a cycle of trying to give a reductive analysis or necessary and sufficient conditions for something that just doesn't admit of that kind of analysis. We might not be able to give a reductive definition that tells us exactly what makes something a

---

[11] See especially Thomasson (2019), Dutilh Novaes (2020), and Diaz Leon (2018) (as well as Haslanger's (2018) reply). Depending on what is meant by "conceptual engineering," ameliorative projects are sometimes viewed as this type of enterprise (as outlined in Burgess and Plunkett (2013)).

[12] Though see Nguyen (2020).

table, but this doesn't mean there's any deep philosophical mystery about the nature of tables.

What I'm suggesting is that sometimes we run into a philosophical impasse that is different in kind from these more familiar forms of confusion. We're theorizing something that's real. And moreover, we're theorizing something for which we need an account of *that very thing* as a unified whole, not just a collection of individual realizers or family resemblances. And yet there are intractable tensions, confusions, and inconsistencies for any unified explanation we might try to give. We thus find ourselves in a situation where we need to provide an illuminating answer to the "what is x?" question, and yet we can't.

It is this type of take on the "what is x?" question that I am calling *ameliorative skepticism*. Unlike the error theorist (more on this shortly), ameliorative skepticism doesn't ask us to forgo theorizing x or consign x to the flames for its failures of consistency. Rather, it seeks to diagnose tensions and confusions that might be central to understanding what (if anything) x is. Skepticism is typically viewed as destructive—it tells us what's false, what doesn't make sense, what we can't say, but it doesn't provide any positive answers. The gadfly buzzes around and bites you, and that's about it. But the idea of ameliorative skepticism is that a skeptical stance can sometimes aid our understanding and be a positive suggestion for theorizing in its own right. Skepticism, that is, can be palliative rather than destructive. If social reality is at least in part constituted and sustained by our norms, concepts, and collective practices, then we should expect there to be places where it is intractable to coherent explanation. Insofar as our norms, concepts, and practices admit of inconsistencies and tensions—and surely that's to a great extent—we should expect that there can be aspects of our collectively shaped social world that inherit those same inconsistencies and tensions. And so when we're theorizing social reality, it might genuinely aid understanding and guide future theorizing to point out the presence—and the intractability—of such features.

Beliefs, norms, and concepts are, of course, often messy, and this is not news to anyone. Moreover, often what philosophers try to do is to clean up the mess—to offer a theory of what precise thing(s) in the world a muddled concept is in fact (perhaps surprisingly) attempting to pick out, or to suggest a cleaned-up concept as an improved alternative. But sometimes, I'm arguing, the most helpful thing to do is to, in effect, put up a warning sign that says "wet floors—danger of slipping."

To explain this idea further, let's return to the particular case of health. As I hope the preceding chapters have illustrated, there are a wide range of distinct roles that we ask health to play. We expect it to be a category that is biologically significant and interesting. The difference between something that harms your health and something that merely upsets you is, at least in part, a biological difference. And we expect pathology to be, at least in most cases, something we can investigate via the methodology of the natural sciences. But we also expect health to explain something significant about our own subjective experiences—

how we feel, our embodied sensations, our sense of what we can do. Likewise, we expect health to explain something significant about our social capacities—how we function in our communities, our capacities, and the roles we can perform.

And we expect all of these roles to feed into health's distinctive normative significance. If you lose health, that is something we think *by itself* matters—even if you are still happy, and even if you can still do the things you want to do. More strongly, a loss of health is normatively significant even if it, on the whole, doesn't lead to a net increase in the amount that you are suffering. Yet a huge part of why we care about loss of health is the suffering it produces. And if you are in chronic pain, or if you are experiencing ongoing fatigue, this is morally important even if (as far as anyone can tell) your biological systems are functioning normally and the tissue damage of an initial injury has healed.

My skeptical contention is this: nothing can do all this work for us. That is, nothing can play all the roles that we ask health to play. It's not just that the roles are too different; it's that they're actively in tension with one another. We cannot fully allow for the subjective distinctness of health while still maintaining its status as biologically and scientifically interesting. Nor can we maintain a close link between health and functional capacity while still maintaining that major aspects of health are subjective, and will be experienced differently from person to person. And so on. There is no healthy cake that we can both have and eat.

And yet we need to talk about health as a whole, not just as its individual components or specific realizers. The things we're tracking when we talk about health are interrelated, and in many cases even interdependent on each other. Stress, for example, is partly an emotional phenomenon, and is deeply influenced by its social context and our interpretations of that context. But it's also partly a biological process, and that biological process has a profound impact both on our overall biological functioning and on our subjective experiences. It's not just that we can't separate "stress" into its mental, physical, and emotional components. It's that we need an understanding of how they're all interwoven in the complex whole that is our health. And at times, of course, we also need to understand how these various factors—whether biological, phenomenological, or normative—can pull against each other. One person, though otherwise biologically normal, says he cannot function due to feeling overwhelmed by stress, a result of "burnout" from his job at a tech start-up; another is undergoing chemotherapy for an aggressive cancer but says he's feeling optimistic and coping very well, thank you for asking. Who is experiencing more "stress" in their life? Do we weigh the impact of that on their overall health equally? What do we owe to each? When resources are scarce, who do we prioritize?

We need a theory of health—health as a unified, real thing that matters to our lives in all sorts of distinctive ways (biological, phenomenological, normative, political). And yet, I'm arguing, there's no account of health that can do the legitimate work we want and need it to do. We need health to play distinctive roles—to be something that explains the distinctive biological, phenomenological, political, and normative significance we think health carries. And we also need to explain how these roles are enmeshed and interdependent, yet can also pull against each other. And there's nothing that can do all this work for us. Health is real, on this view, but there's no consistent, adequate theory of health we can give.

You can think of ameliorative skepticism about some thing x as having two key components: a Metaphysical Claim and an Explanatory Claim. The Metaphysical Claim goes something like this: we're tracking a range of things when we talk about x that are real and that are interdependent. The Explanatory Claim is then: there's no precise, coherent, non-messy way of characterizing how the things we're tracking when we talk about x are related to x. The upshot of the Metaphysical Claim and the Explanatory Claim is a distinctive type of skepticism. Talking about x is important—it's not enough (nor does it allay confusion or solve inconsistency) just to focus on the things we're tracking in isolation, since they're constitutively interdependent in distinctive ways. But there's no way of clearly and coherently explaining what x is. Many of our questions about x are thus unanswerable.

So the *skeptical* half of ameliorative skepticism is that there are questions we want and need to answer—things we want and need to know—but that we fundamentally can't answer and can't know. And this is not because we need to refine our concept or ask better question; it's because of the nature of the thing we're trying to explain.

But how is something this pessimistic *ameliorative*? We'll return to this issue in more detail in section 7, but the gist is this. Ameliorative projects have at their core the idea that we need to change the way we think about the question "what is x?" and we need to do so partly in view of our (legitimate) political and social goals. What I'm arguing is that sometimes the best and most helpful change we can make is to understand that we face intractable dilemmas and confusions—dilemmas and confusions that won't be solved by different, better, or more nuanced attempts at precision. They are, in a sense, baked into the very thing we're trying to understand. Likewise, sometimes what counts as our "legitimate political and social goals"—the values that guide us and the work we need an understanding of health to help us do—can be conflicting and in tension precisely because of the nature of the subject matter. Understanding all this, I argue, can be helpful. It can explain why we feel pulled in so many different theoretical directions, why no answer seems satisfactory, why so much in the area is baffling. It can prevent us from going down unhelpful theoretical rabbit holes and from getting lost in patch-and-puncture cycles of debate. And it can also shape the way we approach questions of how to talk about health, how to measure health, and so on. (More on this in Chapter 6.)

To further articulate what I mean by ameliorative skepticism, I'm now going to proceed to argue against—and contrast the view with—several other approaches to thinking about this type of conceptual messiness. Obviously, I won't be able to address each and every alternative option. The goal is simply to illustrate my own view by comparison and to highlight what I take to be its advantages.

3. Why Not Error Theory?

When faced with the skeptical case I've laid out, some philosophers might suggest that the solution is *error theory* (or eliminativism, or fictionalism, or similar). If our understanding of health is such a mess, we should simply say that there is no such thing as health and move on. There are no such things as witches. In order to be a witch, one needed to be a woman who gained supernatural powers due to a bargain with the devil, and as it turns out nothing in the world actually plays that role. So there are no witches. There is no such thing as phlogiston. To be phlogiston, an entity would have to be a substance (as posited by eighteenth-century chemists) that exists in all flammable bodies and is responsible for combustion. There is, as it happens, no such substance. So there is no such thing as phlogiston.

I do not think, however, that it is a viable option to simply say that there is no such thing as health. Our understanding of health is deeply embedded in our moral lives, our politics, and our practical reasoning. Major health-related events can bring the world to a standstill. Of course health is real. And of course we aren't simply going to stop caring about it because philosophers are dissatisfied with its conceptual clarity. "Philosopher argues there is no such thing as health" is a bad headline at any time, but in the wake of a global pandemic it borders on parody.

Here, it might be useful to contrast health with a morally important concept that still admits of error theory. The idea of immaterial, immortal souls has, in certain times and places, been deeply embedded in our moral reasoning. But that isn't a good reason to say that we really do have such immaterial essences or that—despite beliefs otherwise—our souls just are our bodies. Rather, it seems perfectly appropriate to give an error theory of souls, even if—and maybe partly because—the idea of souls is embedded in certain kinds of moral reasoning.

I take it that error theory is an attractive option for souls if we think that, in reality, there is nothing that exists that comes close to corresponding to the idea of an immaterial, immortal thing that is somehow the essence of personhood. And so insofar as our moral reasoning might hold that idea to be central, that moral reasoning is based on a mistake. With health, though, it's not that simple. It's not just that the idea of health is in fact entrenched in the way we reason. It's that in thinking and talking about health we're tracking things that are real—biological function, pain and suffering, the way the functioning of our bodies and minds shapes our social lives and social capacities. And in talking about these things as a unified whole—interconnected and strongly correlated— we're doing important work that we couldn't easily replace by talking about something else.

We don't really have formless, immaterial essences. But if you have cancer, your health really is compromised. And that loss of health really does matter, politically and morally. We aren't making a wholesale mistake when we're talking about health—there's something(s) real and genuine that we're trying to describe. Likewise, it's important that we recognize the normative and political significance of health, and we aren't making a mistake or engaging with a fiction when we do so. But at the same time, the ways we think and speak about health are intractably in tension.

So the main way in which the skepticism I'm offering about health differs from error theory, fictionalism, or eliminativism is simply that the things we're talking about when we talk about health are real. They have a significant impact on our lives. In trying to fix on something more than basic physiological functioning, but less than overall wellbeing, we focus on a range of features—vital signs and physiological functioning, homeostasis, lifespan and mortality, functional limita- tion, pain, fatigue, quality of life, relationship to the institutions of medicine and to medicalization, and so on. All of these features are real, and they all have distinctive (health-related) importance to us. It's just that there's no stable, precise way of specifying how those features are interrelated to form our health.

Consider, instead, what a moral fictionalist thinks about morality.[13] They think moral discourse is important—it's embedded in our lives, and it's of pragmatic significance to us that we continue to think and speak this way. But when it comes to moral reality, the moral fictionalist thinks that there's basically nothing there.[14] We speak as though we're tracking moral facts, and there aren't any moral facts. So we're engaged in an elaborate fiction—an important fiction that plays a useful role in our lives, but one that isn't tracking anything real.

On the picture I'm defending, there are many—overlapping, interrelated, interdependent—things that we're tracking when we talk about health. It's not that there's pragmatic usefulness but no ontology to underwrite that that prag- matic usefulness (as in the case of fictionalism), but rather that there's, in a sense, *too much ontology*. We're tracking many different things—some of which can at

[13] See, e.g., Nolan, Restall, and West (2005).
[14] A slight contrast to this is the kind of error theory defended by, e.g., Anthony Appiah (1994) about race. Appiah argues that the thing we take ourselves to be tracking when we talk about race—a biological or natural kind—doesn't exist, but that there's still something morally and politically important—racial identity—that we're often tracking. In the case, I don't think we can make a case like this in either direction; we aren't tracking things that aren't really there, and there isn't a neat "substitution hypothesis" we can offer for something that can offer a better explanation.

times be in tension with each other—and what we lack is a coherent story of how they fit together to give us a consistent, stable understanding of health.

It's at this juncture that I think ameliorative skepticism becomes an appropriate response. Sometimes, the things we're trying to theorize are real, and theorizing about them does important normative and practical work for us that we couldn't easily replace by talking about something else. But they also involve ineliminable tension, unclarity, or inconsistency.

The combination of these two features—reality and messiness—is what motivates the kind of skeptical stance I'm adopting. We're not going to stop talking about something like health, or caring about it, or reasoning about it. Nor should we. And yet our reasoning is invariably vexed and pulled in different directions. We need to talk about health as part of what Haslanger might call "our legitimate political and social goals." And yet the answer to the question "what do we want health to be?" is: something it cannot be.

This is where, I contend, there's a useful ameliorative role for skepticism. A modest skepticism can help to explain the confusion we often find ourselves in when talking about health—explain why no definition works, why no measure avoids severe problems, why comparative judgments are so fraught with difficulties. And it can perhaps help provide tools to reason about and through invariable inconsistencies.

Again, skepticism is often thought of as a negative enterprise—telling us what views are wrong without attempting to tell us what might be right. But what I'm arguing is that skepticism in cases like this can be constructive. We are all stuck up that proverbial creek when it comes to thinking about health; an ameliorative skepticism just tries to hand us a paddle.

4. Why Not Non-Reductive Realism?

When faced with unclarity, imprecision, or confusion over something that seems both obviously real and obviously important, another familiar option is to adopt some form of non-reductive realism. A non-reductive realist approach to theorizing x tries, roughly speaking, to point out what thing(s) in the world we're tracking when we think, speak, and care about x. Then it proceeds to give a theory of those things and how they explain the gist of what we think, speak, and care about when we think, speak, and care about x, even if they aren't exactly a theory of x per se, or don't perfectly mesh with "common sense" about x.

Two prime candidates, in the case of health, would be role-functional realism and cluster concept realism. In explaining why error theory is a bad option, I emphasized that the things we're tracking when we care about health are real—the functioning of our basic biological systems, homeostasis, our physical and mental capacities, pain, fatigue, and so on. It's all real (even if some of it depends on our subjective experiences and judgments). So we could, for example, argue that we should separate out the various roles we ask health to play, and then see which of these things we're tracking best explains these individual roles (role-functional realism).[15] Alternatively, we could say that health is a matter of having some sufficient amount of some sufficient number of the things we care about (cluster concept realism).[16] Let's look at these options in turn.

We might, for example, argue that when we're focused on the biological significance of health, what plays the health role are objective indicators of physiological functioning (and perhaps more generally, something like normal species function). Whereas, in contrast, when we're focused on the normative significance of health, what plays the health role are more socially embedded and subjective factors, like how we feel and how we function in our communities. If we tease out these separate roles, we can focus on the different realizers for these roles, and move on from there. Health in general would just be some loose, contextually determined combination of the various roles.

The problem with this approach is that neither the roles nor the realizers in question are separable in this way. In the case of the realizers, we've already seen (Chapter 3) how strongly interconnected and interdependent the subjective and objective features of our health are. We cannot, for example, fully understand the basic biological components of health without assessing the ways that highly subjective, difficult-to-quantify experiences like "stress" impact our physical functioning—raising the likelihood of complications from hypertension, of heart attacks, of relapses of multiple sclerosis, and so on. We also can't understand the biological components of health without addressing the ways in which they are affected by "the social determinants of health"—how people are situated in their communities, the resources they have to, their experiences of discrimination and trauma. And the connection here is more than simple causal interaction—it's, in many cases, closer to constitutive dependence. Your experience of pain, for example, is a part of your overall biological functioning, but it's partly determined and constituted by your emotional reactions and the meaning you assign to it. (More on this in section 6.)

Likewise, the various roles we ask health to play aren't separable—conceptually or otherwise. We can't understand the normative significance of health, for example, without discussing its biological underpinnings. Part of why it was so important to recognize that epilepsy is a genuine disease and not a sign of spiritual interference was that the biological reality of epilepsy is part of what explains what

---

[15] There are many different ways to sort out the details of such an approach, and they don't matter all that much for what I have to say here. But for an illustrative model, see something like Clarke's (2013) approach to the definition of "organism" in biology.

[16] See especially Boyd (1991) and (1999) for the classic presentation of natural kinds as "homeostatic property clusters." Another relevant comparison would be something like Griffiths' (1994) notion of a cladistic kind.

we owe to people with the condition. Likewise, suppose two friends break a promise to us at the last minute because they say they are in pain and not feeling capable of following through on the promise. But one has a stubbed toe and the other has a broken leg. The basic biological reality of their respective conditions is part of how we assess whether their promise-breaking was *reasonable* or *acceptable*.

Similarly, part of how we understand the biological significance of health is tied to its normative and phenomenological significance. As we discussed at length in Chapter 1, we can't fully articulate the sense of "normal function" that seems relevant to us in assessing the difference between abnormality and pathology without (often implicitly) reaching for something normative—pathology is what *harms* us, pathology is when something goes *wrong*. But this assessment often depends both on things like our social arrangements, our goals and values, and how we feel. Moreover, we can't fully understand the biological significance of various objective aspects of our health—levels of thyroid, dopamine, adrenaline—without talking about the subjective dimensions of how they make us feel. Elevated cortisol can make us feel anxious and on edge, for example—feelings that themselves can then contribute to a further rise in cortisol, as well as affecting things like heart rate and blood pressure. Our emotions and feelings are always realized in our physiological functioning, all of which is embedded in, and partly determined by, our social environment.

The roles we ask health to play are invariably interconnected and interdependent. But they are also, invariably, in tension with each other. We want to say that health is normatively significant partly because of the role it plays in determining our wellbeing—except in the cases where health and wellbeing widely diverge. We want to say that health has objective biological significance, while also maintaining that a person's subjective experiences are part of what determine their health. And yet we still want to allow that things get complicated in cases where we judge people to somehow be overreacting or having subjective reactions which are somehow *unwarranted*. We lack a coherent story about how the things we care about when we care about health are related to each other. And we also lack a coherent story about how we should balance them when they—inevitably—pull against each other.

So why not just adopt a fairly straightforward cluster concept analysis of health? There are various features that can determine whether you are healthy. To be healthy, you have to have sufficiently many of them to a sufficient degree. But none of them is itself sufficient (and maybe none is necessary), even though having some combination of them is jointly sufficient. Likewise, there's no neat story to be told about exactly which combinations of these features will tell the difference between health and disease, and there may even be borderline cases for which features are included in the cluster. There is, thus, no simple, reductive story about what it is to be healthy (or what it is to have pathology). But there is also no deep mystery. There's just a bundle of stuff—stuff that tends to group together, but for which there's no precise, context-independent story of how it groups together to determine health in any particular case.[17]

I take myself to be arguing for something substantially stronger—and more skeptical—than a basic cluster concept view. Just as with role functionalism, the difficulty here is not simply that there are many different things we are tracking when we talk about health. The difficulty, rather, is that the things we are tracking are often entangled in ways that mean we can't simply parcel them out into separate roles and realizers. The upshot, for something like a cluster concept approach, is that there isn't really a bundle of discrete things here that we can list as separate, distinct components of a cluster. Likewise, there isn't a coherent story about how these things relate to each other in a way that ultimately determines our health.

Consider, again, the case of pain. (See Chapter 3 section 3, as well as Chapter 3 section 4.3 for more discussion of the case of pain.) Pain is biologically important—both as a protective mechanism and as an indicator of damage or dysfunction. Likewise, pain is typically realized in objective neurobiological processes. But pain is also, in part, a subjective, emotional process. Pain is distressing. We don't like it. And pain is able to play the biological role it does in part because of this subjective element of distress— pain is a good warning mechanism because it distresses us. But how distressing pain is for a particular individual is highly subjective. And there's often no way of disentangling *how much* pain a person is feeling from *how bad* they find that experience of pain to be. Likewise, experiences of pain are strongly influenced by social context, by emotions, and by the meaning we associate with the pain. Pain is often limiting—this is a simple biological fact and part of the biological function of pain. But how limiting, and in what ways, someone's pain may be is once again highly subjective, and mediated by complex social and psychological factors. It's not just that there's a causal connection—that physical processes can cause emotional processes and that emotional processes can cause physical processes. Rather, it's that there's something like constitutive interdependence. Pain is *both* a physiological and an emotional process. We can't understand what pain is without understanding both its objective and its subjective elements, as well as the ways in which they interact with and depend on each other.

A further worry for a cluster concept approach is that, in specifying which things belong in or out of the cluster, we need an account of why some features of the world are part of what determines health and others are not.[18] Which particular things are relevant in assessing health might depend on the context, and our on our background aims, but the basic idea of health constrains what type of things can ever be considered an appropriate part of an assessment of health. Your experience of fatigue might not matter much at all when we are assessing

---

[17] For a view roughly in this area about the nature of health and disease, see Keil and Stoecker (2016).

[18] For a cluster concept to be explanatory, we also arguably need, in Boyd's (1999) sense, something like an underlying mechanism to explain their co-occurrence.

your health in the ICU—where we are focused on keeping you alive—but matter a great deal when you are assessing your health in the context of going about your everyday life. But in neither case is your income an appropriate part of an assessment of your health status. Income is, sadly, an excellent predictor of your long-term health outcomes. But causal factors that contribute to health aren't the same thing as health itself. And for health to have the importance that it does, we have to be able to single out a specific set of factors, distinct from the wider range of features that might affect our wellbeing or our quality of life.

If your job is boring and frustrating, that might affect your wellbeing. If your job is harming your health, that might also affect your wellbeing. Part of the distinctive importance of health, though, is that we think it matters in particular ways, and with particular significance, if something is harming your health. Health has a distinctive role to play in what we care about and how we care about it. A cluster concept analysis, while purporting to explain the nature of health, leaves this distinctiveness unexplained and ungrounded. It can't tell us why the health-related factors are health-related, and why other things—things that are nevertheless causally related to health, or that are strongly associated with wellbeing and quality of life—are not. For a cluster concept to be interesting or informative, we need to have some account of why it is that the things in the cluster are related to each other in a distinctive way—that is, why the features like fatigue and heart rate are things that determine health (though to different extents in different context), but income is not. And the overarching story, in the case of health, is a confused one.

The core objection, then, to both the role-functional and cluster concept approaches is that they try to clean up conceptual messiness by shifting to something more precise and distinct in the world (the realizers of a specific role, the cluster of features we're tracking, etc.). But part of what I'm arguing is that, in the case of health, there is messiness all the way down. The things that we are tracking when we care about health are real, but they are also interdependent, entangled, and sometimes conflict with each other in ways that resist stable explanation. We can't neatly separate the biological from the social, or the objective from the subjective, not just because of conceptual confusion but also because of the nature of the things we're trying to explain—how we feel, our ability to function, our embodied experiences of distress.

And so, although the we're tracking real features of the world when we talk about health, the problems we encounter when talking about health can't be solved simply by narrowing our focus to those more specific features. Many of those features are themselves embedded with tensions and inconsistencies, and perhaps more significantly the central work we need health to do is to explain the complex ways in which they both depend on and pull against each other. My skeptical contention is that there is nothing—not a specific thing, not a cluster concept, not a disjunctive kind—that can do all this work for us. Nothing can fill the role we ask health to play, quite simply because it's an unfillable role.[19]

5. Why Not a Vague or Inconsistent Object?

The metaphysician reader (hello, I see you) may at this point be wondering: what, exactly, is the ontology of the view being defended? Is it that there is a precise reality and a messy concept, such that the concept doesn't apply neatly to the underlying (precise) reality? Or is it that reality itself is imprecise, indeterminate, or inconsistent in some way? And if it's the latter, why is this *skepticism*, per se? Why isn't the view just that health is a metaphysically vague[20] or even metaphysically inconsistent[21] thing?

By way of analogy, consider vagueness. Most people think that facts about physical shape—including the physical shape that determines height—are precise and settled. That is, the underlying reality that we're tracking when we consider whether something is tall (and which at least partly determines whether something is tall) is itself settled. Things have a specific size, and nothing about that is confusing. The concepts we use to talk about and categorize physical shape, however, are a different story. Concepts like "tall" are, themselves, imprecise— there isn't a sharp boundary between the tall and the not tall, people can faultlessly disagree on which things are tall in a context, there can be borderline cases, and so on. But this is generally thought to be a case in which an imprecise concept fails to map perfectly onto a precise world. All the messiness, that is, is at the level of the concept.

In contrast, some (including me) have defended the idea that in specific cases, it might be the underlying reality itself that is unsettled, and which (at least partly) gives rise to the resulting imprecision in whether and how to precisely apply a concept. There might, for example, be some cases in which its indeterminate whether two objects compose a further object—and this isn't (or isn't only) because of imprecision in our use of concepts like "compose," or "part" but rather is due to genuine indeterminacy in the world. For cases like these, that is,

---

[19] One option here might be to adopt the kind of stance that, e.g., Frank Jackson (1998) takes on free will: there is nothing that will give us everything we want, so we should settle for a "second best" account—one that comes closest to giving us everything we want. This is, I take it, sometimes what proponents of normal function theories seem to suggest. Nothing will quite capture all the various roles that we ask health to play, but normal function comes the closest, and prioritizes physiological function, which seems to be the most important aspect of health. My claim here, though, is that we can't adequately separate the subjective, objective, and social components of health sufficient to arrive even at a "second best" account of health. This will, obviously, be a question of degrees—but my argument is that the kind of skepticism I'm defending here does a better job of explaining health as we encounter it than does any particular precisification.

[20] As in Barnes (2010) and Barnes and Williams (2011).   [21] As in Priest (1987).

there might simply be no fact of the matter as to whether there is a third object composed by the two smaller objects—and that's because of what reality itself is like, not because of how we think about or represent reality.

What, then, is the corresponding view I'm defending here? Is it that there's a clear underlying reality—there are clear and precise things we're trying to track when we talk about health—but messiness because our concept doesn't apply well to this basic reality? Or is it that reality itself—the nature of human health itself—is somehow indeterminate, inconsistent, or otherwise unsettled?

I think that this is a place where social ontology is unique, and needs to be evaluated somewhat differently than other cases. When doing social ontology, we are dealing with aspects of the world that are at least partially created by—and which have the nature they do at least partly because of—our collective norms, concepts, and practices. On the view I am defending, some of the things we are tracking when we talk about health are themselves indeterminate, imprecise, or otherwise unsettled. Things like how functionally limited a person actually is (versus how limited they perceive themselves to be), distinctions like how much pain a person is in versus how distressed they are by their pain, or the boundary between distressing and limiting experience of negative emotion that is nevertheless normal and genuine mental health disorders, may admit of unsettledness because of what social reality itself is like. Likewise, there may sometimes simply be no fact of the matter, for example, how much of someone's functional limitation or suffering is due to their health and how much is due to their broader belief set and their wider social circumstances.

But this unsettledness, on the view I'm defending, arises primarily because health is a part of social reality—a part of the reality that our aims, values, collective practices, and beliefs have shaped. Thus, it's at least partly because our aims, values, and collective practices are messy that social reality is messy. But it doesn't follow that we could simply clean up those aims, values, or practices and thereby eliminate the mess. This messiness, I'm arguing, is part of the way the world is. But it's part of the way the world is because it's the way we've created it. Social reality inherits the tensions and unclarity of our norms and practices in a way that fundamentally shapes the social world.[22]

I can imagine the insistent metaphysician, at this point, protesting—yes, but is health *real*? Is it that you're committed to an ontology of health, but that ontology is unsettled, or is it that you don't think our idea of health corresponds to anything

---

[22] Richardson (2022) has recently argued that this type of indeterminacy inherited from our norms can sometimes be a matter of injustice—that is, social reality can inherit a kind of imprecision or indeterminacy from our failing to have thought about or included certain groups (e.g., trans women, non-binary people). Cameron (2022) argues more generally that, because of the way it is shaped by our norms and practices, we should consider social reality as a potential locus for true contradictions. See also Brouwer (2022) for a defense of inconsistent social ontology.

in the world? The former is something like simple metaphysical vagueness and the latter is closer to standard error theory. My view is something in between. There are many aspects of social reality that we're tracking when we talk about health. They're real. That's why this isn't an error theory. And many of them are unsettled or otherwise imprecise, because of how we've shaped them to be, so perhaps there is also an element of metaphysical indeterminacy (or even metaphysical inconsistency, if you prefer). These factors are all relevant to health. But none of them *are* health. That's why this isn't the view that there is some thing in the world that is health, but that that thing is imprecise or unsettled.

And we can't explain what health is simply by saying it's a specific thing, but that it's an indeterminate or inconsistent thing. Part of what we need to explain— but ultimately can't explain in a stable way—is the way that the various aspects of health both connect with and pull against each other in bizarre ways. Invoking something like metaphysical indeterminacy or inconsistency could be part of the story, but it won't be anything like a full explanation.

6. Why Not Pluralism or Pragmatism?

Overall, I've been arguing that there are fundamental tensions in how we think and speak about health. But if Chapter 1 was attempting to make the case that extant theories of health are wrong—and ill equipped to handle these tensions— an interesting feature of many such theories is that they also get *something* right. That is, while I think that most extant views of health fail as attempts to tell us what health is, they succeed in giving important insights into significant aspects of health. So given everything I've been saying, why not simply adopt a *pluralist* approach to theorizing health?

On the view I'm defending, we can't give a coherent, extensionally adequate account of what health is. But there are, nevertheless, important things we are tracking when we talk about health. And many of the most salient of those are highlighted in extant theories of health. A pluralist could then argue that what we need are many different theories of health—theories that can account for the different aspects of health or the different goals we might have in talking about health.

Arguably, this is the gist of the response favored by so-called "hybrid" approaches, as discussed in Chapter 1 (section 6). And the problem for these approaches, which reemerges for the skeptical case here, is that it doesn't look like you can get out of the worries we encounter simply by combining different aspects of what we care about when we care about health. There is, again, messiness all the way down—the different things we're tracking are themselves messy, in addition to being interrelated in complicated ways. A hybrid approach merely inherits the problems of the things it's combining and smushes them together.

A pluralist approach—at least a pluralism of a pragmatist stripe—is a slightly different, though closely related, strategy.[23] Pluralism argues that we simply need different theoretical approaches to account for these different aspects of health. We can then select the understanding that best fits our goals and purposes in a given context. Problem solved.

As will be evident in Chapter 6, there's a lot to the pragmatist approach that I'm sympathetic to—especially to the idea that, in talking about what health is, we need to consider our specific goals and interests.[24] And when it comes to things like deploying specific measures of health—where we obviously have to pick a specific metric in order to do the work we need to do—I'll argue for a limited type of pluralism. But what I think the pluralist account misses is the thorough-going confusion we encounter in trying to understand health.

Again, the idea I'm pushing here is not simply that there is messiness, but that this messiness infects all aspects of a theory of health—both conceptual and ontological. So, just as with cluster concept or role-functional accounts, we can't eliminate the mess by narrowing our focus (in this case, via our norms or contextual saliency). That is, we can't simply single out the right approach to health in a context, or the approach to health that best suits "our goals and purposes" in that context, and thus avoid the skeptical worries. Suppose we're in a context where we care most about how people feel and how they function in their communities. Then consider again the comparative cases from Chapter 3— Ama and Brynn, Chris and Dev. (As a refresher, these are cases in which complicated social factors seem to influence how much pain people report, and how limited they say they are in response to that pain.) We still don't have an answer—and we're still pulled in multiple, conflicting directions—about whose health is more compromised, who we should prioritize for the allocation of medical treatment and accommodation, who is in more immediate need of social support, and so on. And yet we have to act. Nothing about pluralism resolves this tension. Saying that there are multiple ways we could view health doesn't overcome the epistemic (and perhaps metaphysical) quagmire of some-thing like pain, and it doesn't tell us what to do. Nor can we solve the problem by zeroing in on what goals or values should guide our reasoning, since those goals and values are *themselves* bound up in our confusion about health. (Are we trying to reduce pain or to reduce embodied distress and suffering that's related to pain? And is there a difference? And if there were a difference, would it matter normatively? Should we think there's a moral difference between func-tional limitation that's biologically inevitable and functional limitation that's

---

[23] Some examples of broadly pluralist approaches along the lines I'm considering here include Haueis (2021) and Lilienfeld and Marino (1995).

[24] As Anderson (1995) persuasively argues, the way we answer a question—and what counts as a good answer to that question—will often be guided by the reasons we have for asking the question.

mediated by psychological or social factors? If someone is coping well, do we owe them less? And so on.)

In arguing for ameliorative skepticism, I'm arguing that appreciating these tensions—and appreciating their ineliminability—has to be at the core of how we reason about health. Just giving various distinct, overlapping theories[25] doesn't get us out of this intractable confusion, nor does it give us an account of the distinctive ways in which the various things we care about when we care about health both depend on and, at times, pull against each other. And so, while the approach I'm defending has a lot of methodological commonalties with pluralism, I take it to be committed to something significantly stronger—and more metaphysically robust—than pluralism typically involves. Likewise—and partly as a result— the line I'm taking about what solutions we can provide is, I take it, also somewhat more skeptical than what's typically offered by pluralism or pragmatism.

7. Amelioration and Messiness

Thus far, I have laid out what I take to be the central case for skepticism about health. What we need health to be is something in between physiological functioning and wellbeing. But there is no stable ground in the middle—no "medium place" to situate our understanding of health. We either prioritize our physical functioning, at the cost of discounting the psychological, emotional, and social aspects of health, or we emphasize the psychological, emotional, and social aspects of health, at the risk both of rendering health too subjective and of collapsing the distinction between health and wellbeing. The foundations of our understanding of health are, I'm arguing, built on quicksand.

We are thus left with questions that I don't think are mistaken or based on false presuppositions—we aren't asking who the Queen of America is or how much a soul weighs—and yet lack fully informative answers. It's not just that it's hard to figure out who is healthier or who has greater health needs and so on in some of the contrast cases I've been talking about; it's that the answers are unknowable— and they're unknowable because of the nature of health.

So far, so skeptical. In what sense, then, is this skepticism *ameliorative*? Again, I think the project of ameliorative skepticism is apt in cases where we don't want to abandon a concept—where error theory or eliminativism are too strong—but where we're faced with ineliminable tensions. Health is real. Health matters. Measuring, assessing, and comparing health is a crucial part of many of our most important social endeavors. And yet there is no way for us to stably fix on

---

[25] This specific form of pluralism is sometimes called a "patchwork concept" or "Rorschach concept" approach; see, e.g., Lilienfeld and Marino (1995).

what we're trying to target when we talk about health. That is, there's no way for us to stably isolate something that is more than the functioning of the body but less than the wellbeing of the person.

My (admittedly optimistic) contention is that this kind of skepticism can be useful in what it explains. We have many different ways of measuring health and many different tools for assessing health outcomes—and they are all imperfect, and they all seem subject to significant flaws. We have all witnessed public debates about health policies—whether schools should reopen during a pandemic, whether governments should fund treatments that make people feel and function better but that appear to be placebos—in which the parties involved all seem to be saying true things but are working with different understandings of health. And especially in murky cases like these, the parties involved can easily be advocating very different approaches, without either party having said anything that's obviously false or inappropriate—they simply prioritize different things in their understanding of what health is and why it matters. The tensions and difficulties in understanding health are as ingrained as the need to talk about it.

Skepticism is useful, then, insofar as it allows us to diagnose these tensions—and perhaps even allows us to chart a way forward. (I'll flesh this idea out in Chapter 6.) An ameliorative approach looks at the legitimate political and social goals we might enhance by theorizing a particular social kind. There are, arguably, many true things we could say about gender, depending on the context. Haslanger is not saying that hers is the only definition we could give, or that her definition gives deep insight into the nature of the people who are in fact women. Rather, I take Haslanger to be arguing that by pointing to the definition she identifies, we are bringing attention to a real structure that exists in the world—and, furthermore, that drawing attention to that structure is part of how we pursue the goal of gender justice. Perhaps because the structure is often hidden from view, and so we too easily view gender categories as natural or inevitable.[26] Or perhaps because understanding the inherently hierarchical nature of our current gender categories might be part of how we can make progress in dismantling them.[27]

In the case of health, I'm arguing that we can't say anything this specific or informative. Furthermore, I'm arguing that this fact is itself crucial to our understanding of health. It is diagnostic of the swamp of confusion we find ourselves in when we try to theorize health. The legitimate work we need from a theory of health is also work that can't adequately be done. And my hopeful contention is that we'll make progress, in theorizing, measuring, and comparing health, if we can appreciate the ineliminability of this confusion, rather than getting mired in a search for "the best" or "the correct" understanding of health.

---

[26] As in Haslanger (2012c).   [27] As in Haslanger (2012b, d).

Ameliorative approaches are salient primarily in the realm of social ontology. Haslanger would not, I suspect, endorse an ameliorative approach to the question "what are electrons?" or "is there a privileged present?".[28] When we are discussing social ontology, we are discussing parts of the world that are real but that are the way they are in part because of how we've made them—their nature is shaped by our collective norms, beliefs, and practices. They thus have a certain amount of social and political significance baked into them, which makes an ameliorative approach apt.

But, precisely because they are shaped by our collective norms, beliefs, and practices, they can also have tensions and inconsistencies baked into them. After all, our norms, beliefs, and practices are shot through with inconsistency, vagueness, and imprecision. If we take seriously the idea that those norms, beliefs, and practices can shape the social world, then we should take seriously the idea that some parts of the social world will themselves be inconsistent, vague, or imprecise.[29]

The distinction—if any—between the mind and the body. The relationship between physical pain and emotional pain. The difference between what makes us live longer and what makes us live *better*. The variation in what different people can endure and how different people cope and function. These are all things we struggle to understand. And they're also central to understanding health. It makes sense, then, that our deep-set confusion on basic issues like these, which in turn shape our understanding of health, would bleed into fundamental tensions about the nature of health itself.

What I'm arguing is that understanding these inherent tensions—a project that involves an inevitable aspect of skepticism—is crucial to understanding both the distinctiveness of the ways in which health matters to us and the confusions we so often find ourselves left with.

8. A Model for Ameliorative Skepticism

To illustrate ameliorative skepticism a bit more clearly, it might be helpful to consider how it would work in a case other than health. As an example, let's return to Haslanger's original ameliorative case and consider gender. To be clear, I'm *not endorsing* ameliorative skepticism about gender. I'm just using gender as a way of explaining what it would look like to apply ameliorative skepticism to a case other than health.

It's certainly true that contemporary discussions of gender are messy, and that people seem to mean very different things by—and want very different things

---

[28] In whatever sense the present time might or might not be privileged, it's not *that kind* of privilege.
[29] Again, see Cameron (2022), Richardson (forthcoming), and Brouwer (forthcoming).

from—an account of gender. Many extant views of gender try to do the typical philosophical work of cleaning up this mess—by telling us that we should understand gender as a type of social position,[30] or as a set of dispositions,[31] or as a type of self-identification,[32] or as a direct correlate of biological sex.[33] Such accounts, though, are often pulled into multiple different directions. We want a theory of gender to explain the connection between gender and sex—to explain how they are closely connected, but (at least according to most contemporary theories) not the same thing. We want a theory of gender to explain the normative and political significance of gender—why recognizing someone's gender is something we owe to them and misgendering them is a way of harming them, how gender is linked to social oppression and inequality, or why gender norms are pervasive and entrenched And we want a theory of gender to be able to account for the personal aspects of gender—its connection to a private, internally felt sense of self, its connection to our gender expression and our desires for how others perceive us.

Within social and feminist philosophy, a recurring problem is that theories of gender seem to inevitably run into trouble trying to do all this work. A social position account of gender, for example, will emphasize the public and political importance of gender but face problems accounting for the ways in which someone's internally felt sense of gender can come apart from their public gender role/gender position. An identity-based account will be able to fully capture the role of gender identity but will struggle to explain some of the public, political significance of gender. At this juncture, some will argue that we simply need a disjunctive account[34] or a cluster concept account.[35] Others will argue that we should be error-theorists[36] or fictionalists.[37] And yet others will argue for pluralism, deploying different understandings of gender depending on what best suits a particular context.[38]

An ameliorative skeptic, in contrast, will say that we should abandon the idea that we can give a fully informative, non-messy answer to the question "what is gender?". But she'll also say that this isn't the same thing as being an error theorist or eliminativist. Instead, she'll say that the things we're tracking when we talk about gender are real. Perhaps, when talking about gender, we're tracking some complex mix of: sex-based social position, gender identity, gender expression and performance, sex itself, and so on. These are all real, and they're all significant. More strongly, the ameliorative skeptic will say, they're also interconnected. We can't understand why gender identity is *gender* identity without understanding both sex-based social position and perhaps sex categories themselves.

---

[30] As in Haslanger (2012b) and Alcoff (2005).   [31] As in McKitrick (2015).
[32] As in Bettcher (2009) and (2013).   [33] See Byrne (2020)
[34] As in Jenkins (2016), although she's since revised her position in her (forthcoming).
[35] As in McKitrick (2015).   [36] As in Labrada (2016).
[37] As in Logue (2022)   [38] As in Saul (2006) and Jenkins (2022).

Likewise, we can't understand how it is that gender norms shape our society in ways that constrain the actions of individuals (to form a distinct social position) without understanding the ways in which our internalization of norms is influenced by our self-identification. And so on. So we can't simply toss out the idea of "gender" and focus on the realizers in isolation, and nor can we, contra a view like Jenkins (2016), offer a disjunctive account.[39] Similarly, we can't merely default to the idea that we can somehow precisely define the underlying things we're tracking and switch between them depending on our goals and values in a context. Our goals and values might themselves be unstable. (How do we balance the public and the private? How do we understand the interplay between a person's sense of their own social position and the way others treat and react to them within society?) Likewise, we might be equally unable to offer clear accounts of things like gender identity, gender expression, or gendered social position, at least in part because of the way they all seem to be tangled together.

Instead, the ameliorative skeptic will say that we're left with the following picture. Gender is real: there are real, important, and interconnected things in the world that we're tracking when we talk about gender. It's important that we continue to talk about gender (and not just the various things that we're tracking as separate entities) because it's important that we focus on the ways in which these things are entangled and interconnected. But there is no coherent, non-messy way to do this.

Perhaps the core of the skepticism in a case like gender could be something like this: we want gender to be both deeply private/personal (a matter of self-identification and self-expression) and strongly political and public (a matter of rights, a target for political intervention, an axis of oppression). We also want to it have *something* to do with biological sex, and yet be meaningfully distinct and separate. We often need to have accounts of this type of difference between the natural and the social, and yet in doing so rely on a distinction between the natural and social that is artificial at best, pure fiction at worst. We are, as a part of our nature, social organisms. Part of the nature of a species like ours is shaped by our dimorphic sex-based reproductive process—and yet that process is deeply, inherently social. We want gender to be "the social meaning of sex," to be about personal self-identification, to be about how we are socially positioned in the world and about contingent, unjust realities of material oppression.

And so, says the ameliorative skeptic: there is nothing that can do all this work for us. There is no way, for example, of balancing between the political and the personal aspects of gender. We're simply pulled in conflicting (but real) directions. We want a theory of gender that captures both its public and its private

---

[39] Jenkins' account, to be fair, isn't simply disjunctive, since she argues that gender identity in some sense depends on gendered social position (but not vice versa)—gender identity, for Jenkins (2016), is having a "mental map" for navigating social situations based on a particular social position.

significance, but there's no "medium place" between these two that can give us stable ground to say what gender is.

If we take seriously the idea of social construction, then categories like gender are arguably something that we created. They're real aspects of the world, but the world is this way because of our thoughts, norms, beliefs, and practices. And insofar as those are messy, the ameliorative skeptic says that gender itself is also messy.

An inclination in analytic philosophy is to say that when things seem inconsistent, or when we can't adequately define or theorize them, we need to do away with them. Often we do this by saying that the things in question aren't metaphysically significant—reality is what's "fundamental," and everything else is just loose convention. We might also do this by "conceptual engineering"—by replacing our messy concepts with something more precise, more coherent, more suited for our purposes.

The ameliorative skeptic says we should resist these options. An ameliorative skeptic about gender would say that gender is real because the things we're tracking when we talk about gender are real, and also interdependent in complex ways. But she'd further say that there's no single, coherent theory of gender that can explain the ways in which the various things we care about when we talk about gender interact with each other across the wide range of contexts in which we want to talk about gender.

Again, I'm not endorsing this as a view about gender—I'm just using gender as an illustration of what an ameliorative skeptical approach would look like. (It seems like an apt example case, since it's Haslanger's most famous application of her original ameliorative project.) But one could reasonably argue that we have more choice and political agency in what genders are, and what we understand genders to be, than we do in the case of health. Both gender and health have obviously social components, but gender could arguably be seen as much more thoroughly social, and more determined by collective social norms, than health.

To reiterate the main point, then: the two key components to ameliorative skepticism about some thing x can be seen as a Metaphysical Claim and an Explanatory Claim.[40] The Metaphysical Claim is something like this: we're tracking a range of things when we talk about x that are real and that are interdependent. The Explanatory Claim then becomes something like: there's no coherent, non-messy way of characterizing the way in which the things we're tracking when

---

[40] It's worth noting that, in the gender case, we can also see prospects for these two claims coming apart. I see them as mutually reinforcing, and at least in the case of health I think the arguments for each are closely related. But Jenkins (2023) outlines a position on the nature of gender that seems to endorse something like the Metaphysical Claim, while remaining agnostic on the explanatory claim. Likewise, someone could in principle endorse the Explanatory Claim without taking a view on the Metaphysical Claim—which seems to be at least somewhere in the neighborhood of the view put forward in Lopez De Sa (in progress).

we talk about x are related to x. The net result is that it makes sense to continue asking questions about the nature of x. And yet some of those questions will inevitably be unanswerable.

9. The Important, Impossible Role of Health

In summary, ameliorative skepticism about health sits as a kind of middle ground between deflationary forms of realism (cluster concept accounts, realizer-based accounts, etc.) on the one hand and eliminativism, fictionalism, and error theory on the other. Contra fictionalism or error theory, we're tracking real parts of the social and natural world when we talk about health—and we're often tracking them in systematic ways. And contra the eliminativist, there's important work to be done in talking about these features of the world distinctively as health-related features. We wouldn't be better off if we replaced talk of health with something more specific or more fine-grained. Part of why we need to talk about health—and insist that health is real—is that the things we're tracking when we talk about health (our biological function, our subjective experience of our minds and bodies, our social capacities, etc.) are entangled with and interdependent on each other in distinctive ways.

But contra more standard versions of realism, we can't give a clear, consistent theory of what health is. Nor can we make progress simply by theorizing the distinct realizers of health, or by giving a cluster concept analysis. Again, the problem is that the features of the world we're tracking are bound up with each other in complex ways, and we need an account of both how they're connected and how they're connected such that they're also related to health (but other things that causally impact health—like income or social status—aren't themselves part of our health). And there's no coherent, precise story we can tell here, at least in part because of the nature of what we're trying to explain. We're trying to theorize something that's less than overall wellbeing but more than physiological function, and that in-between space—although important—is inherently unstable.

Our understanding of health is thus inevitably messy and in tension. Indeed, talk of health might be most straightforward when applied to things like oak trees or penguins—living organisms that can be thriving instances of their species without the complex psychological and social variables we encounter for humans. Humans, perhaps uniquely, struggle with the relationship between our minds and our bodies, between what is natural for us and what is good for us, between our behavioral drives and our sincere desires. And so it would make sense if health for human beings, compared to other living organisms, is an especially murky thing.

Why, then, is talking about health still important? Why not just get rid of the idea that anything unifies the various features in the cluster—subjective experience, physical function, social capacity, relationship to the institutions of medicine—and move on? We could then talk about each individual feature, and be clearer and less confused.

Again, the problem is that these various features relate to each other—and shape the values we care about—in distinctive and often confusing ways. If we want to understand the distinctive moral and political importance of the prevalence of diabetes, for example, we need to understand not just its social determinants, not just the things it can keep people from doing in their everyday lives, not just the way it makes people feel, and not just its impact on major organ systems. We need to understand how all of these form a complex, interdependent whole. This entangled web is why we need a concept of health.

We need, that is, a way of talking about how these various dimensions of health are unified—how they are related to each other and how they form the complex whole that is health. And we need this to explain why the interrelation of these features is distinctively important—biologically important but also normatively, politically, and phenomenologically important.

Ameliorative skepticism about health says: this is work we need to be able to do, but there is no coherent, non-messy way of doing it.

6

Ameliorative Skepticism, Shifting Standards, and the Measure of Health

In Chapter 5, I introduced and defended a moderate skepticism about health—the view I'm calling *ameliorative skepticism*. In defending that view, I made a claim about the skepticism involved: skepticism can sometimes be ameliorative, rather than cynical or destructive, because of the ways in which it can help us. Much that is of central importance to us centers around how we think, speak, and reason about health. And my contention is that we'll be in a better position to understand what we value so much—and in a better position to work toward promoting and achieving it—if we give up on the idea that there's any coherent, settled thing that health is.

The crux of the kind of skepticism I'm defending is this: the things we're tracking when we talk about health are real and important, but there's no way of understanding their connections, interdependencies, and tensions that will tell us what health "really" is or give us everything we want from a theory of health. The goal of this chapter is to sketch, using the work of Delia Graff Fara, a model for two interrelated claims: (i) it can still be useful and informative to talk about health, even if there's no coherent, settled thing that health is; (ii) acknowledging that there's no coherent, settled thing that health is can help us communicate more effectively when talking about health.

In Chapter 4, I argued for a form of contextual shiftiness about the relationship between disability and health, inspired by David Lewis' work on counterparts and similarity. Here, I pick back up on that same idea of shifting standards, but expand it more broadly to talking about health in general. Lewis' and Fara's approaches to context dependence have much in common, and they both provide useful models for how we can understand—and communicate about—entrenched tensions and apparent inconsistencies. I am not a philosopher of language, and the goal of this chapter is not to defend a particular view within philosophy of language, or to articulate specific technical details of such a view. I'm also not trying to defend a story about what the word "health" means. Rather, the goal is to articulate a way of approaching cases where we think it's important to continue discussing and focusing on something, and we think much of what we're talking about is real and important, and yet there's irresolvable unclarity in what we're talking about. That is, cases like health—or so I've been arguing. More strongly, I argue that sometimes the unclarity or imprecision that we're stuck with can actually help us

Health Problems: Philosophical Puzzles about the Nature of Health. Elizabeth Barnes, Oxford University Press.
© Elizabeth Barnes 2023. DOI: 10.1093/oso/9780192883476.003.0007

to communicate in the ways that we want to communicate.[1] Just as in Chapter 5, I argue here that trying to clean up the mess might (at least in some cases) be missing the point.

1. Health Judgments and Hard Cases

There are various questions about health we might be interested in asking. Is some person, x, healthy? Is a person, x, healthier than some other person, y? Is a condition, C, had by some person, x, pathological? Does a condition, C, tend to reduce health when had by an individual, x? And so on.

On the kind of view of health I'm defending, the differences between these questions are not particularly substantial, and none of them is the most important or perspicuous for an analysis of health. And I've been shifting loosely between them. But for the purposes here, I am going to focus on questions like:

(i) is x healthy?
(ii) how healthy is x?

where x is a particular individual under consideration in a context.

What I've been arguing so far is that there is no consistent, correct way of addressing these questions because there is no coherent, unified thing that health is. That doesn't mean that there can't be clear or objectively correct answers in many cases. Someone with advanced lung cancer is not healthy—indeed they are very unhealthy—and it's never correct to say otherwise. We have, of course, no trouble at all understanding that someone with lung cancer is not healthy (or is very unhealthy, and considerably less healthy than someone without substantial pathology). Likewise, we have no trouble understanding that someone on a ventilator in an ICU is not healthy, or that someone with a seriously debilitating, fatal disease like amyotrophic lateral sclerosis is not healthy.

Judgments in cases like these are often straightforward because the multiple dimensions of health converge. The person's biological function is compromised; that decrease in biological function is closely linked to reduced ability to function socially, to subjective experiences of pain and discomfort, to the need for complex medical care, and so on. In other clear cases, the various aspects of health might not be well correlated but the person's biological function is so severely compromised that it allows for an obvious verdict. Aneurysms, tumors, blood clots, and arterial blockages can all be symptomless and yet their immediate risk of death or

---

[1] For different arguments that lead to an interestingly similar conclusion, see Nguyen (2021) and (2020)

severe illness are enough to allow us to say that persons with such significant pathologies are unhealthy.

The trouble comes when we need to make comparative judgments, or when we need to make judgments about more ordinary and less acute cases. And things get especially murky when the various dimensions of health conflict to some degree, but the basic biological reality isn't clear enough to deliver a verdict (e.g., the person isn't about to die).[2]

Let's consider a range of cases where judgments might reasonably conflict over whether, and to what extent, the person in question is healthy. For the sake of simplicity, let's hold fixed some broad social variables and suppose that the people in question are all middle-class, securely employed Americans in their thirties. (Comparisons, unsurprisingly, get astonishingly more complicated once we introduce variations in age, socioeconomic class, etc.)

Joe works in tech. There is nothing obviously wrong with Joe, and his body functions fairly normally for someone of his age. If you ask Joe how he feels—either physically or psychologically—he'll probably say, "Fine, I guess?". However, Joe works very long hours, his job is sedentary, and he gets very little physical exercise. He doesn't sleep all that well, and he's recently taken up occasional smoking to help manage the stress from his job.

Kiko is an elite athlete currently recovering from a concussion. Let's say her sport is skiing—so it involves substantial amounts of impact, risk, and wear on the body. Kiko is in incredible physical shape, she's highly energetic, and she's very scrupulous about nutrition and stress management. She works regularly with a trainer, a physical therapist, and a sports medicine doctor to help keep her in her best shape. This is her fourth traumatic brain injury in the last ten years.

Mika is an academic (with the stable employment and flexible working hours of a secure academic job) diagnosed with lupus several years ago. She considered her diagnosis of lupus a "wake-up call" and has made major changes to her life

---

[2] The discussion in this chapter isn't about the meaning of the word "health", but the discussion of "health" from philosophy of language is nevertheless informative. The linguistic aspect of this shiftiness is a phenomenon that is well documented. Indeed, health is often given as a paradigm example of what linguists call a "multidimensional adjective" (see especially Kamp (1975)). Gradable adjectives like "tall" have a single dimension of comparison, and we can easily construct a comparative ordering of how tall individuals are. But as Kamp points out, "such adjectives are rare. Even 'large' is not one of them. For what precisely makes an object large? Its height? or its volume? or its surface? or a combination of some of these?" (p. 141). Instead, most adjectives have multiple dimensions of comparison along which we are evaluating them, and we have to consider the balance of these various (sometimes competing) dimensions when using them.

Here again is Kamp: "Suppose for example that Smith, though less quick-witted than Jones, is much better at solving mathematical problems. Is Smith cleverer? ... This is perhaps not clear, for we usually regard quick-wittedness and problem-solving facility as indications of cleverness, without a canon for weighing these criteria against each other when they suggest different answers. When faced with the need to decide the issue, various options may be open to us."

since. She's started to prioritize getting regular exercise, maintaining a more comfortable work–life balance, and making sure she gets enough sleep. She often says that she wouldn't be nearly as healthy as she is today if she didn't have lupus.

Jess is also an academic, in a similar social situation to Mika. Jess struggles with moderate depression, and has been especially burnt out at work in recent years. Lately, she's been feeling constantly run down and tired, and she has been experiencing a lot of muscle aches and stomach problems. She's been to multiple doctors, but all her tests come back normal and her doctors reassure her that she appears to be "very healthy."

How healthy are Joe, Kiko, Mika, and Jess, respectively? How unhealthy are they? Who is healthier than whom?

Questions like these should, I hope, appear fairly intractable. Joe feels fine, and there's nothing wrong with him, but his lifestyle involves various major risk factors for poor long-term health outcomes and reduced life expectancy. Kiko is fit and energetic, engages in a lot of positive health behaviors, and would assure you that she feels very healthy; but the long-term toll her sport has taken on her body is substantial. Mika is grappling with a serious disease that requires ongoing medical intervention; but most days, she has the capacity to do what she wants to do, and she feels enthusiastic about her own health and health management. Jess, in contrast, really struggles to do the things she wants to do, and doesn't feel well at all, although there's no obvious biomedical pathology that's causing her to feel this way.

Questions such as who is healthy, or who is healthier, quickly become murky in cases like these. And on the view of health I'm defending, there is no single best answer to them. They are, nevertheless, questions we often *have* to ask when deciding how to make judgments or allocate resources. Moreover, they can still be useful questions to ask and they can still have useful (and correct) answers in particular contexts. What I want to argue is that skepticism about health can help promote, rather than obscure, good answers to (inevitable) questions like these.

2. Health and Shifting Standards

We seem to very naturally employ shifting standards when talking about health. So, for example, the following (paraphrased) conversation didn't sound jarring:

MY SISTER: I've been running more and more lately—it's the main way I stay healthy, so I always want to make time for it.
ME: Have you had any trouble with injuries?

My sister: Ugh, yeah, I have shin splints again. I have to go back to PT. It's so
hard to stay healthy when you're a runner!

Anyone in a conversation like this easily recognizes that slightly different standards are being applied to our understanding of health as our focus shifts.
Sometimes, as in my sister's first claim, we can focus on things like overall
cardiovascular fitness and stress relief. But we can then easily shift the focus to
things like injury and pain.[3]

It is, of course, consistent that one and the same thing (in this case, running)
can increase your fitness and promote stress relief but also increase your pain and
cause injury. So her two statements are fully compatible with each other.

We can—and often do—apply very different standards of health, depending on
the context. Consider the example cases from section 1. We might say of Kiko, for
example, that she is finally healthy this season—meaning that she is able to return
to competition. And then we might immediately follow up this claim with the
statement that the long-term physical toll of competitive skiing makes it the case
that athletes are rarely healthy—meaning, in this context, that things like the
traumatic brain injury associated with skiing have serious impacts on life expectancy, and neurological function. We can make this shift easily, and people will
understand what we mean.[4]

In talking about health-related behaviors and fitness—the kinds of behaviors
that predict long-term health complications, chronic lifestyle related diseases, and
life expectancy—we might say that Mika is healthy. Indeed, we can plausibly say
that she's healthier than she would've been had she not gotten lupus. And we
might favorably contrast her health to Joe's. Joe's smoking, for example, is, at least
statistically, a significantly greater threat to his long-term health outcomes than
Mika's lupus, provided she can get good medical care.

But if Joe and Mika are both at the doctor, it makes sense for Joe to be
described as healthy. There's nothing wrong with Joe—no active pathologies or
disease processes—and all his major organ systems function within normal
limits. He is not in any immediate risk and doesn't require any active medical
intervention. Mika, in contrast, wouldn't be viewed as healthy in this context. She
has a serious disease that needs regular monitoring and regular treatment. Her
long-term functioning and survival depends on this maintenance, and the presence of an autoimmune disease like lupus means she's always at risk, even if the

---

[3] Perhaps unsurprisingly, then, the word "healthy" is often a paradigm example of the phenomenon
of *polysemy*—where a single word has distinct but related meanings that it can shift between.
[4] Again, Kamp (1975) is instructive here: "Often it is the context of use which indicates how the
criteria should be integrated ... Exclusive preoccupation with shape, for example, can be evident to both
speaker and audience ... think of a session about shape during a conference on industrial design ...
preoccupation with shape will last throughout the session."

risk is small, of serious medical complications that are much less salient for someone like Joe.[5]

We might, in many ordinary contexts, describe Jess as someone who isn't healthy. She's struggling with her mental health, she doesn't feel well, she isn't able to do what she wants to do. But in the context of something like the COVID-19 pandemic, it might make sense to say that Jess is healthy—she isn't obviously at increased risk of medical complications, she isn't immunocompromised, she's basically physically normal for an adult of her age, and so on.

Jess is also someone who might easily find herself employing different standards in evaluating health than others in the same conversation. Suppose Jess seeks out medical care for her pain and stomach issues. The doctor runs lots of tests, and those tests are normal. From the doctor's perspective, Jess' major biological systems are working normally, and she's not—unlike Mika—in need of regular monitoring or major medical intervention to prevent serious sequelae of a disease process. So the doctor, seeing that she is anxious and in distress, attempts to reassure her by saying that she appears to be healthy.

But to Jess, this isn't reassuring at all—it feels dismissive. Jess is considering her health by the standards of how she feels and the impact on her everyday life, and from that perspective she is *not* healthy. So when the doctor tells her she is healthy, she feels like she isn't being heard or taken seriously.

The doctor isn't wrong, though. If we evaluate Jess by the standards most salient in a medical practice, Jess is healthy. She isn't at significant risk of major medical complications and she doesn't require major medical intervention to stay that way. Moreover, the kinds of physical symptoms Jess is seeking medical care for—while very real—aren't obviously the kinds of problems that would benefit from the treatment modalities that the doctor has to offer. It makes sense that the doctor is employing very different standards of health from Jess. They have different goals, and they prioritize different things.

It also makes sense, then, how a conversation about health between Jess and the doctor can so easily go off the rails. They're both employing standards of health that it makes sense for them to employ, and neither is making a mistake. Yet they might both leave an encounter like this baffled and frustrated.

---

[5] Sassoon (2012) makes similar observations in her extensive discussion of health as a multidimensional adjective. Again, I not giving an account of what the word "healthy" means, but the comparisons are useful. She argues that there are different dimensions to both "healthy" and "healthier," and that context restrictions play a significant role in our evaluation of health: "On a scenario whereby a man has high blood pressure, while his dying wife has normal blood pressure, intuitively, the husband is healthier despite him doing less well in one respect." Sassoon's overall analysis is a conjunctive view of "healthy," according to which someone is healthy if they are healthy in every respect. I'm pushing against a view like that here but sympathetic to much of what she says about multiple dimensions and the role of context.

3. Health as a Useful Shorthand

In Chapter 5, I argued that there isn't any coherent, settled thing that health is, at least for human beings. Rather, there are a range of features we're tracking when we talk about health, and those features are interrelated and interdependent in complex ways. There's no account, though, of how they are interrelated that gives us a consistent, extensionally adequate theory of health. But this doesn't mean that health isn't real, or that we should abandon talk of health. Of course health is real, and of course we need to talk about it. In this section, I want to sketch a general picture for how this might work. With this general picture in place, we can begin to make some progress on the type of evaluative puzzles from the example cases.

The German economy is healthy. The Yellowstone ecosystem is healthy. My marriage is healthy.

It's true of all of these things that they are healthy. But we don't typically infer, when talking about health in these contexts, that there's a single thing that it is for an economy to be healthy, an ecosystem to be healthy, or a marriage to be healthy. Likewise, we don't typically think we can give necessary and sufficient conditions for the health of marriages or ecosystems, and many philosophers at least would be skeptical that there's one single thing that is economic health. And yet, it often makes sense and is useful for us to speak about health in these contexts.[6]

When evaluating health for economies and ecosystems, we're often engaging in a type of useful shorthand. Economies are many things that matter to us—they are stable, diversified, growing. We can communicate useful information about economies by calling them healthy without needing a precise or reductive account of what health-for-economies means. Indeed, a pedant might come along and insist that the only things that are ever *really* healthy or diseased are living organisms—things like economies or ecosystems are healthy only insofar as they bear some kind of loose family-resemblance relationship to healthy living organisms. Suppose the pedant is right (I have no particular view on this either way)—it would still be useful and informative to talk about healthy economies.

Similarly, we can say that economies are flourishing or thriving, but insofar as an economy is an excellent instance of its kind or a well-functioning instance of its kind, it gets these accolades from how it suits our goals and intentions. There is nothing that it is for the existence of a German economy to be going well *for the German economy*. There is nothing that it is for the Germany economy to be an exemplar of an economy other than the goals we have in establishing economies— it's not that these goals are good *for the economy itself*.[7]

---

[6] Again, this is what philosophers often have in mind when they talk about "analogical concepts" or "family-resemblance" concepts, and why health is often a paradigm example in these cases.

[7] Ecosystems might be different—depending on whether you think environmental value is at least partly objective.

There are various features an economy can have: consumer confidence and consumer spending, GDP and domestic growth, rates of (un)employment, and so on. Moreover, some of these factors are themselves pretty messy (what exactly is "consumer confidence"? who should count as "unemployed"?) and can be interwoven in complicated ways. Saying that an economy is healthy is a way of speaking loosely about these various factors, and the complex relationships between them. We can misconstrue the health of an economy by focusing too myopically on any one aspect (e.g., paying sole attention to the stock exchange averages and not looking at the financial status of individuals within the economy). And we can be wrong about whether an economy is healthy. Just because everyone in 2005 was buying homes and interest rates were low and stock prices were high didn't mean the US economy in 2005 was healthy. But there's no single thing that it is for an economy to be healthy, and no universally correct, reductive analysis of health for economies. After all, whether an economy is healthy will depend partly on what you care about when you're assessing it. Is the presence of multi-billionaires an indicator of health or dysfunction? Is substantial economic inequality by itself an indicator of an economic problem? Is growth always good? The answers to these questions will depend on what you think economies are for, and what purposes you think they serve. And nothing about the nature of economies themselves will answer those questions.

When we talk about healthy economies, we're loosely gesturing at a range of features. We can be wrong about which economies are healthy and wrong about what features factor into an assessment of economic health. But any answer to a general question like "is the German economy healthy?" will depend on more than just the basic metrics of economic health—it will also depend on the context of the question, on what we're trying to accomplish by talking about economic health, by what we think economies are for and how we think they ought to be structured. Any answer, that is, requires more than a simple assessment of baseline economic factors themselves.

Of course, that doesn't mean talk of economic health is totally unconstrained or subjective. Again, we can be straightforwardly wrong about whether an economy is healthy. And, for example, number of border collies is never a good indicator of economic health, even if it's something we happen to care about *very strongly*. There are norms that govern which things—such as growth, inflation, and employment—are part of the overall story of economic health. These norms allow us to communicate effectively when talking about economic health, and they track something real and significant about the social world. But it doesn't follow that there's something specific that economic health *really* is, or that we can give necessary and sufficient conditions for when an economy is healthy.

So when we say that an economy is healthy, we're in effect communicating via shorthand. In describing them as healthy, we're loosely referring to a wide range of features economies can have. Which features matter most, and how we interpret their relationship to an economy's health, however, will depend on our what our purposes are in talking about economic health. Or, to frame it in Elizabeth Anderson's terms: contextual interests will often determine the "goal of inquiry," such that they'll shape what question we're in fact asking, and what the best way of answering that question (and, indeed, the objectively correct answer to that question) might be.[8]

When we're speaking loosely about economies being healthy or sick, of course, we're often trading on analogy to the ways that living organisms can be healthy or sick. But part of the skepticism I'm defending here is that there isn't any clear, coherent thing that it is to be healthy in the case of human beings. Rather, health for human beings is actually quite similar to health for economies. There are a range of features—real features, important features—which we are loosely tracking with our talk of health. But some of those features are themselves quite messy, as are the relationships they have to each other. And which features are most salient or which of them matter most to determining and assessing health will depend on our interests and purposes.

We often talk about health, for human organisms, as though there is one single, unified thing—a property or feature of being healthy. But on the view I'm defending, there isn't any one thing we're tracking when we talk about health. There are many interrelated things—some of them really messy, and some of them at times actively in tension with each other. And which of them matters most, in a particular contextual assessment of health, can depend on a wide range of factors.

Nevertheless, it can often be useful to speak about health *as though* it is a single, coherent thing. Health can serve as a kind of useful shorthand, and we can often say true things about the world by talking about health in this way. But it would be a mistake to infer from that we can give necessary and sufficient conditions for being healthy, or that there's a coherent story we can tell about what exactly health is.

4. Delia Graff Fara on Shifting Standards

Since we are discussing philosophical messiness, let's return again to the case of *vagueness*. Notoriously, many predicates admit of a distinctive type of paradoxical reasoning, captured by the Sorites Paradox (or the paradox of vagueness). Take, for example, "is tall." Let's suppose that someone who is 6'3 is definitely tall. Suppose, then, that I subtracted 1 mm from their height. A single millimeter surely shouldn't make a difference to whether someone is tall. So if the person is tall at

---

[8] See especially Anderson (1995).

6'3, they're tall at 6'3 (-1 mm). But now I apply the same reasoning: a single millimeter shouldn't make a difference to whether someone is tall. And I shave off another millimeter. If a person who is 6'3 (-1 mm) is tall—and we just established that they are—then a person who is 6'3 (-2 mm) is tall. And now we are off to the races. I can keep applying the same reasoning, impeccably, until I reach the conclusion that someone who is 5'3 is tall, which is absurd.

The crucial problem of vagueness is this: it looks like small changes *shouldn't* make a difference, that large changes *must* make a difference, but that if you add up enough small changes, you'll inevitably get a large change. Imagine that I line up for you a series of men, ranging in height from 6'3 to 5'3, in order of tallest to shortest, with only a millimeter height difference between each. The tallest is clearly tall. The shortest is clearly not tall. If I reason that a millimeter shouldn't make a difference to whether or not a person is tall, I have an argument that they are either all tall or all not tall. Either conclusion is obviously wrong. Somewhere, a change occurs: some of the men are clearly tall, and some of the men are clearly not tall. But it looks wrong to locate the change in any particular place. A millimeter's difference in height doesn't determine whether or not a man is tall.

To be clear, I don't want to argue that the philosophical problems we encounter with health are simply problems of vagueness, or that we can address any of these problems simply by treating "is healthy" as a vague predicate. There has been an ocean of philosophical ink spilled on the problem of vagueness, and I'm not going to wade into that debate. Rather, I want to argue that some key insights from thinking about vagueness might help us make progress in other cases of philosophical complexity and unsettledness. It is probably true that "is healthy" is semantically vague. It is also probably not very interesting, and not much hangs on it. Most predicates are semantically vague. The philosophical problems we encounter in the nature of health (not just in the word "health")—where unclear norms, practices, and beliefs shape and in turn are shaped by a messy social reality—extend far beyond this ordinary phenomenon of semantic vagueness. What I'm suggesting, though, is that some models for thinking about imprecision and unclarity from literature on vagueness might give us a useful frame for thinking about the kinds of complexity we find ourselves landed with when we try to talk about health.

Contextualism, very simplistically, is the view that terms like "is tall" can change their meaning[9] depending on the context in which they are used. Most straightforwardly, this shift in meaning is explained as change in *comparison class*. The man who is 6'3 is clearly tall if we're at a philosophy conference, comparing him to the other attendees. He's also clearly tall if we're comparing him to the average American population. But he's not clearly tall if we're comparing him to NBA

---

[9] For technically minded readers: pretty much everywhere I'm about to say "meaning" you should sub in "extension."

players. Steph Curry is tall at the grocery store. But he isn't tall on the basketball court. The meaning (or at least the extension) of "is tall" changes depending on the context, and often that's because the relevant comparison class changes.

This kind of shift in meaning can also, sometimes, give rise the phenomenon called "faultless disagreement." Suppose you and I spot Steph Curry at the grocery store. Not following basketball and having no idea who he is, I say, "Wow, that guy is tall." When I say that, I mean something like "tall for a random person walking around the grocery store." You recognize immediately that it is Steph Curry, and you reply "What? No, he's not tall!" Here, you mean by "tall" something like "tall for a professional basketball player." We appear to be disagreeing with each other—we're pointing to the same individual and disputing whether he is tall. But in fact, we simply mean different things by "is tall."

In her highly influential work on vagueness, Delia Graff Fara (2000) developed an alternative approach. According to Fara, terms like "tall" can shift not only because of variation in the comparison class but also because of variation in the interests, aims, and purposes of the person using the term.[10] Fara argues that we often have vague or otherwise imprecise purposes that drive our use of terms like "tall," and that imprecision can affect what words like "tall" mean. More strikingly, she also argues that often this imprecision in our aims and purposes is a useful part of how we communicate rather than something that we would ideally precisify away.[11]

Suppose I am considering whether to plant a cherry tree in my garden, and I ask you whether cherry trees are tall. Part of what I mean by "tall," in this context, will be determined by what my purposes are in planting a cherry tree. Maybe I want privacy screening from the neighbors. Maybe I want to know whether I'll be able to see it from my kitchen window. Maybe I want to know whether it will block too much light to my other plants as it grows. These different purposes might give me importantly different aims in asking whether cherry trees are tall, and according to Fara will involve me employing different standards of assessment and evaluation in deciding whether a tree counts as tall. A tree that's tall enough to be seen from my kitchen window might not be nearly tall enough to be a privacy screen, for example. And crucially, according to Fara, it isn't just that I'm varying the comparison class (although in some cases I might be). It's that my aims and purposes shape my standards for "tall." And given that my aims and purposes can change—I might say of the very same tree that it's tall when deciding to prune it,

---

[10] To be clear, Fara was not the only person interested in the role that our aims and practical interests play in our use of vague predicates—and more traditional contextualist approaches also sometimes emphasized these features. See especially Raffman (1994).

[11] This is, I think, something particularly interesting and distinctive about Fara's approach to vagueness. Contextualists have also emphasized the idea that vagueness might be ineliminable—see, e.g., Kennedy (1999) and (2007)—but it's closer to the idea of verisimilitude in scientific theorizing: we may never fully precisify but the closer we can get, the better. Fara's view, in contrast, is that sometimes precisification is helpful but sometimes it obscures as much or more than it clarifies—and that sometimes imprecision is actively helpful to us. See her (2008) for elaboration.

but that it's not tall when talking about how much light it lets through—so too can the extension of the term "tall."[12]

It's not my goal here to explain Fara's solution to the Sorites Paradox—although I encourage any philosophy aficionado to read it, as it's a thing of beauty. Fara defends a view known as interest-relative invariantism, and her approach to the Sorites Paradox is most technically distinctive (and varies from a contextualist approach most specifically) insofar as she argues that whether an object has a particular (vague) property can depend on our interests. Rather than arguing (as the traditional contextualist does) that a word like "tall" will pick out different properties in different contexts, she is arguing that it always picks out the same property, but that whether an object—such as a particular cherry tree—has that property in a context depends on our aims and purposes in evaluating height. The details of this solution—and the broader debate between contextualists and interest-relative invariantists—is not something I'm focused on here, and not something I want to take a view on.[13]

Rather, I want to highlight what I take to be some of the key points of Fara's broader approach to vague predicates like "tall," since those same insights are—I suggest—equally applicable to how we speak and reason about conceptually messy things like health. Again, some of these points aren't at all distinct to Fara. But the overall approach she defends—especially her focus on the idea that vagueness is often a helpful tool for communication rather than an impediment that should be expunged whenever possible—is what I take to be distinctive, and what I want to focus on.

4.1 First Key Point: Terms Can Inherit the Imprecision of Our Aims and Purposes

More often than not, we don't use—and aren't trying to use—terms like "tall" in a way that allows us to make millimeter-based discriminations. We have some sort

---

[12] It is worth being clear on this technical detail: Fara's view is not that the meaning of a word like "tall" changes depending on the standards used when employing it, but rather that which things fall in the extension of "tall" shifts depending on the standards we employ in our use of the term, and that this is part of what fixes the meaning of "tall."

[13] Fara's (2008) reply to criticisms raised by Jason Stanley (2005) is particularly instructive on this point, for those interested. Fara (p. 200) writes that: "It is possible that the predicate could express the same property from occasion to occasion, and that the reason the extension may change as the heights of things do not change is that the property expressed context-invariantly by 'tall' is a property which is such that whether a thing has it depends not only on heights, but on other things as well ... there is much less context-dependence than one might have initially thought would be a consequence of the bare-bones view." See Fara (2000), p. 64.

Whether one wants to commit to full-throated interest-relative invariantism—which is a fairly strong view—is somewhat orthogonal to the point I'm most interested in extracting from Fara's work. Just as you don't need to be a die-hard counterpart theorist to go along with most of what I said in Chapter 4, you don't need to accept interest-relative properties to accept most of what I say here.

of rough, not-quite-specified intention to single out people (or trees, or buildings, etc.) that are to some significant degree above average height for the kind of thing they are. And we can use "tall" in that way just fine, and it accomplishes that sometimes necessary but relatively underspecified purpose perfectly well without needing added precision. But Fara argues that *partly for this reason*, we encounter the problem of vagueness. The purposes for which we apply a term like "tall" to people just don't require us to make distinctions at the level of millimeters. And so when we're asked how many millimeters we could subtract from the height of a tall person in order to make him no longer tall, or where, in a series of people lined up in height, with only a millimeter's difference between them, the cut-off between the tall and the not-tall is, we have no good answers. We have no good answers— and we find ourselves prone to paradoxical reasoning—because, according to Fara, our normal use of a term like "tall" isn't intended to, and doesn't require us to, make judgments like that.

4.2 Second Key Point: We Can Make the Application of a Term More Precise by Making Our Aims and Goals More Precise

Fara also points out that terms like "tall" can inherit precision, just as they can inherit vagueness or unclarity. Suppose that rather than a gardening enthusiast, I am a botanist. (Reader, I am neither.) I am studying cherry trees, and am particularly interested in the effect of soil types on height. Suppose I know the average height for cherry trees in the region, and consider any cherry tree that's 25 percent or more above the average to be tall for a cherry tree in this region. I now set about measuring, down to the millimeter, the cherry trees growing in a particular soil type, in order to assess how many of them are tall. My use of "tall for a cherry tree" now means something *much more precise* than any of the gardening usages. And, as a result, it's much less prone to vagueness-style reasoning. In a case like this, the meaning of "tall" is more precise—and less vague—simply because my aims and purposes are much more precise.

4.3 Third Key Point: We Often Benefit from Having Imprecise Aims and Goals, and It Would Not Always Be Better for Us to Make Our Aims and Goals More Precise

Crucially, though, Fara doesn't argue that, as a result, we should all be more like the botanist and less like the gardener in our use of terms like "tall for a cherry tree." The botanist is using "tall" in a way that's more precise than the gardener. But the gardener doesn't need to use "tall" in that precise a way. It's not worth her time and effort to make the kinds of fine-grained discriminations about height

that the botanist is engaged in. She doesn't need to. Millimeter-based discriminations in cherry tree height are relevant to the botanist's aims and goals but not to
the gardener's. The gardener is thus *better off* using "tall" in a way that's less
precise. It's a better fit for her purposes. Her purposes themselves are somewhat
less specified than the botanist's, but they don't need the same level of specification
in order to fulfill the gardener's goals. Trying to use the term in the way the
botanist does would be frustrating for her, and get in the way of what she's trying
to do. A less precise use of the term is a better fit for a less precise goals—the
flexibility and underspecification can be genuinely helpful in communicating
when we have flexible and underspecified projects. And plenty of legitimate and
ordinary goals just aren't that precise (and wouldn't be improved by adding
precision).

4.4 Fourth Key Point: Sometimes We Can Disagree with Each Other about the Application of a Term Because We Have Different Aims and Goals

Suppose that you and I are discussing whether Joe is tall. I say that Joe is tall, and
you say he's not tall. Let's suppose that we both agree that Joe is 6 ft, so we're not
actually disagreeing about his height. One way that we could disagree with each
other is via the standard contextualist story: we're comparing Joe to different
groups, and thus each mean something slightly different by "tall." But Fara
proposes that we can also disagree with each other by having different purposes
and aims in our use of the word "tall." Suppose we are both considering the men at
the conference, and so our comparison class is the same—the men at the conference, and maybe "the average man" as well. But we're implicitly thinking about
different projects when bringing up the subject of Joe's height. You're thinking
about how hot it is in the room, and wondering whether anyone around is tall
enough to open the window. I'm thinking about my single friend at the conference, who I officiously keep setting up, and who I know likes tall guys. I say that
I think Joe is tall, thinking that 6 ft is definitely tall enough to pass muster for my
friend. You say that Joe is not tall, thinking that it might be a stretch for him to get
to that window. We disagree about whether Joe is tall in this context—and
disagree, more fundamentally, about what the standards are for tallness—because
we each have different practical purposes in considering whether Joe is tall.

5. A Fara-Style Approach to Shifting Standards for Health

Let's now examine what it might look like to apply the key insights I've picked out
from Fara's work to the case of health. Again, the point is not to say that we can

solve the problems we encounter in reasoning about health by saying that "is healthy" is a vague predicate or that "is healthy" is context dependent—both of which seem true, but neither of which seems particularly surprising or interesting. Rather, the claim is that, taken together, some of the central points of Fara's approach to vagueness can be usefully applied to other cases of philosophical messiness, including the ameliorative skepticism I'm endorsing for health.

The case of health, at least as I'm presenting it, will have important differences from a case like tallness. For tallness, we all agree that there's a single underlying factor we're tracking: height. It's just that there are different standards we might employ for how much height something has to have in order to be a tall instance of its kind. For health, in contrast, we're tracking many different factors—factors that are interrelated, interdependent, and at times even in tension with each other. Likewise, in the case of tallness, the underlying reality itself is not messy. Things have the height they have, and in normal circumstances what height they are typically isn't something we're confused about or something that gives rise to paradoxical reasoning. The puzzling question is whether the height they have is sufficient for being tall, not anything about their height itself. In contrast, I've argued that parts of the underlying reality we're tracking when we talk about health are itself messy. When we talk about health, at least part of what we're talking about are aspects of social reality, and some of those things can themselves be unsettled.

Moreover, my analogy to Fara doesn't incorporate all the technical details of her approach, nor is it focused on what the word "health" means. Rather, I'm using Fara's model as a springboard to explain how it can still be useful to focus on, talk about, evaluate, and care about health as a though it's a specific and unified thing, even though there's not a consistent and fully coherent story to be told about what health is. But I'm not, in doing this, focused on the semantic details of how we use terms like "health" and "healthy".

Despite these differences, however, I think the insights from Fara's model can provide a useful framework for the shifting ways we evaluate health. And, more generally, they can provide an explanation for how, in talking about health, we can communicate clear and important things about our lives, even if there's no consistent, non-messy story to be told about what health is.

5.1 We Inherit the Imprecision of Our Aims and Purposes—Health Redux

Fara argues that the application of our terms can often inherit imprecision from our aims and purposes. Likewise, our aims and purposes, when we focus on health, are often very imprecise. We want to talk about a feature of our lives that is somewhat more than the simple functioning of our major organ systems, but somewhat less than our overall wellbeing. We want it to include our current physical and mental status (any active disease processes or injuries, for example) but also to encompass things like our life expectancy and major long-term risk factors. And we also want it to include subjective information about how we feel (if we're in pain, for example) and to what extent we feel capable of pursuing our "activities of daily living." We want to bring attention to all these things in some loose and not particularly specified way. We thus typically don't have precise aims or purposes when we're talking about health. And our discussion of health can, as Fara argues, easily inherit the lack of specificity of our purposes.

Return to our example cases. Is Joe healthy? For some purposes we might have in talking about health—life expectancy, long-term risk factors, and so on—Joe isn't healthy. But for other purposes—how he feels and what he can do day to day, his current physiological function and his immediate medical risk—is makes sense to count Joe as healthy. In many contexts, though, we just aren't clear about which of these aims we have in asking whether Joe is healthy, or how such aims should be balanced and weighed against each other. It's hard for us to say whether Joe is healthy because, in many cases, the different aims and purposes we might have in asking about Joe's health pull against each other, and nothing really settles which aims are most salient.

Who is healthier, Jess or Mika? Again, often when asking a question like this, we don't have particularly specific aims. Do we want to know who is at more risk of major medical complications? Do we want to know who feels better, and who feels themselves to be capable of more normal daily activities? Are we evaluating long-term health outcomes? More often than not, we take ourselves to be doing a little bit of all of this, but without a specific sense of how we're combining or weighing these various factors.

And so, our understanding of health in cases like these can often be unclear, muddled, or otherwise confusing. What Fara's model explains is the way in which unclarity in our aims and purposes can be a major contributor to this muddle.

5.2 We Can Make Our Aims and Goals More Precise—Health Redux

Similarly to our use of predicates like "tall," we can also make discussions of health clearer and more precise by making our purposes clearer and more specific. Let's consider the question of whether Kiko is healthy. A sports commentator might declare "Kiko is finally healthy again!", and in the particular context in which she says this, it might be clearly true. And that's because talking about health in this context has a very specific purpose: we're singling out which athletes are currently able to compete at their typical levels of fitness for their sport. Kiko may be at risk for serious long-term complications of a traumatic brain injury, but that doesn't make her any less healthy for this very specific purpose of talking about health.

In contrast, when she visits her neurologist, Kiko may be sternly told "you need to stop this—skiing has seriously damaged your health." In talking about health, the neurologist likewise has specific aims. He's assessing Kiko's current and long-term risk for major medical complications, especially neurological complications. Now suppose, for the sake of the example, that Kiko adores skiing, and she would say that the training she's put into has helped her build a positive relationship with her body and work through some significant personal challenges in her life. All of this might contribute to aspects of her health, according to various ways we might talk about health and various things we might legitimately factor into an assessment of health. But when she's sitting in the neurologist's office, those aren't the relevant ways in which we care about health. Rather, the neurologist has the specific aim of evaluating and treating Kiko's neurological function, and his discussion of health is guided by that aim. He is thus evaluating health in a narrower—and correspondingly less unclear—way.

5.3 We Often Benefit from Having Imprecise Aims and Goals, and It Would Not Always Be Better for Us to Make Our Aims and Goals More Precise—Health Redux

Crucially, though, just as in the case of vague predicates, more detailed and specified purposes aren't always better. And the solution to the problems we encounter when talking about health isn't simply to make our purposes hyper-specific. The invocations of health by the sports commentator and the neurologist are in many ways clearer. But they're also, for the same reasons, less adaptable and in many cases less useful. Often, we *want* our aims and purposes to be relatively non-specific when talking about health. As discussed in section 3, we want a loose shorthand—a flexible frame from which we can discuss a wide range of intersecting features that matter to us in a distinctive way. Talking about health can do this for us—it can be flexible, loose, and adaptable—in part because the goals we have in discussing health are often fairly imprecise and open-ended.

Fara argues that this is true for many vague predicates such as "tall" and "bald", even when the underlying reality (height, arrangement and number of hairs on a head, etc.) is perfectly precise. It's even more salient, I suggest, when the underlying reality is itself messy. Very often, our aims and purposes, when we talk about health, are fairly non-specific and unsettled—and very often, it's useful for them to be unsettled, at least in part, because the social reality we're talking about *is itself unsettled*. We have the underspecified intention of singling out some aspect of our flourishing that isn't the same thing as our wellbeing, but isn't just our physical health. We target a range of features with this intention—our physiological and psychological functioning, what we are capable of doing in our everyday environment, how we feel, how long we can expect to live, and what diseases we might be most at risk for over the course of our lives—but those features are entangled, interdependent, and in tension with each other in complex ways. The social reality of health, that is, is messy. And so, very often, imprecise purposes are in fact fairly well adapted to suit imprecise reality.

The ways that the sports commentator or the neurologist evaluate health are useful for them in their highly specific contexts. But this specification can obscure as much as it clarifies. It makes sense, at the start of the season, for a commentator to say Kiko is healthy. But if we rely on that understanding of health too much, we miss the significance of the current and long-term effects of her injuries. Likewise, it makes sense for a neurologist, in a clinical setting, to say she's not healthy. But if we rely on that usage too much, we over-prioritize physiological function and disease and don't give enough priority to the mental and social aspects of health, and to the way someone like Kiko's long-term health will be shaped not just by her injuries but also by her fitness and diet, her health literacy, and her general love of what she does.

Specific purposes are extremely useful in specific contexts. But the solution to the quandaries we find ourselves in when talking about health isn't to adopt them in all contexts. Often, our aims and goals in talking about health aren't just non-specific—they're *appropriately* non-specific.

5.4 Sometimes We Can Disagree with Each Other Because We Have Different Aims and Goals—Health Redux

And finally, just as in the case of vague predicates, disagreement can arise when individuals in the same context have different—sometimes very different—aims or goals when talking about health. Consider again the case of Jess. Jess has particular aims and goals in discussing her health—she wants to feel better, to experience fewer distressing physical symptoms, and to have the sense that she can do more in her everyday life without being so run down. Jess' doctor also has particular goals—she wants to identify and treat significant biomedical pathology, she wants to make sure that she does what she can do prevent serious medical complications, she wants to avoid the harms of over-treatment. Sometimes, these aims and goals can happily overlap. But sometimes, as in the case of Jess, they don't. Jess considers herself very unhealthy. And given her aims and goals in talking about her health, this judgment is correct. But given the doctor's aims and goals, it's correct to say that Jess is healthy.

We can leave open the question of whether either, or both, Jess and the doctor should have *different* goals than they in fact do. Often, disagreement in these cases may be as much about which standards of health we ought to apply as they are about whether a particular person is healthy.[14] The point is simply that part of the disagreement that arises in a case like this can be understood as a divergence in purposes. Jess feels dismissed and unheard—and the doctor feels frustrated and baffled when Jess isn't reassured—at least in part because they have different goals when they speak about health.

They may, to be clear, be having a more specific factual disagreement as well. Jess may believe that the only thing that would explain her symptoms of pain and gastrointestinal distress is some kind of serious biomedical pathology, and so she may think the doctor is simply wrong—and perhaps not listening to her—when the doctor says that she likely doesn't have such pathology.[15] The point here is that some aspects of disagreement—including some of the most entrenched—can arise simply because of a disagreement in purposes.

To circle back to the skepticism defended in Chapter 5, though, some of their disagreement in aims and purposes might, at least in part, be a disagreement about the best understanding of health. Perhaps the doctor thinks that we're better off if we give more weight to biological dysfunction and observable disease processes when considering the special normative and political importance of health. Perhaps Jess, in contrast, thinks that what matters most is how people experience their bodies and live their lives. They are, in that case, partly having a clash of aims over the way we should understand health. Moreover, this is a case in which the various ways we might understand health—and the various things we might care about when we care about health—pull against each other. There will sometimes be, on the picture I'm defending, no best or clearly correct way of adjudicating such disputes.

6. Messy Times, Messy Measures

I opened this book with a philosophical sales pitch: the puzzles of health are especially pressing because we have immediate and practical need for a theory of health. We can be mystified as to what tables are really, but face no obstacles in using them. We can be unsure about the nature of free will, and still make choices.

---

[14] See, e.g., Plunkett (2015) for discussion of this phenomenon, sometimes called "metalinguistic negotiation." My claim here is that Jess and the doctor aren't just disagreeing over what "healthy" should mean in the context, but are—more fundamentally—experiencing a clash of aims and purposes. They have, as Anderson (1995) would put it, different goals of inquiry when asking how healthy Jess is.

[15] Carel and Kidd (2017) argue that the doctor–patient relationship is a frequent locus of epistemic injustice—and given the power differentials involved, this is likely true. But a curious example case they give is when physicians refuse to believe that somatic symptom presentations have physiological rather than psychological origins. It's not at all obvious that patients have special first-person authority over the causal origin of their symptoms, however, and there's a significant risk that we undermine the seriousness of psychological illness by construing psychologically mediated pathology as someone "not real" or "all in the head." I tread into these waters a little in my (2020a), but see especially Suzanne O'Sullivan's (2021) exploration of psychologically mediated illness, *The Sleeping Beauties*. See also Barnes (forthcoming).

And the philosophy of these issues is, to be clear, no less interesting as a result. But one reason to look askance at the comparative lack of attention that health has received in philosophical theorizing is that health is one of the places where theory collides with practice.

We need to be able to measure, quantify, and compare health. We need to be able to assess the health of different populations, to evaluate the success of a particular public policy intervention or compare the effectiveness of two competing treatments, to evaluate the impact on health of socioeconomic inequality or social status. And yet measuring health is notoriously difficult, and our current measures of health tend to be plagued with problems.

If what I've argued is correct, though, how is it of any use? That is, how does it cash in on the original sales pitch? And how is the skepticism here in any sense *ameliorative*? Many of the political and social goals we have in asking "what is health?" involve our ability to quantify, measure, and assess health outcomes. If a philosopher then comes along and says that we can't give a consistent, coherent theory of health, it's hard to see what the upshot is supposed to be for those goals.

Taking a Fara-style approach, however, allows us to see particular ways in which the kind of skepticism I'm defending her can be useful—can be ameliorative—rather than simply destructive. As discussed in Chapter 3, many of our extant measures of health inherit the problems and tensions we have in reasoning about health. Self-reported health assessment, for example—in which individuals evaluate their own health status, their satisfaction with a particular health intervention, their current functional capacity—highlight the subjective experience of individuals. Such measures, contrasted to those that tend to focus on more objective features of pathology, often do a better job of assessing which health interventions actually make improvements in people's daily lives, which features of a person's health status are most important to them, and how limited individuals are in their everyday activities. They're plagued, though, with the major pitfalls of subjective evaluation. They often track a person's expectations and beliefs more closely than they track significant functional change, and responses are mediated by many confounding personal and social factors.

Suppose we wanted, for example, to assess the health of Population A in comparison to Population B. Maybe one of the main reasons we're interested in this question is that A is in general much wealthier, more educated, and has much more access to advanced healthcare than B. And so we want to know whether, and to what extent, these differences are correlated with health disparity. If we simply ask members of A and B to assess their own health and health outcomes, however, we might not get a full picture of what's going on.[16] Because of the relative comfort

---

[16] Dowd and Zajacova (2010), e.g., found that individuals from different socioeconomic statuses (SES) interpreted self-rated health questionnaires quite differently, and that more educated individuals showed higher levels of objective biomarkers for health for the same level of self-rated health when

and ease in which people in A live their lives, and because of the ready availability of state-of-the-art care, people in A might have very high expectations for their health, and treat pathology that is not life-threatening (but is still uncomfortable or otherwise distressing to them) as very serious. In contrast, because of the different standards of living, people in B might have very different expectations of what's normal, what counts as healthy, and what counts as a serious compromise to health.[17] Similar reports of distress, limitation, or compromise in health from A and B don't necessarily mean relatively equal health between the two groups. Someone in A, given his expectations, might rate his back pain to be as distressing and limiting as someone in B rates his malnutrition. We risk obscuring the genuine health disparities between these populations—and the way in which those health disparities are shaped by socioeconomic inequality—if we just look at subjective assessment of health.

In contrast, measures that focus on objectively verifiable aspects of health (life expectancy, incidence of disease, rates of major medical complications, etc.) often miss important aspects of how people live, what they're able to do, and how they feel. Suppose that we compare A and B in terms of life expectancy. If the life expectancy of both groups is relatively similar, we might be led to conclude that the groups are similarly healthy. Likewise, we could look at the incidence of major life-limiting diseases and conclude from similar rates that both groups are similarly healthy. But again, such measures can obscure as much as they illuminate. The differences between A and B could mean that, although they live for a similar amount of time, members of B are more likely to spend a longer amount of time feeling worse and being much more limited in what they can do in their everyday lives.[18] By simply looking at objective rates of mortality and major disease, we risk missing much of what's most important to us in promoting health.

The first major upshot of the kind of the skepticism I'm defending here is that there will never be a measure of health that solves these problems. When we talk about health in many ordinary contexts, we're making use of its flexibility and imprecision—we're employing it as a kind of useful shorthand, and we both want and need it to be flexible and adaptable. Any way of measuring health,

---

compared to poorer and less educated individuals. (Interestingly, self-rated health more predictably tracked mortality risk for the higher SES group, suggesting that lower SES individuals might be more likely to downplay or misconstrue their own health risk.)

[17] From Dowd and Zajacova (2010): "If one group has consistently lower standards for what is considered 'excellent' health for instance, they will report systematically better health than other groups. For example, individuals with less education may compare themselves to their relatively less healthy peers, leading [them to] report better health" (p. 748).

[18] Women have often been reported as healthier than men, for example, in virtue of having a longer life expectancy and being less likely to die from major pathologies such as heart attack, stroke, and cancer. But some research suggests that, if we look at the effect of various chronic conditions on pain, functioning, and activities of daily living, women end up spending more time in poorer health than men. See especially Luy and Minagawa (2014).

though, imposes specific restrictions—it gives a precise metric of how health will be understood for the purposes of that measure.

Recall Fara's insight from the case of tallness. Very often, we encounter the paradox of vagueness in cases like tallness at least in part because our aims and purposes in talking about tallness are themselves imprecise. Nor is that always (or even often) a bad feature—a concept like "tall" works well for us at least partly *because* it's fairly non-specific—because we can apply it in different ways, and it doesn't typically require precise judgments from us.

We can, of course, apply more precise standards of tallness—we can say, for example, that a man is tall if and only if he's 6'1 or taller. But in applying precise standards like this, we can sometimes obscure as much as we clarify. Using this standard of tallness, we'll no longer be able to point to the 6-ft man standing among 5'8 men and refer to him as "the tall one," for example. We'd no longer be able to describe a man as tall because he, well, just *seems* kind of tall (we'd need to get the tape measure out first). We'll make our meaning of "is tall" more precise, but we'll lose some of its utility, and fail to capture important aspects of how people ordinarily use the term.

Whether this is worth it, to take Fara's insight, depends largely on our aims and purposes. Sometimes, we have very specific goals in talking about health, and correspondingly specific goals in measuring health. If we want, for example, to assess whether providing free prenatal care improves maternal health, we might specifically measure things like medical complications during pregnancy, miscarriage, premature delivery, or birth and labor complications. In this very specific context, we have very specific goals and purposes—and those goals and purposes will merit the use of a very specific health measure. Just as with the scientist interested in cherry tree heights, the aims and goals will guide the application of a particular standard, and it's important that we pick a standard that suits those aims and goals. Just as the scientist and the gardener mean something different by "tall for a cherry tree," though, it's important to realize that these specific aims and goals—and their corresponding measures—won't generalize.

In many contexts, though, we need to measure health, but our aims and purposes are relatively non-specific. We want to assess health disparities between richer and poorer countries, or quantify the health benefits of major public policy initiatives. Here, our purposes in talking about health—though important—aren't all that specific. We want, in general, to get a sense of how people are faring in that medium place between basic physiological function and wellbeing, except we can't say exactly what this medium place is or where we want to locate it. In these cases, any measure of health will, by making things more precise, also *leave things out*. The only way to clarify and quantify some aspects of our health, in order to measure it, involves obscuring others.

The second major upshot, then, is that rather than aiming for the ultimate definition of health, or the best measure of health, we should instead aim for many different, overlapping, and contrasting measures of health. There is no measure of health that can give us everything we want, just as there is no theory of health that can give us everything we want. Every measure will leave out something important, and get some wonky results. But if the skeptical picture I'm defending is correct, the best way forward is to be pluralistic about our approaches to measuring and assessing health—to get as much information as we can, and to accept that all the information that we're getting is incomplete and imperfect.

Note that this isn't to endorse pluralism about health, or about health measures. Sometimes, there will be a measure or measures that are the best fit for a context. Sometimes you just want to know if an athlete can play a season, or if a person has an infectious disease, just like sometimes all you want to know is if a cherry tree will block the light from your kitchen window. But those contexts are hyper-specific, and relatively rare. More often than not, we want to know quite a lot of things—some of which are themselves messy, some of which can pull against each other, and none of which have neat stories about how they fit together.

Recall our original example cases. If we ask people to rate their own health, Kiko will come out as very healthy, Jess as quite unhealthy, Joe and Mika likely somewhere in between. Such a ranking captures something important, but it doesn't give us the whole picture. If instead we look at objective long-term medical risk factors, Kiko and Joe will be much less healthy than Mika, who is herself less healthy than Jess. Again, this tells us something important, but it doesn't tell us everything. And if we look at who's currently most in need of medical intervention, and currently most at risk of major medical complications, Mika, the only one with a serious disease, is the least healthy of the bunch. Once again, this ranking tells us something that matters a lot to us in assessing health, but it misses things as well.

What I'm suggesting is that we can't rely on any one of these rankings to tell us the full story about health. We also won't make progress by trying to come up with a single rating that somehow "gets it right"—that way lies the jungle. Nor can we simply combine these rankings (since we don't have a good account of what to do when the rankings conflict) into one wild patchwork, or shrug and say "pragmatism" and assume values will tell us which ranking to employ in which context—since our values, aims, and goals are themselves caught up in this confusion.

What we can do, instead, is appreciate the mess we find ourselves in. The best that we can do is be clear that there are many different ways of assessing health, many different things those measures are assessing, and many different results given—each getting at something important, but each missing out on important information as well. This gives us good reason to both use as many different measures as pragmatic constraints allow and use any particular measure with a note of caution—to be aware that precision can often obscure as much as it clarifies.

If we're assessing health in the loose and general sense, the questions "how healthy are these people?" and "who is healthier?" don't have clear or definite answers. In more specific contexts, where we have more specific aims and purposes in talking about health, they might have clear answers. But in more general contexts, the best we can do is point to the complexity—show the various aspects of health, and the various ways in which the health of these individuals diverges along those aspects. Having varied and diverging ways of assessing health allows us to do that, and minimizes the pitfalls of relying on any single interpretation of health. We still won't be able to definitively answer the question of who is healthy or who is healthier. But part of the skepticism I'm defending here is that those questions often don't have definite answers.

7. The Distinctive Significance of Health

I have relied, throughout this book, on the claim that health has distinctive types of significance. It's not just that health matters to us, but that heath matters in distinctive ways, and that any successful account of health must provide an explanation of how it is that health can play these roles. That is, any successful theory of health must provide a grounding for the biological, normative, political, and phenomenological significance of health.

But I've also argued that these roles are inevitably in tension with each other. When we emphasize the phenomenological significance of health, it's hard to maintain its biological significance and distinctiveness. If we zero in on the biological distinctiveness of health, we quickly lose track of its political significance, and the way that, for example, how we function in our social environments is part of what determines whether we are healthy. And I've ended up with a view that suggests there's no way to square this circle—there's no theory that will give us everything we want from health, at least all at once. But I'm still giving a theory of health, at least in some sense. So you might reasonably wonder whether my theory succeeds by my own standards.[19] If there's no one thing that health is, and communicating about health is invariably shifty and often underspecified, does health really have the significance that it ought to have?

On the model I'm offering, health has special significance because all of the things we're tracking when we talk about health have special significance. The homeostasis and standard species functioning of the human organism matter in special ways—they're distinctively connected to how long we can live, what we can do, and how we feel. Our experiences of pain and distress matter in special ways—they're distinctively connected to our quality of life, to our social capacity, to how we feel. And so on. More strongly, though, the ways in which all the various things we're tracking when we talk about health are connected to each other—and depend on each other—matter in distinctive ways. That's why there's a special

---

[19] Particular thanks to Sean Aas on this point.

importance to theorizing and talking about *health* (and not just focusing on the particular realizers).

But on the view I'm offering, the work that a theory of health can do is inherently limited. Consider the case of two people arguing over prolonged school closures in the context of the COVID-19 pandemic. There are many things they might be debating about, but let's suppose that they are debating which option is best for overall health outcomes. One points to the risk of the spread of infectious disease, the immediate potential for illness and death, the long-term impacts of the vaguely defined "long COVID". The other points to the over-whelming impact of educational disparities on long-term health outcomes, the link between social isolation and increased mental health problems, the rising rates of suicide in young people. They are both insistent that their preferred option is better for the community's health, and that the other option will do serious damage to health. They are both convinced their interlocutor is misunderstanding health. They each have studies and statistics and expert opinion to bolster their side.

Such a dispute, I'm suggesting, is not the kind of dispute that's going to be settled by figuring out *the truth about health*—there's no study, no statistic, and (especially important for our purposes here) no correct understanding of health that will resolve this. They are each applying different standards when assessing health, and prioritizing different things. Neither standard is obviously incorrect. They just have different aims and purposes in talking about health: one is especially concerned with immediate risk of infectious disease; the other is especially concerned with the social determinants of health and long-term health outcomes over a lifetime. Both of these can be reasonable standards to apply in a conversation about health, depending on the context.

Often, in a dispute like this, both parties want to insist that their stance is what's *really* best for health or what will *really* lead to the best health outcomes. If the view I'm defending is correct, however, there is no fact of the matter about this. Instead, there is just a question of which goals, aims, and purposes are the best to employ given the situation we find ourselves in. We won't solve a dispute like this by figuring out the true nature of health. Instead, we have to figure out which aspects of health we care about in a context, and which it's morally or politically best for us to prioritize. This is not to say that one party in the dispute isn't right and another isn't wrong. It could absolutely be the case that one is clearly right and the other clearly wrong. But they aren't right or wrong, in a case like this, because they've somehow found the true understanding of health.[20] Rather, they're right or wrong because they have better or more apt goals and purposes.

---

[20] Again, this also isn't to say that there couldn't be cases in which someone is simply misapplying "health" or wrong about health outcomes. If two parties are debating which option is best for health— vaccines or horse dewormers that you bought on Amazon because a friend read some good reviews on

There might be cases, though, where we simply have competing norms and purposes, and where there's no best answer for how to proceed. I argued in Chapter 4 that disputes can sometimes become intractable because both parties are applying good standards—"good" both in the sense of reasonable and in the sense of morally and politically appropriate—and there's simply no master stand-point from which we can say which standard is the best standard to apply. Sometimes, this might be a case of ignorance—we just don't know which standards would be the best to employ, given all the things we care about. But sometimes, it might be a case of genuine value incommensurability or moral indeterminacy.[21] There might, for example, sometimes simply be genuine moral unclarity in whether it's better to prioritize the way individuals subjectively *feel*, or to prioritize the risks and realities of biomedical pathology. We often can't prioritize both. Prioritizing either has costs and benefits. They're both legitimate goals. And in some contexts, deciding which is the better standard of health to apply might represent a genuine moral dilemma.

Health, in all of these conversations, has distinctive importance (biological, normative, political, and phenomenological). But the importance it has is, in a sense, inherited from the complex web of things we're tracking when we talk about health, and the complex ways in which they're related. Because we can reasonably apply different standards and aims when talking about health, health can also have different types of importance in different contexts. In an emergency triage setting, the biological significance of health is what matters most to us, and the rest can be placed on the back burner. In a meeting about public health policy, the political significance of health is as important as its biological significance, if not more so. And so on. This flexibility, rather than diminishing the significance of talking about health, is—I've been arguing—part of why health as a concept is so valuable to us. The inherent flexibility is a feature, not a bug. The reality we are trying to describe is messy, and we need a level of shiftiness and imprecision in order to communicate about this messiness.

It can be tempting, when dealing with something as morally and politically important as health, to think that we can only make progress by coming up with the single correct understanding—the best definition, the best metric, the most accurate measure. The kind of skepticism I'm defending here says that this is a mistake. It's a mistake to think we can arrive at a definition or measure that will give us everything we want, and a mistake to think that all our varied definitions and measures are somehow assessing the same thing simply because they fall under the broad label "health."

Reddit—then one party really is making a mistake about what is best for health. The point is not that people can never be wrong but rather that there can often be disputes in which both parties are using reasonable standards, but those standards conflict.

[21] See, e.g., Schoenfield (2016) and Chang (2002).

On a practical level, I'm arguing for something like a "more is better" pluralism for evaluating health outcomes. The more ways we have of measuring health, the more ways we have of assessing aspects of health that we know are important to us, and the more ways we have of talking about health, the closer we can get to seeing just how messy the reality of health is, and to capturing the diverse panoply of things we care about when we talk about health. No measure, definition, or standard will give us what we want, and they'll often obscure as much as they clarify. But my—optimistically skeptical—contention is that the more we can pursue layers of imperfection, the better off we'll be, at least insofar as we'll be able to more fully appreciate the complexities involved.

Health itself remains elusive. We don't know how best to define it. We don't know how best to measure it. And I'm arguing that this isn't a simple fact of ignorance. It's due to the nature of what we're trying to understand. Health is a mess. Health is an *intractable* mess, and any artificial precision we layer on top of it will obscure as much as it illuminates.

But if we accept this, we can make progress by embracing the mess.

Afterword: Parkinson's Disease

In the final stages of completing this manuscript, I was diagnosed with young-onset Parkinson's disease. I didn't write the material in this book with Parkinson's disease in mind, nor did I know much of anything about Parkinson's during the process of writing it. But because my primary love language is complex technical jargon, I've since gone on an information deep dive. I am now the proud owner of a brain with fewer living cells in the substantia nigra than might be hoped, but brimming with trivia about this curious condition. And in a strangely harmonious way, this one condition—which I almost certainly had while I wrote most of this book, without knowing it—serves as a useful example of many of the key points in the text. The goal of this Afterword is to review those key points, using this one condition—Parkinson's—as an illustration.

In doing this, I don't in any way wish to make it seem like these arguments are grounded primarily in my own "lived experience," or that in disagreeing with them one would be somehow discounting that experience. They're just arguments, I'm probably getting things wrong, and I'm not claiming special insight into any of this. But hey, I've got this information bubbling around in my head, and there's a certain kind of elegance to summarizing all the key points with a single example. More than that, though, I've found analyzing Parkinson's in these terms to be strangely comforting. (No, maybe *you* need therapy.)

To me, philosophy is beautiful. It's an exercise in exploring the mystery of the world, of pointing at ordinary things and asking why and then refusing to be satisfied with simple answers. And so, odd as it may seem, viewing my disease in this light is one small part of making it beautiful. Let's go.

1. The Multiple Roles of Health

Health matters to us. And it matters to us in specific, distinctive ways. It has distinctive biological significance, distinctive normative significance, distinctive political significance, and distinctive phenomenological significance. To explain health, we need to explain not just these axes of significance but also how they interact with and depend on each other.

If you have Parkinson's disease, the cells in a central portion of your brain known as the substantia nigra begin to die off. Because many of these cells are dopamine producers, the brains of people with Parkinson's disease have

Health Problems: Philosophical Puzzles about the Nature of Health. Elizabeth Barnes, Oxford University Press.

less available dopamine. Many parts of the nervous system are affected as a result—most famously and obviously the motor system, but nearly all aspects of brain function can be affected. There's a basic biological reality to the disease that's central to understanding everything else about it.

If you have Parkinson's disease, you are harmed in specific ways. You lose things of value. As you lose motor function, your projects and goals can be interrupted. Your freedom and autonomy can be compromised. Your life may even be shortened. (Technically speaking, Parkinson's itself is not fatal, but people with Parkinson's very often die of Parkinson's-related complications.) This type of loss of health has distinctive normative significance—loss of health matters to us, and it matters in unique ways.

If you have Parkinson's disease, there are also distinctive things we might need to ensure you have access to—certain types of healthcare and assistance—and distinctive ways we might need to make sure you're able to function in your community. If, as is the case, the number of people with Parkinson's is increasing rapidly, this might affect how we allocate resources of healthcare, including healthcare research. It might prompt us to reexamine how we handle elder care. It might raise thorny questions of vulnerability and prioritization in something like a vaccine rollout (people with Parkinson's typically aren't immunocompromised, but hospitalization is often especially risky for them, for example).

And finally, if you have Parkinson's disease, it will have a profound impact on how you feel. There is no single "what it's like" for a condition as varied as Parkinson's. A seventy-five-year-old man has gradually become somewhat depressed and apathetic, although he insists that he is fine; his family is concerned that his personality is changing, especially as they begin to realize that he is seeing things that are not there. A thirty-two-year-old woman begins having unusual cramps and problems with coordination in her right leg while training for a marathon. These are both cases that could characterize the onset of Parkinson's. There is probably no distinctive experience that unites them. But variation notwithstanding, a neurodegenerative disorder will dramatically affect how you feel—your movement, your energy levels, your sense of control over your own body.

When we care about health, we care about all these things. And, crucially, these various aspects of distinctive significance are often interrelated. We can't understand the normative significance of Parkinson's without also understanding its biological significance. Part of what we owe to people with Parkinson's, part of the sense in which they are harmed, is related to the objective biological dysfunction. But likewise, we can't fully understand the biological significance of Parkinson's without understanding its normative significance. Part of why movement disorders affect creatures like us in the way they do, and part of their progression trajectory, is determined by our social arrangements—what goals we pursue, what we expect from our bodies on an everyday basis, what supportive relationships and social networks we have, and so on. Understanding this interplay is part of understanding why and how Parkinson's affects us in the way it does, and why it sometimes affects different people in different ways.

We need to understand the interconnection of these various roles to understand what matters to us when we care about our health. And yet, inevitably, these axes of significance sometimes end up in tension with each other or pull us in different directions.

2. Health and Wellbeing

You don't want to get Parkinson's disease. And ostensibly you don't want to get Parkinson's disease not because of some intrinsic attachment to the cells in your substantia nigra—lovely though they are—but because of the impact it would have on your life. We care about health, the thought goes, insofar as it affects our quality of life—our ability to do what we want to do and be the people we want to be. Perhaps it's too strong to say that health and wellbeing are the same thing, or that they are perfectly correlated. But health is valuable to us (at least as individuals, leaving aside its public or civic value) insofar as it promotes our wellbeing, and loss of health is disvaluable to us insofar as it detracts from our wellbeing.

This basic idea is incredibly popular, and has a lot going for it. But it's also, I suggest, far too simple to capture the often vexed "it's complicated" relationship between health and wellbeing. You don't want to get Parkinson's disease. But many people with Parkinson's disease lead rich, rewarding lives, and the relationship between Parkinson's disease and overall wellbeing is mediated by all sorts of factors—from social circumstances and previous life experience to personality.

Health inevitably declines as Parkinson's progresses. The nature of a neurodegenerative disorder is that the course of the disease worsens over time. And yet a common theme in many individual testimonials and support groups is some form of "it gets better"—that is, people with Parkinson's often report that their mood, their quality of life, and their ability to cope with the disease improves over time.[1] This experience is, of course, far from universal. But at least for some people, the experience of Parkinson's is, in part, a skill—they have to learn how to adapt their daily life, learn how to manage symptoms, learn how to handle fluctuations. And it's a skill they become more proficient in over time. So, perhaps unsurprisingly, they become better able to deal with greater setbacks in health, and their wellbeing adjusts even as their health continues to decline. This doesn't mean, of course, that their health is somehow less valuable than it used to be. But it does mean it's less correlated to their wellbeing than it used to be.

---

[1] See, for an example, Dr. Soania Mathur's beautiful "Letter to My Newly Diagnosed Self": www.unshakeablemd.com/post/to-my-newly-diagnosed-self-1.

There may also be more systematic decoupling of health and wellbeing in the context of Parkinson's. Some research suggests that women, for example, report higher life satisfaction in the context of Parkinson's disease than men at similar stages of the disease.[2] It's important not to overstate findings like this—sample sizes are small, and much more research needs to be done. But prima facie, such a result wouldn't be surprising—norms surrounding masculinity and physical dependence can sometimes make it less stigmatizing for women to accept physical caretaking or to adapt to social roles that involve a certain level of dependence. Suppose this were true in the case of Parkinson's. We wouldn't want to conclude, as a result, that women simply lose something of less value as their health declines, or that their health simply matters less to them.

Similarly, individual adaptation strategies and sense of self-efficacy have been shown to be major predictors of life satisfaction and self-assessed wellbeing in the context of Parkinson's.[3] But again, we shouldn't say that the person who adapts especially well—who accepts their condition, makes positive changes in their life, rebuilds for their new normal—has simply lost something of lesser value as their health inevitably declines. Making positive adaptive changes in the face of a neurodegenerative disorder can be part of what helps you live well—and thus a huge part of what determines your wellbeing. But it doesn't mean that what you lose, as you adapt to loss of health, is less valuable.

Health and wellbeing are intimately connected. We care about conditions like Parkinson's because of how they impact our lives, not just our brains. And it would be callous in the extreme to ignore the impact that a condition like Parkinson's has on wellbeing. But we also can't measure the disvalue of a condition like Parkinson's—even to individuals—just by looking at its impact on wellbeing. People with Parkinson's lose something of great value—the inevitable chipping away at their freedom of movement, at the functioning of their nervous system, is loss. But we can't measure the value of that loss by looking at the overall flourishing of their lives.

3. Health and Subjectivity

As the cells in your substantia nigra die off, you lose motor function. Your movement becomes slower (bradykinesia) and some of your movement becomes uncontrollable and involuntary (tremor). This is an objective biological process. Tremors, like facts, don't care about your feelings.

Except they do. A well-known aspect of Parkinsonism is that symptoms worsen with stress, distraction, or emotional upset. This is partly, again, an objective

---

[2] See, e.g., Rosqvist et al. (2017).   [3] Rosengren et al. (2016) and (2021).

biological process. Parkinsonian symptoms are caused by reduced levels of dopamine. Dopamine is also, famously, the "feel-good" neurotransmitter. Things that put strain on the available levels of dopamine in the brain will worsen Parkinsonian symptoms. And so feelings of stress, mental distraction, and the like can all make something like a tremor worse, and that's part of the objective reality of the disease.

But it's not, of course, *purely* objective. How stressful or how upsetting a particular situation is will always be, in part, a matter of interpretation. Your interpretation of events will be part of what shapes the way your brain regulates dopamine, but your brain's regulation of dopamine is also part of *what it is* to experience certain responses.

Partly for this reason, people with Parkinson's are notoriously susceptible to placebo responses. The expectation that a treatment will do you some good is often enough to generate brain changes that are, by themselves, the kind of thing that will have a positive impact on the symptoms of Parkinson's. Such responses aren't imaginary, and they're also not purely subjective—they're objectively measurable changes to objectively measurable symptoms. But they're caused by the subjective expectation that a treatment will be beneficial. Those subjective expectations will only take you so far, though. Placebo reactions are often strong, but they're also temporary. You can't positive-think your way out of a condition like Parkinson's, even though mood and expectation play a major role in the experience of the disease.

There is an objective reality to the disease process that marches on—inexorably—and demands attention. But curiously, we can't simply observe the objective disease severity of someone's condition and then make conclusions about their level of functioning. One of the biggest predictors of overall functioning—as well as life satisfaction—in the context of Parkinson's is depression.[4] Depression is sometimes a symptom of Parkinson's—again, when you have less dopamine to play with, that affects how you feel, including your mood, and your mood in turn affects, directly, what you are capable of doing. But the relationship between depression and Parkinson's doesn't appear to be fully explained by loss of dopamine due to Parkinsonian changes. This isn't surprising—Parkinson's is something that happens to people, and depression is independently common in people. Likewise, Parkinson's can cause major changes to work, life, and relationships in a way that can precipitate depression—depression that's in some sense causally related to having Parkinson's, but isn't a *symptom* of Parkinson's in the way slow movement is.

If someone is dealing with depression, it will quite literally make their Parkinson's worse. Lower mood will often mean less dopamine production, and

---

[4] See Schrag (2006).

reduced dopamine is at the root of many aspects of Parkinson's. Lower mood can also make it more difficult to do some of the things that seem to have a substantial positive impact on Parkinson's over time—regular exercise, daily routine, good sleep, and so on.

But a person's mood and outlook might also, of course, worsen their own subjective assessment of their Parkinson's. And if someone is dealing with depression, that can influence how they interpret and quantify the impact of Parkinson's on their life. Suppose two people with Parkinson's are in roughly the same social circumstances and at roughly similar stages of the disease. But one is struggling with depression (something they dealt with long before Parkinson's). They're both dealing with postural instability, and their doctor tells them that walking every day is the best thing they can do for themselves. The person unaffected by depression gets out and walks every day. The person affected by depression insists they can't do this, it's too much, that even the thought is overwhelming.

A difference like this will make a substantial impact on physical capacity over time. But what do we say about physical capacity in the moment? Is it that the person with depression could walk, but is simply choosing not to? That seems harsh. Is it that the person without depression is less affected, and has more physical capacity? That seems to misdescribe the physical reality of the disease process. It's not clear how to describe the interplay between subjective and objective factors in a case like this, and they pull us in somewhat different directions.

This "pulled in different directions" phenomenon only increases when we consider, once again, the curious case of the functional disorder. People with Parkinson's disease have tremors and difficulty with movement—as well as a huge range of other issues—because cells in a specific region of their brain are dying. But there is also a well-described phenomenon in which people have similar symptoms, but don't have the associated neurological changes. In so-called "functional movement disorders," people will have a wide range of neurological symptoms[5] that are very distressing and impairing for them. But they are physiologically normal, and their symptoms, though very real, are primarily psychogenic—caused by things like heightened awareness of bodily changes, or illness anxiety.

One fascinating study compared quality-of-life assessments for a group of people suffering from functional movement disorders to a group suffering

---

[5] It's worth noting that "functional disorder," at least in this context, isn't simply (as it's sometimes described) a diagnosis of exclusion when no cause can be found for symptoms—though it sometimes can be used in that way. The symptoms of functional neurological disorders are *similar*, but they tend to have recognizable features that mark them out as having a primarily psychological/functional cause. In the case of tremor, for example, Parkinsonian tremors tend to have a specific amplitude, and they typically get worse when a person is distracted (e.g., when a patient is asked to do some math in their head). In contrast, the frequency of functional tremors is often variable, and they typically go away when the person is distracted or asked to do specific tasks with another part of the body.

from Parkinson's.[6] The functional disorder group reported levels of impairment and self-assessed health similar to the Parkinson's group. They also reported substantially lower levels of emotional wellbeing (the Parkinson's group reported basically normal emotional wellbeing), and were more distressed by their symptoms.

What are we to make of a comparison like this? On the one hand, it's telling us something significant—and often neglected—about the impact of functional neurological disorders. These people are suffering. Their suffering is every bit as real as any other kind of suffering, and having a cause that is primarily psychological in origin doesn't make it any less real. The functional disorder group was also vulnerable in all the ways you might expect such a group to be. They were more likely to be women, more likely to have lower socioeconomic status, more likely to have past experiences of trauma. *Of course* such a group might be more likely to be distressed by symptoms, or to rate their own functioning as compromised.

But the cells in the brains of the people with Parkinson's are dying; the cells in the brains of those with functional disorders are not. The neurological function of those with Parkinson's will inevitably decline, over time, until ultimately their bodies do not have enough control over movement to survive. They will lose function, piece by piece, until they die, and that is the biological reality—and the biological inevitability—of their disease.

Whose disease is worse? Who is more impaired? In conditions of scarce resources, what treatment and research do we prioritize, or who do we see as having greater need when it comes to accommodations and social benefits? (Perhaps you think the answer to questions like these is "Revolution!" That's fine, but we still have to act in the here and now, with the resources we currently have.)

The subjective and objective aspects of health depend on each other in complicated ways, such that at times they can't be fully separated. Is it that someone's mood affects how they interpret their own physical capacity, or is it that understanding the relationship between mood and function in the context of an ongoing illness is part of what it is to understand physical capacity for something like Parkinson's? There might, at times, be no good answer to a question like this.

And to make matters worse, the subjective and objective aspects of health—as intimately connected as they are—can also be in tension with each other. Whose health is more compromised, someone with a functional disorder who is physiologically normal but feels that she is too sick to do even the most basic tasks, or someone with Parkinson's who is walking marathons as part of their effort to stay fit? Again, there might be no good answer to a question like this.

---

[6] Anderson et al. (2007); see also Gendre et al. (2019).

4. Health and Disability

There is a flat-footed interpretation of disability-positive views that assumes something like the following. If you are saying that disability is mere difference, then you must think it's not in any way bad or harmful. Since disability is associated with—and partly constituted by—significant loss of health, you are also thereby saying that loss of health is mere difference. You're probably committed to thinking that having Parkinson's is just like having an unusual hair color. You probably think it's bad to look for cures for Parkinson's. You probably think people with Parkinson's are excited about having Parkinson's, and if they aren't its false consciousness.

For the record, although a defender of a mere-difference take on disability, I do not think any of these things. I am not, in fact, a sociopath.

But it requires some nuance to see why these kinds of connections don't hold. And perhaps for that reason, critics—and also, at times, defenders—of mere-difference-style views often construe it as stronger than it really is, or as committing to more than it needs to.

What any sensible mere-difference position needs is the ability to say that loss of health is not mere difference, even though loss of health and disability are strongly, inevitably correlated. To begin with, it's important not to overly medicalize disability. Parkinson's is a specific disease process, characterized by a specific set of symptoms. But you cannot understand what it is to be disabled because of the disease process of Parkinson's just by understanding those symptoms, or the specific functional effects they have on the body. One reason why a mere-difference view, when formulated carefully, doesn't entail that Parkinson's disease is mere difference is simply that the entity "Parkinson's disease" isn't, by itself, the kind of thing people are talking about when they talk about disability in this context. Being physically disabled is a matter of being in a bodily state that makes you subject to certain norms and stigmas, that requires certain adaptations in the way you navigate spaces and social situations, and so on. Parkinson's disease can be the reason you're disabled, but it's never the whole story.

Some theorists push this idea even further. On some versions of the Social Model of disability, there's nothing biomedical involved in being disabled. The death of cells in your substantia nigra might create an *impairment*—a biological condition—but having a disability is social, not biological. It's about living in a world that isn't set up for, and often unjustly structured against, people with impairments. Disability, on such views, is the unjust social disadvantage caused by lack of accommodation for the biological difference of impairment.

The experience of living with Parkinson's, though, is one of many that put pressure on any kind of firm distinction between the social and biological. Every day I make coffee in my AeroPress. Almost every day, as I navigate through the process with the unruly meatpaws attached to my arms, I spill something. Is this social? Yes. I make coffee in the AeroPress because I'm an aging millennial fond of vaguely pretentious, overly complicated food rituals in a culture in which coffee drinking is common. Is this biological? Yes. I have less fine motor control than I used to (and let's be honest, I always ran toward clumsiness) because of a neurodegenerative process in my brain. Can we separate which elements of this should be classified as biological, and which as social? Not really. Nor is it clear whether any of the difficulties we might class as social are related to injustice. The AeroPress might not be easy for someone like me to navigate, but that doesn't make it a site of Coffee Injustice. And I don't *have* to make the coffee that way. I just like to, meatpaws notwithstanding. But it still frustrates me when I can't, even though it's no one's fault.

Such an example is trivial, of course. (Although the net effect of many such examples added up may not be—people with motor difficulties take longer to do ordinary tasks and have to concentrate harder to do them. The impact of each ordinary task is small, but add up all the ordinary tasks of a day and you've got a significant impact.) The point is that for many people's experience of disability, there's no neat separation between the social and the biological—we're social animals. And likewise, many of the harms associated with disability/impairment, even if they're partly social in origin, aren't obviously the kinds of things that are rooted in injustice or that could be solved by making the world a better place. Even in an ideal society, lack of motor control might still be—legitimately—frustrating.

While we can't separate the social from the biological, though, we can sometimes view the very same thing—the same condition of the body—from different perspectives. And viewed from different perspectives, the same state of the body might have different things truly said of it. When we're viewing people with Parkinson's, we sometimes consider them from a "whole-life" perspective—we look at the richness of their experiences, the way that the difference in their physical capacities is interwoven into so many things they care about. Many people—certainly not all, but many—report leading good and rich lives. And more than that, they talk about ways in which having Parkinson's has enriched their lives—how the slower pace imposed has actually been valuable to them, how it's brought them hobbies, connections, and opportunities they never would have had, how it's changed their relationship to their bodies in positive ways. From a socially embedded perspective, in which we're considering their overall flourishing and quality of life, it's not at all obvious that such people would be better off without Parkinson's. (Again, this isn't true for everyone, by any means. But it definitely seems true for some.)[7]

---

[7] Again, Soania Mathur's "Letter to My Newly Diagnosed Self" provides an apt example: "To my newly diagnosed self. You may not feel it at the moment but Parkinson's will be your ultimate blessing, not the curse you see it as... That it will force you to live in the moment, to take nothing for granted. That it will not defeat you as you may now believe, but instead be the very element that causes you to thrive."

But for the very same people, when we view them specifically in a biomedical context, all of the following is still true: they are undergoing a process of neuro-degeneration that will slowly kill them, they are losing function, they are losing motor control, their brain cells are dying. All of that, viewed from a biomedical context, is harmful to them. All of it should be minimized; prevented if possible. These perspectives—and many others, including the perspective of caregivers, of public health officials, and so on—are all important, legitimate perspectives we can take on the embodied state of people with Parkinson's. But they can, at times, say conflicting things.

The actor Michael J. Fox (you didn't think you were getting all the way through this without hearing about him, did you?) has called his experience of Parkinson's an incredible gift—but a "gift that keeps on taking."[8] In saying this, Fox is articulating a sense of ambivalence that's common among disabled people, espe-cially people with medically complex or degenerative conditions. Fox often claims that his life has been enriched and positively changed by the experience of having Parkinson's—that he is, in the context of considering his overall wellbeing, better off for having Parkinson's. More strongly, he makes this case not only because of the instrumental goods associated with Parkinson's—that it's taught him humility or "what really matters," perhaps—but also because of experiences directly bound up with (and perhaps partly constitutive of) having Parkinson's.[9] And yet he is frank about the struggles, the difficulties, and the grief of his disease—especially as he faces his own mortality and the likelihood of a substantially shortened life-span.[10] His own experience of Parkinson's, at least as he describes it, is a mixture of positive overall wellbeing and tremendous loss—a gift, but a gift that keeps on taking. And to do justice to this experience, we need to be able to capture not only both aspects but also the connection between them.

We make a mistake if we overly medicalize an experience like Fox's—if we conceptualize his experience of disability as mere reduction in health, or a collection of symptoms. But we also make a mistake if we fail to account for the biomedical reality of that experience—the way that a gradual process of neuro-degeneration affects his body's motor control and autonomic systems in a way that radically alters his functional abilities and will ultimately (barring some other accident or illness) take his life.

What I have been arguing is that there is no single, God's-eye view from which we can say everything we want to say—or everything that is true—about

---

[8] See his *Lucky Man* and *No Time Like the Future*.

[9] "I couldn't be still until I couldn't keep still" is how he sometimes elaborates it—the idea being that the imposed changes in his movement removed some of the frenetic rushing from thing to thing and place to place that had previously marked his life, and in ways that impeded his happiness. See his *Lucky Man*, as well as his interview on the Mike Birbiglia podcast, *Working It Out* (https://static1. squarespace.com/static/5c12c27796e76ffc6466e182/t/6294548d9cad707a59c0a4bf/1653888141407/ WIO_MichaelJFox_Transcript.pdf).

[10] See especially his recent memoir *No Time Like the Future*.

experiences like these. And this is why we are often pushed toward saying things that seem contradictory. Some people are grateful to have Parkinson's disease, and find value in their experiences of Parkinson's disease, but at the same time they seek to minimize the effects of Parkinson's disease on their lives and slow its progression as far as possible. Some people will both speak openly and meaningfully about the valuable aspects of Parkinson's disease and devote much of their time and resources to preventing others from getting it in the future, and to developing a cure. These conflicting attitudes, I'm arguing, make sense; and it makes sense that they exist alongside each other. But they are also, invariably, somewhat in tension with each other, and they will always pull us in different directions.

5. Ameliorative Skepticism

The intuitive gloss we're aiming at when we talk about health is something like thriving or flourishing. A healthy maple tree is a thriving maple tree; a healthy hawk is a thriving hawk. But something in the neighborhood of flourishing or thriving is also what we're aiming to capture when we talk about wellbeing. Your life is going well for you if you're thriving. But health and wellbeing aren't the same thing—you can live a good life with reduced health, or have good health without living a good life.

Can we just say that health is *physical* thriving? No, at least not for more complex apes like us. To evaluate the health of someone with Parkinson's, we have to consider their mental state, the amount of stress they're under, the positive experiences and supportive relationships in their lives, and so on. Such factors will play a direct, determinative role in their motor function. There isn't anything like a neat separation, for brain disorders, between "physical health" and "mental health," even if some paradigm instances do come apart.

So perhaps, instead, we aim for Philippa Foot's idea of the flourishing of the human organism. Health is the thriving—in body and mind—of the kind of organism that we are, but not the thriving of the *person*. This is, I think, the idea we're aiming at. But it's also an idea that's ultimately unworkable. There is no stable middle ground that's more expansive than simple physiology but narrower than overall wellbeing.

To evaluate the health of someone with Parkinson's, we have to consider how she can train her brain, and how she can regulate dopamine. And so we have to consider her daily activities and hobbies—how stable her environment is, what her living situation is like. And we have to consider her mood, how often she can do new and interesting things, how much stress she's under in her everyday commitments, whether she can take time for self-care. Basic aspects of her personality (is she optimistic? does she tend to have a strong sense of self-efficacy?) will also be part of what determines how much dopamine she has access to, and thus how much motor function she has. Factors like these aren't just causally related to her health. They are partly *constitutive* of her health—part of what determines her functional capacity given the state of her brain. But they're also, quite obviously, part of what determines her overall quality of life and wellbeing, and that make her the person she is. Disorders of the brain give us perhaps the clearest example of why there isn't a neat separation between the person and the organism, at least when it comes to health.

When theorizing health, we want something that can play multiple roles at once. We want to explain the biological significance of dying brain cells, the normative significance of loss of functional capacity, the political significance of changes in opportunity and access and healthcare needs, the phenomenological significance of a body that has forgotten how to move. And we want to explain—we need to explain—how these various factors are interconnected. Part of Parkinson's normative and political significance is determined by its biological significance: at least some of why Parkinson's matters and what we owe to people with Parkinson's is related to the objective changes in their brains. Part of Parkinson's biological significance is determined by its distinctive normative and phenomenological significance: you can't understand the disease process itself without understanding the complex relationship between how it makes people feel, the impact it has on how people function in their communities, and the interaction that such factors have with the function of the brain. And so on.

We need to theorize these roles as a unified whole, to explain how they all fit together into the complex tangle that is our health. And yet, I suggest, we can't. There is no consistent, coherent explanation of health we can offer that gives us everything we need from a theory of health. For someone with Parkinson's, we can't consistently, stably explain the difference between their health and their quality of life, between their health-related limitations and their own decisions or choices, between symptoms and personality. We have lines we need to draw, but there is no good way to draw them.

This isn't the same as suggesting error theory or eliminativism. The things we're talking about when we talk about health are real. The things we care about are real. And we need to be able to explain how they're unified into a complex whole. It's just that there's no consistent, explanatorily adequate way to do this.

Sally Haslanger has argued for the idea of ameliorative analysis—that in asking a question "what is x?" we should something be asking "what do we want x to be?". How does answering the question "what is x?" best suit our legitimate political and social goals? What I'm suggesting is that sometimes the most helpful answer to the question "what is x?" is a skeptical answer. What is health? We don't know. We can't know. It's a question we can't give a consistent, adequate answer to, even though it's a question we need to ask.

Skepticism like this, I'm arguing, can sometimes be therapeutic rather than destructive. It can help us explain why some questions are unanswerable questions, and why some disputes feel utterly intractable. It can help diagnose confusion, and point the way toward places where we might be able to make more progress.

6. Health in Context

Our understanding of health is messy. Consider three people with Parkinson's. One has uncontrolled tremor and very limited movement on the entire left side of his body. But he's active, happy, and able to pursue his goals. One has relatively minor symptoms, but since his diagnosis has felt overwhelmed by depression. He rarely leaves his house and has withdrawn from most of his previous interests and relationships. One has symptoms that are very well controlled by medication. He's feeling cheerful and optimistic, and is working full-time. But as a side effect of that medication, he has developed a compulsive gambling addiction that led to the dissolution of his marriage.

Who is healthier? Whose health has been more severely affected by Parkinson's? There aren't, I suggest, definitive answers to such questions. But this doesn't mean that talking about health is unhelpful or that we always need to precisify in order to communicate. Rather, sometimes it is the very messiness that is useful to us in communication. Our lives are messy. Our aims and purposes are often unclear, imprecise, or otherwise underspecified. When we talk about health, we're gesturing at a range of things we care about, without any specific story about how those things fit together, or how to balance them when they're in tension with each other. And often, that level of messiness is fit for purpose—it allows us a loose, adaptable form of communication that works for us.

But sometimes, our aims and goals are more precise, and we can make our understanding of health more precise to fit those purposes. At the psychologist's office, the man with the gambling addiction has seriously compromised health. But at the vaccine clinic, he's fairly healthy. At the movement disorder specialist, the man with pronounced tremor has significantly compromised health. But when he's out hiking with a friend and they ask him how he's feeling, he can reply, "Good! There are challenges, but I feel strong, I feel healthy!" and be understood without difficulty.

When our aims and purposes are more precise, we can mean something more precise when we talk about health. But it doesn't follow that the more precise meaning is *better*. By narrowing our focus, we clarify, but at the same time we also obscure. (In the vaccine clinic, we're missing something important about the addict's health, even though it makes sense to narrow our focus in that way in that context.)

We will, inevitably, be faced with circumstances where we have to weigh different aims and purposes against each other. What research do you fund, the research for treatment of tremor of the research for treatment of the compulsive disorders sometimes associated with Parkinson's medication? Whose request for accommodation is of more immediate need, the man with severe tremor who reports high quality of life or the man with minimal symptoms and early-stage disease who is struggling emotionally? When faced with such questions, part of what we need to be able to do is quantify and compare health. And no theory of health, no correct understanding of health, no nuanced pluralistic approach to health will solve intractable problems we face in trying to do this. But appreciating that there are different ways we can legitimately measure health, that will weight these competing needs differently, is at least a part of appreciating the complexity of the picture. It's also diagnostic, at least to some extent, of why we often feel so stuck.

7. Conclusion

I've intended this Afterword as a little guided tour of some of the key ideas presented in this book, using Parkinson's as a central example throughout. It's not, by any means, an exhaustive summary, and if you're looking for a Cliff Notes version of the arguments in this book, this isn't that. Rather, using this single condition I now know an unexpected amount about, I've tried to draw together some central themes.

But if there are two key takeaways from this book that I hope will stick, it's these: (1) health is distinctively philosophically interesting, and deserves more mainstream philosophical attention as such; (2) as much as philosophy loves to bring clarification and precision, we also need to leave room for uncertainty and messiness. Some things resist neat, coherent explanations. And I think part of what's especially interesting about health is that it's one of those things.

I don't think I have any special new insight into the nature of health in virtue of having Parkinson's. I wish I could say that what I lose in health I gain in extra special philosophy skills, but we all know that's false. What I have, perhaps, is a deeper appreciation of the complicated ways in which health matters to us, and a deeper dissatisfaction with overly simplistic analyses of that. It would be foolish—perhaps hubristic, even—to suggest that life experience has offered me answers. But what I think it *has* offered me is a growing sense of fascination and wonder. And I hope this book can be an invitation to share in that fascination. Health is weird. Weird can be good.

Bibliography

Abrams, J. A., Hill, A., & Maxwell, M. (2019). Underneath the mask of the strong Black woman schema: Disentangling influences of strength and self-silencing on depressive symptoms among US Black women. *Sex Roles*, *80*(9), 517–26.

Abrantes-Pais, F. de N., Friedman, J. K., Lovallo, W. R., & Ross, E. D. (2007). Psychological or physiological: Why are tetraplegic patients content? *Neurology*, *69*(3), 261–7. DOI: 10.1212/01.wnl.0000262763.66023.be.

Aidoo, M., Terlouw, D. J., Kolczak, M. S., McElroy, P. D., ter Kuile, F. O., Kariuki, S., Nahlen, B. L., Lal, A. A., & Udhayakumar, V. (2002). Protective effects of the sickle cell gene against malaria morbidity and mortality. *The Lancet*, *359*(9314), 1311–12. https://doi.org/10.1016/S0140-6736(02)08273-9.

Aiello, L., & Dean, C. (1990). *An introduction to human evolutionary anatomy*. Oxford: Elsevier Academic Press.

Aiken-Morgan, A. T., Bichsel, J., Savla, J., Edwards, C. L., & Whitfield, K. E. (2014). Associations between self-rated health and personality. *Ethnicity & Disease*, *24*(4), 418–22.

Alcoff, L. (2005). *Visible identities: Race, gender, and the self*. New York: Oxford University Press.

Anderson, E. (1995). Knowledge, human interests, and objectivity in feminist epistemology. *Philosophical Topics*, *23*(2), 27–58.

Anderson, K. E., Gruber-Baldini, A. L., Vaughan, C. G., Reich, S. G., Fishman, P. S., Weiner, W. J., & Shulman, L. M. (2007). Impact of psychogenic movement disorders versus Parkinson's on disability, quality of life, and psychopathology. *Movement Disorders: Official Journal of the Movement Disorder Society*, *22*(15), 2204–9.

Andrykowski, M. A., Donovan, K. A., & Jacobsen, P. B. (2009). Magnitude and correlates of response shift in fatigue ratings in women undergoing adjuvant therapy for breast cancer. *Journal of Pain and Symptom Management*, *37*(3), 341–51.

Anie, K. A., Steptoe, A., & Bevan, D. H. (2002). Sickle cell disease: Pain, coping and quality of life in a study of adults in the UK. *British Journal of Health Psychology*, *7*(3), 331–44.

Appiah, A. (1994). *Race, culture, identity: Misunderstood connections*. Tanner Lectures on Human Values.

Arpaly, N. (2005). How it is not "just like diabetes": Mental disorders and the moral psychologist. *Philosophical Issues*, *15*(1), 282–98.

Australian Institute of Health and Welfare. (2009). *The geography of disability and economic disadvantage in Australian capital cities*. Canberra: AIHW.

Bäckström, D., Granåsen, G., Domellöf, M. E., Linder, J., Mo, S. J., Riklund, K., Zetterberg, H., Blennow, K., & Forsgren, L. (2018). Early predictors of mortality in parkinsonism and Parkinson disease: A population-based study. *Neurology*, *91*(22), e2045–56. DOI: 10.1212/WNL.0000000000006576.

Badredlin, N., Grobman, W. A., & Yee, L. M. (2019). Racial disparities in postpartum pain management. *Obstetrics and Gynecology*, *134*(6), 1147.

Bagenstos, S. R., & Schlanger, M. (2007). Hedonic damages, hedonic adaptation, and disability. *V and. L. Rev.*, *60*, 745.

Baker, D. W. (2017). *The Joint Commission's pain standards: Origins and evolution*. The Joint Commission, Oak Brook Terrace, IL.

Baker, L. R. (1997). Why constitution is not identity. *Journal of Philosophy*, 94, 599–621.

Baker, T. A., Buchanan, N. T., Small, B. J., Hines, R. D., & Whitfield, K. E. (2011). Identifying the relationship between chronic pain, depression, and life satisfaction in older African Americans. *Research on Aging*, 33(4), 426–43. https://doi.org/10.1177/0164027511403159.

Barclay, R., & Tate, R. B. (2014). Response shift recalibration and reprioritization in health-related quality of life was identified prospectively in older men with and without stroke. *Journal of Clinical Epidemiology*, 67(5), 500–7.

Barclay-Goddard, R., King, J., Dubouloz, C. J., Schwartz, C. E., & Response Shift Think Tank Working Group. (2012). Building on transformative learning and response shift theory to investigate health-related quality of life changes over time in individuals with chronic health conditions and disability. *Archives of Physical Medicine and Rehabilitation*, 93(2), 214–20.

Barf, H. A., Post, M. W. M., Verhoef, M., Jennekens-Schinkel, A., Gooskens, R. H. J. M., & Prevo, A. J. H. (2007). Life satisfaction of young adults with spina bifida. *Developmental Medicine and Child Neurology*, 49(6), 458–63. https://doi.org/10.1111/j.1469-8749.2007.00458.x.

Barnes, E. (2010). Ontic vagueness: A guide for the perplexed 1. *Noûs*, 44(4), 601–27.

Barnes, E. (2014). Valuing disability, causing disability. *Ethics*, 125(1), 88–113.

Barnes, E. (2016). *The Minority Body*. Oxford: Oxford University Press.

Barnes, E. (2017). Realism and social structure. *Philosophical Studies*, 174(10), 2417–33.

Barnes, E. (2018). Against impairment: Replies to Aas, Howard, and Francis. *Philosophical Studies* 175(5), 1151–62.

Barnes, E. (2020a, June 26). The hysteria accusation. *Aeon Magazine*. https://aeon.co/essays/womens-pain-it-seems-is-hysterical-until-proven-otherwise.

Barnes, E. (2020b). Feminist metametaphysics. In R. Bliss & J. T. M. Miller (Eds.), *The Routledge handbook of metametaphysics* (pp. 300–11). London: Routledge.

Barnes, E. (Forthcoming). Trust, distrust, and 'medical gaslighting'. *Philosophical Quarterly*.

Barnes, E., & Williams, J. R. G. (2011). A theory of metaphysical indeterminacy. *Oxford Studies in Metaphysics*, 6.

Bassett, M. T., & Galea, S. (2020). Reparations as a public health priority: A strategy for ending black–white health disparities. *New England Journal of Medicine*, 383(22), 2101–3.

Becker, N., Thomsen, A. B., Olsen, A. K., Sjøgren, P., Bech, P., & Eriksen, J. (1997). Pain epidemiology and health related quality of life in chronic non-malignant pain patients referred to a Danish multidisciplinary pain center. *Pain*, 73(3), 393–400. https://doi.org/10.1016/S0304-3959(97)00126-7.

Benony, H., Daloz, L., Bungener, C., Chahraoui, K., Frenay, C., & Auvin, J. (2002). Emotional factors and subjective quality of life in subjects with spinal cord injuries. *The American Journal of Physical Medicine & Rehabilitation*, 81(6), 437–45.

Bettcher, T. M. (2009). Trans identities and first-person authority. In L. Shrage (Ed.), *You've changed: Sex reassignment and personal identity*. Oxford: Oxford University Press.

Bettcher, T. M. (2013). Trans women and the meaning of "woman". In A. Soble, N. Power, & R. Halwani (Eds.), *Philosophy of sex: Contemporary readings*, sixth edition (pp. 233–50). Lanham, MD: Rowman & Littlefield.

Birtane, M., Uzunca, K., Taştekin, N., & Tuna, H. (2007). The evaluation of quality of life in fibromyalgia syndrome: A comparison with rheumatoid arthritis by using SF-36 Health Survey. *Clinical Rheumatology, 26*(5), 679–84. https://doi.org/10.1007/s10067-006-0359-2.

Bleiweis, R., Boesch, D., & Gaines, A. C. (2020, August 3). The basic facts about women in poverty. *American Progress.* www.americanprogress.org/article/basic-facts-women-poverty.

Boakye, M., Leigh, B. C., & Skelly, A. C. (2012). Quality of life in persons with spinal cord injury: Comparisons with other populations. *Journal of Neurosurgery: Spine SPI, 17* (Suppl1), 29–37. https://doi.org/10.3171/2012.6.AOSPINE1252.

Bogart, K. R., Rottenstein, A., Lund, E. M., & Bouchard, L. (2017). Who self-identifies as disabled? An examination of impairment and contextual predictors. *Rehabilitation Psychology, 62*(4), 553–62. https://doi.org/10.1037/rep0000132.

Bognar, G., & Hirose, I. (2014). *The ethics of health care rationing: An introduction.* London: Routledge.

Bombak, A. E. (2013). Self-rated health and public health: A critical perspective. *Frontiers in Public Health, 1*, 15. https://doi.org/10.3389/fpubh.2013.00015.

Bonathan, C., Hearn, L., & Williams, A. C de C. (2013). Socioeconomic status and the course and consequences of chronic pain. *Pain Management, 3*(3), 159–62. https://doi.org/10.2217/pmt.13.18.

Boonstra, A. M., Reneman, M. F., Stewart, R. E., Post, M. W., & Schiphorst Preuper, H. R. (2013). Life satisfaction in patients with chronic musculoskeletal pain and its predictors. *Quality of Life Research 22*, 93–101. https://doi.org/10.1007/s11136-012-0132-8.

Boorse, C. (1977). Health as a theoretical concept. *Philosophy of Science, 44*(4), 542–73.

Boorse, C. (1987). Concepts of health. In D. VanDeVeer & T. Regan (Eds.), *Health care ethics: An introduction* (pp. 377–7). Philadelphia, PA: Temple University Press.

Boorse, C. (1997). A rebuttal on health. In J. M. Humber & R. F. Almeder (Eds.), *What is disease?* (pp. 1–134). Totowa, NJ: Humana Press.

Boorse, C. (2011). Concepts of health and disease. In F. Gifford (Ed.), *Philosophy of medicine* (pp. 16–13). New York: Elsevier.

Boorse, C. (2014). A second rebuttal on health. *Journal of Medicine and Philosophy, 39*(6), 683–724.

Boyce, C. J., & Wood, A. M. (2011). Personality prior to disability determines adaptation: Agreeable individuals recover lost life satisfaction faster and more completely. *Psychological Science, 22*(11), 1397–1402.

Boyd, R. (1991). Realism, anti-foundationalism and the enthusiasm for natural kinds. *Philosophical Studies, 61*(1), 127–48.

Boyd, R. (1999). Kinds, complexity and multiple realization. *Philosophical Studies, 95*(1), 67–98.

Brekke, M., Hjortdahl, P., & Kvien, T. K. (2002). Severity of musculoskeletal pain: relations to socioeconomic inequality. *Social Science & Medicine, 54*(2), 221–8. https://doi.org/10.1016/S0277-9536(01)00018-1.

Broadbent, A. (2019). Health as a secondary property. *British Journal for the Philosophy of Science 70*, 609–27.

Bronsteen, J., Buccafusco, C. J., & Masur, J. S. (2008). Hedonic adaptation and the settlement of civil lawsuits. *Columbia Law Review, 108*(6), 1516–50.

Broome, J. (1988). Good, Fairness and QALYs. *Royal Institute of Philosophy Lectures, 23*(1), 57–73.

Broome, J. (2002). Measuring the burden of disease by aggregating well- being. In C. Murray, J. Salomon, C. Mathers, & A. Lopez (Eds.), *Summary measures of population health: Concepts, ethics, measurement and applications* (pp. 91–113). Geneva: World Health Organization.

Broome, J. (2004a). *Weighing lives*. Oxford: Oxford University Press.

Broome, J. (2004b). The value of living longer. In S. Anand, F. Peter, & A. Sen (Eds.), *Public health, ethics, and equity* (pp. 243–60). Oxford: Oxford University Press.

Brouwer, T. (2022). Social inconsistency. *Ergo: An Open Access Journal of Philosophy*, 9.

Buczak-Stec, E. W., König, H. H., & Hajek, A. (2018). Impact of incident Parkinson's disease on satisfaction with life. *Frontiers in Neurology*, 9(589). DOI: 10.3389/fneur.2018.00589.

Burckhardt, C. S., Clark, S. R., & Bennett, R. M. (1993). Fibromyalgia and quality of life: A comparative analysis. *The Journal of Rheumatology*, 20(3), 475–9. https://europepmc.org/article/med/8478854.

Burgess, A., & Plunkett, D. (2013). Conceptual ethics I&II. *Philosophy Compass*, 8(12), 1091–1101.

Burke, M. (1992). Copper statues and pieces of copper: A challenge to the standard account. *Analysis*, 52, 12–17.

Burke, M. (1994). Preserving the principle of one object to a place: A novel count of the relations among objects, sorts, sortals and persistence conditions. *Philosophy and Phenomenological Research*, 54, 591–662.

Byrne, A. (2020). Are women adult human females?. *Philosophical Studies*, 177(12), 3783–803.

Cadogan, G. (2016, July 6). *Walking while black*. LitHub..

Cameron, R. (2022). *Chains of being*. Oxford: Oxford University Press.

Canavan, C., West, J., & Card, T. (2014). The epidemiology of irritable bowel syndrome. *Clinical Epidemiology*, 6, 71–80. https://doi.org/10.2147/CLEP.S40245.

Canguilhem, G. (1991). *The normal and the pathological*. New York: Zone Books.

Cappelen, H., Plunkett, D., & Burgess, A. (Eds.). (2019). *Conceptual engineering and conceptual ethics*. New York: Oxford University Press.

Carel, H. (2008). *Illness*. London: Routledge.

Carel, H. (2010). Phenomenology and its application in medicine. *Theoretical Medicine and Bioethics* 32(1), 33–46.

Carel, H. (2018). *The phenomenology of illness*. Oxford: Oxford University Press.

Carel, H., & Kidd, I. J. (2017). Epistemic injustice and illness. *Journal of Applied Philosophy*, 34(2), 172–90.

Carr, A. J., Gibson, B., & Robinson, P. G. (2001). Is quality of life determined by expectations or experience? *BMJ*, 322, 1240. DOI: 10.1136/bmj.322.7296.1240.

Chang, R. (2002). The possibility of parity. *Ethics*, 112(4), 659–88.

Chang, R. (2004). All things considered. *Philosophical Perspectives*, 18, 1–22.

Chang, R. (2012). Are hard choices cases of incomparability? *Philosophical Issues*, 22, 106–26.

Chang, R. (2017). Hard choices. *APA Journal of Philosophy*, 92, 586–620.

Chang, E. M., Gillespie, E. F., & Shaverdian, N. (2019). Truthfulness in patient-reported outcomes: factors affecting patients' responses and impact on data quality. *Patient Related Outcome Measures* 10, 171–86. https://doi.org/10.2147/PROM.S178344.

Chen, R., & Crewe, N. (2009). Life satisfaction among people with progressive disabilities. *Journal of Rehabilitation*, 75(2), 50–8. www.researchgate.net/publication/287942122_Life_Satisfaction_among_People_with_Progressive_Disabilities.

Chen, R., Kessler, R. C., Sadikova, E., NeMoyer, A., Sampson, N. A., Alvarez, K., Vilsaint, C. L., Green, J. G., McLaughlin, K. A., Jackson, J. S., Alegría, M., & Williams, D. R. (2019).

Racial and ethnic differences in individual-level and area-based socioeconomic status and 12-month DSM-IV mental disorders. *Journal of Psychiatric Research, 119*, 48–59. https://doi.org/10.1016/j.jpsychires.2019.09.006.

Cholbi, M. (2019). Regret, resilience, and the nature of grief. *Journal of Moral Philosophy, 16*(4), 486–508.

Cholbi, M. (2021). *Grief: A philosophical guide*. Princeton, NJ: Princeton University Press.

Chou, R., Fu, R., Carrino, J. A., & Deyo, R. A. (2009). Imaging strategies for low-back pain: systematic review and meta-analysis. *The Lancet, 373*(9662), 463–72.

Chou, R., Hartung, D., Turner, J., Blazina, I., Chan, B., Levander, X., McDonagh, M., Selph, S., Fu, R., & Pappas, M. (2020). *Opioid treatments for chronic pain: Comparative effectiveness review no. 229.* (Prepared by the Pacific Northwest Evidence-based Practice Center under Contract No. 29-201500009-I.) *AHRQ Publication No. 20-EHC011.* Rockville, MD: Agency for Healthcare Research and Quality. https://doi.org/10.23970/AHRQEPCCER229.

Christensen, D. S., Dich, N., Flensborg-Madsen, T., Garde, E., Hansen, Å. M., & Mortensen, E. L. (2019). Objective and subjective stress, personality, and allostatic load. *Brain and Behavior, 9*(9) e01386. https://doi.org/10.1002/brb3.1386.

Clare, E. (2017). *Brilliant imperfection*. New York: Duke University Press.

Clark, C. W., Yang, J. C., Tsui, S. L., Ng, K. F., & Clark, S. B. (2002). Unidimensional pain rating scales: A multidimensional affect and pain survey (MAPS) analysis of what they really measure. *Pain, 98*(3), 241–7. https://doi.org/10.1016/S0304-3959(01)00474-2.

Clarke, E. (2013). The multiple realizability of biological individuals. *Journal of Philosophy, 110*(8), 413–35.

Cohen, B. J., & Gibor, Y. (1980). Anemia and menstrual blood loss. *Obstetrical and Gynecological Survey*, Oct, *35*(10), 597–618.

Connecticut Coalition against Domestic Violence. (2022). *National statistics about domestic violence.* www.ctcadv.org/information-about-domestic-violence/national-statistics.

Cooper, R. (2002). Disease. *Studies in History and Philosophy of Science Part C: Studies in History and Philosophy of Biological and Biomedical Sciences, 33*(2), 263–82.

Cope, H., McMahon, K., Heise, E., Eubanks, S., Garrett, M., Gregory, S., & Ashley-Koch, A. (2013). Outcome and life satisfaction of adults with myelomeningocele. *Disability and Health Journal, 6*(3), 236–43. https://doi.org/10.1016/j.dhjo.2012.12.003.

Corrigan, J. D., Bogner, J. A., Mysiw, W. J., Clinchot, D., & Fugate, L. (2001). Life satisfaction after traumatic brain injury. *Journal of Head Trauma Rehabilitation, 16*(6), 543–55.

Cox, D. (2017, April 27). *The curse of people who never feel pain*. BBC: Future. the-people-who-never-feel-any-pain.

Crawford, R. (1980). Healthism and the medicalization of everyday life. *International Journal of Health Services: Planning, Administration, Evaluation, 10*(3), 365–88. https://doi.org/10.2190/3H2H-3XJN-3KAY-G9NY.

Cresce, N. D., Davis, S. A., Huang, W. W., & Feldman, S. R. (2014). The quality of life impact of acne and rosacea compared to other major medical conditions. *Journal of Drugs in Dermatology: JDD, 13*(6), 692–7.

Cubukcu, D., Sarsan, A., & Alkan, H. (2012). Relationships between pain, function and radiographic findings in osteoarthritis of the knee: A cross-sectional study. *Arthritis.* Article ID 984060. https://doi.org/10.1155/2012/984060.

Cutler, D. M., & Lleras-Muney, A. (2006). Education and health: Evaluating theories and evidence. *Making Americans Healthier: Social and Economic Policy as Health Policy*, 12352. DOI: 10.3386/w12352.

Dahlhamer, J., Lucas, J., Zelaya, C., Nahin, R., Mackey, S., DeBar, L., Kerns, R., Von Korff, M., Porter, L., & Helmick, C. (2018). Prevalence of chronic pain and high-impact chronic pain among adults—United States, 2016. *Morbidity and Mortality Weekly Report*, 67(36), 1001–6. http://dx.doi.org/10.15585/mmwr.mm6736a2external icon.

Daniels, N. (1985). *Just health care*. Cambridge: Cambridge University Press.

Datablog. (2012, January 18). Disability living allowance: The benefit broken down by condition. *The Guardian*. www.theguardian.com/news/datablog/2012/jan/18/disability-living-allowancedata#:~:text=Arthritis%2C%20with%20over%20500%2C000%20claima nts,than%20one%20million%20DLA%20claimants.

Davis, T. M., Clifford, R. M., Davis, W. A., & Fremantle Diabetes Study (2001). Effect of insulin therapy on quality of life in Type 2 diabetes mellitus: The Fremantle Diabetes Study. *Diabetes Research and Clinical Practice*, 52(1), 63–71. https://doi.org/10.1016/s0168-8227(00)00245-x.

DeGood, D. E., & Tait, R. C. (2001). Assessment of pain beliefs and pain coping. In D. C. Turk & R. Melzack (Eds.), *Handbook of pain assessment* (pp. 320–45). London: The Guilford Press.

Delpierre, C., Lauwers-Cances, V., Datta, G. D., Lang, T., & Berkman, L. (2009). Using self-rated health for analysing social inequalities in health: A risk for underestimating the gap between socioeconomic groups? *Journal of Epidemiology & Community Health*, 63(6), 426–32.

DeSalvo, K. B., Bloser, N., Reynolds, K., He, J., & Muntner, P. (2006). Mortality prediction with a single general self-rated health question: A meta-analysis. *Journal of General Internal Medicine*, 21(3), 267–75. https://doi.org/10.1111/j.1525-1497.2005.00291.x.

Diamond, J. (2003). The double puzzle of diabetes. *Nature*, 423(6940), 599.

Díaz-León, E. (2018). On Haslanger's meta-metaphysics: Social structures and metaphysical deflationism. *Disputatio*, 10(50), 201–16.

Dickinson, H. O., Parkinson, K. N., Ravens-Sieberer, U., Schirripa, G., Thyen, U., Arnaud, C., Beckung, E., Fauconnier, J., McManus, V., Michelsen, S. I., Parkes, J., & Colver, A. F. (2007). Self-reported quality of life of 8–12-year-old children with cerebral palsy: a cross-sectional European study. *The Lancet*, 369(9580), 2171–8. https://doi.org/10.1016/S0140-6736(07)61013-7.

Dieppe, P. A., & Lohmander, L. S. (2005). Pathogenesis and management of pain in osteoarthritis. *The Lancet*, 365(9463), 965–73.

Dijkers, M. P. J. M. (2005). Quality of life of individuals with spinal cord injury: A review of conceptualization, measurement, and research findings. *Journal of Rehabilitation Research and Development: Supplement 1*, 42(3), 87–110.

Din-Dzietham, R., Nembhard, W. N., Collins, R., & Davis, S. K. (2004). Perceived stress following race-based discrimination at work is associated with hypertension in African–Americans. The metro Atlanta heart disease study, 1999–2001. *Social Science & Medicine*, 58(3), 449–61.

Donovan, R. A., & West, L. M. (2015). Stress and mental health: Moderating role of the strong Black woman stereotype. *Journal of Black Psychology*, 41(4), 384–96.

Dorsett, P., & Geraghty, T. (2004). Depression and adjustment after spinal cord injury: A three-year longitudinal study. *Topics in Spinal Cord Injury Rehabilitation*, 9(4), 43–56.

Dowd, J. B., & Zajacova, A. (2010). Does self-rated health mean the same thing across socioeconomic groups? Evidence from biomarker data. *Annals of Epidemiology*, 20(10), 743–9.

Dunn, D. S., & Brody, C. (2008). Defining the good life following acquired physical disability. *Rehabilitation Psychology*, 53(4), 413–25. https://doi.org/10.1037/a0013749.

Dunn, D. S., Uswatte, G., & Elliott, T. R. (2009). Happiness, resilience, and positive growth following physical disability: Issues for understanding, research, and therapeutic intervention. In S. J. Lopez & C. R. Snyder (Eds.), *Oxford handbook of positive psychology* (pp. 651–64). Oxford: Oxford University Press.

Dusenberry, M. (2017). *Doing harm: The truth about how bad medicine and lazy science leave women dismissed, misdiagnosed, and sick.* New York: Harper One.

Dutilh Novaes, C. (2020). Carnapian explication and ameliorative analysis: A systematic comparison. *Synthese, 197*(3), 1011–34.

Easterlin, R. A. (2003). Explaining happiness. *Proceedings of the National Academy of Sciences, 100*(19), 11176–83. https://doi.org/10.1073/pnas.1633144100.

Emond, A., Ridd, M., Sutherland, H., Allsop, L., Alexander, A., & Kyle, J. (2015). The current health of the signing deaf community in the UK compared with the general population: A cross-sectional study. *BMJ Open, 5,* e006668.

Engelhardt, T. (1996). *The foundations of bioethics.* New York: Oxford University Press

Eiesland, N. L. (1994). *The disabled God: Toward a liberatory theology of disability.* Nashville, TN: Abingdon Press.

Fadiman, A. (1997). *The spirit catches you and you fall down.* New York: Farrar, Straus and Giroux.

Fafchamps, M., & Kebede, B. (2012). Subjective well-being, disability and adaptation: A case study from rural Ethiopia. In D. A. Clark (Ed.), *Adaptation, poverty and development* (pp. 161–80). London: Palgrave Macmillan. https://doi.org/10.1057/9781137002778_7.

Fang, J., Madhavan, S., & Alderman, M. H. (2000). Maternal mortality in New York City: excess mortality of black women. *Journal of Urban Health, 77*(4), 735–44.

Fara, D. G. (2000). Shifting sands: An interest-relative theory of vagueness. *Philosophical Topics, 28*(1), 45–81.

Fara, D. G. (2008). Profiling interest relativity. *Analysis, 68*(4), 326–35.

Fara, D. G. (2009). Dear haecceitism. *Erkenntnis, 70*(3), 285–97.

Fenton, J. J., Jerant, A. F., Bertakis, K. D., & Franks, P. (2012). The cost of satisfaction: A national study of patient satisfaction, health care utilization, expenditures, and mortality. *Archives of Internal Medicine, 172*(5), 405–11. https://doi.org/10.1001/archinternmed.2011.1662.

Ferrie, J. E., Virtanen, M., & Kivimaki, M. (2014). The healthy population–high disability paradox. *Occupational and Environmental Medicine, 71,* 232–3. http://dx.doi.org/10.1136/oemed-2013-101945.

Finan, P. H., Buenaver, L. F., Bounds, S. C., Hussain, S., Park, R. J., Haque, U. J., Campbell, C. M., Haythornthwaite, J. A., Edwards, R. R., & Smith, M. T. (2013). Discordance between pain and radiographic severity in knee osteoarthritis: Findings from quantitative sensory testing of central sensitization. *Arthritis and Rheumatism, 65*(2), 363–72. https://doi.org/10.1002/art.34646.

Fishbain, D. A., Cutler, R., Rosomoff, H. L., & Rosomoff, R. S. (1997). Chronic pain-associated depression: Antecedent or consequence of chronic pain? A review. *The Clinical Journal of Pain, 13*(2), 116–37. https://journals.lww.com/clinicalpain/Abstract/1997/06000/Chronic_Pain_Associated_Depression___Antecedent_or.6.aspx

Fletcher, G. (2016). *The philosophy of well-being: An introduction.* Oxford: Routledge.

Foot, Philippa. (2001). *Natural goodness.* Oxford: Oxford University Press.

Forber-Pratt, A. J., Lyew, D. A., Mueller, C., & Samples, L. B. (2017). Disability identity development: A systematic review of the literature. *Rehabilitation Psychology, 62*(2), 198–207. https://doi.org/10.1037/rep0000134.

Fox, Michael J. (2002). *Lucky Man.* Hyperion Press: New York.

Fox, Michael J. (2020). *No Time Like the Future*. Macmillan Press: New York.

Francis, L., & Silvers, A. (2013). Infanticide, moral status and moral reasons: The importance of context. *Journal of Medical Ethics, 39*(5), 289–92.

Fuchs, V. R. (2016). Black gains in life expectancy. *Journal of the American Medical Association, 316*(18), 1869–70.

Fuggle, P., Shand, P. A., Gill, L. J., & Davies, S. C. (1996). Pain, quality of life, and coping in sickle cell disease. *Archives of Disease in Childhood, 75*(3), 199–203.

Fuhrer, M. J., Rintala, D. H., Hart, K. A., Clearman, R., & Young, M. E. (1992). Relationship of life satisfaction to impairment, disability, and handicap among persons with spinal cord injury living in the community. *Archives of Physical Medicine and Rehabilitation, 73*(6), 552–7. https://doi.org/10.5555/uri:pii:0003999392901908.

Gendre, T., Carle, G., Mesrati, F., Hubsch, C., Mauras, T., Roze, E. . . . & Garcin, B. (2019). Quality of life in functional movement disorders is as altered as in organic movement disorders. *Journal of Psychosomatic Research, 116*, 10–16.

Gieler, U., Gieler, T., & Kupfer, J. P. (2015). Acne and quality of life: Impact and management. *Journal of the European Academy of Dermatology and Venereology, 29*, 12–14.

Gillen, R., Tennen, H., McKee, T. E., Gernert-Dott, P., & Affleck, G. (2001). Depressive symptoms and history of depression predict rehabilitation efficiency in stroke patients. *Archives of Physical Medicine and Rehabilitation, 82*(12), 1645–9.

Giltay, E. J., Vollaard, A. M., & Kromhout, D. (2012). Self-rated health and physician-rated health as independent predictors of mortality in elderly men. *Age and Ageing, 41*(2), 165–71. https://doi.org/10.1093/ageing/afr161.

Glackin, S. N. (2016). Three Aristotelian accounts of disease and disability. *Journal of Applied Philosophy, 33*(3), 311–26.

Glackin, S. N. (2019). Grounded disease: Constructing the social from the biological in medicine. *The Philosophical Quarterly, 69*(275), 258–76.

Goetz, C. G. (2011). The history of Parkinson's disease: Early clinical descriptions and neurological therapies. *Cold Spring Harbor Perspectives in Medicine, 1*(1).

Gollust, S. E., Thompson, R. E., Gooding, H. C., & Biesecker, B. B. (2003). Living with achondroplasia in an average-sized world: An assessment of quality of life. *American Journal of Medical Genetics, 120A*(4), 447–58. https://doi.org/10.1002/ajmg.a.20127.

Gooren, L. and Gijs, L. (2015). Medicalization of homosexuality. In A. Bolin and P. Whelehan (Eds.), *The International Encyclopedia of Human Sexuality*. https://doi.org/10.1002/9781118896877.wbiehs296.

Goulia, P., Voulgari, P. V., Tsifetaki, N., Andreoulakis, E., Drosos, A. A., Carvalho, A. F., & Hyphantis, T. (2015). Sense of coherence and self-sacrificing defense style as predictors of psychological distress and quality of life in rheumatoid arthritis: A 5-year prospective study. *Rheumatology International, 35*(4), 691–700.

Graff, D. (2000). Shifting sands: An interest-relative theory of vagueness. *Philosophical Topics, 28*(1), 45–81.

Graupner, C., Kimman, M. L., Mul, S., Slok, A. H. M., Claessens, D., Kleijnen, J., Dirksen, C. D., & Breukink, S. O. (2021). Patient outcomes, patient experiences and process indicators associated with the routine use of patient-reported outcome measures (PROMs) in cancer care: A systematic review. *Supportive Care in Cancer, 29*, 573–93. https://doi.org/10.1007/s00520-020-05695-4.

Green, C. R., & Hart-Johnson, T. (2012). The association between race and neighborhood socioeconomic status in younger black and white adults with chronic pain. *The Journal of Pain, 13*(2), 176–86. https://doi.org/10.1016/j.jpain.2011.10.008.

Griffiths, P. E. (1994). Cladistic classification and functional explanation. *Philosophy of Science*, *61*(2), 206–27.

Griffith, B. N., Lovett, G. D., Pyle, D. N., & Miller, W. C. (2011). Self-rated health in rural Appalachia: Health perceptions are incongruent with health status and health behaviors. *BMC Public Health*, *11*, 229. https://doi.org/10.1186/1471-2458-11-229.

Griffiths, P. E., & Matthewson, J. (2016). Evolution, dysfunction, and disease: A reappraisal. *The British Journal for the Philosophy of Science*, *69*(2), 301–27. www.journals.uchicago.edu/doi/full/10.1093/bjps/axw021.

Grol-Prokopczyk, H., Freese, J., & Hauser, R. M. (2011). Using anchoring vignettes to assess group differences in general self-rated health. *Journal of Health and Social Behavior*, *52*(2), 246–61. https://doi.org/10.1177/0022146510396713.

Gross, R. A. (1992). A brief history of epilepsy and its therapy in the Western Hemisphere. *Epilepsy Research*, *12*(2), 65–74. https://doi.org/10.1016/0920-1211(92)90028-r.

Hall, J. C., Conner, K. O., & Jones, K. (2021). The strong Black woman versus mental health utilization: A qualitative study. *Health & Social Work*, *46*(1), 33–41.

Hamilton, N. A., Zautra, A. J., & Reich, J. W. (2005). Affect and pain in rheumatoid arthritis: Do individual differences in affective regulation and affective intensity predict emotional recovery from pain? *Annals of Behavioral Medicine*, *29*(3), 216–24.

Hammell, K. (2004). Exploring quality of life following high spinal cord injury: A review and critique. *Spinal Cord*, *42*, 491–502. https://doi.org/10.1038/sj.sc.3101636.

Hampton, S. B., Cavalier, J., & Langford, R. (2015). The influence of race and gender on pain management: A systematic literature review. *Pain Management Nursing*, *16*(6), 968–77.

Harrington, A. (2019). *Mind fixers: Psychology's troubled search for the biology of mental illness*. New York: W. W. Norton and Co.

Haslanger, S. (2000). Gender and race: (What) are they? (What) do we want them to be? *Noûs*, *34*(1), 31–55.

Haslanger, S. (2012a). *Resisting reality*. Oxford: Oxford University Press.

Haslanger, S. (2012b). Gender and race: (What) are they? (What) do we want them to be? In *Resisting reality* (pp. 221–47). Oxford: Oxford University Press.

Haslanger, S. (2012c). Social construction: The debunking project. In *Resisting reality* (pp. 113–38). Oxford: Oxford University Press.

Haslanger, S. (2012d). What are we talking about: The semantics and politics of social kinds. In *Resisting reality* (pp. 365–80). Oxford: Oxford University Press.

Haslanger, S. (2012e). What good are our intuitions? Philosophical analysis and social kinds. In *Resisting reality* (pp. 381–405). Oxford: Oxford University Press.

Haslanger, S. (2016). What is a (social) structural explanation? *Philosophical Studies*, *173*(1), 113–30.

Haslanger, S. (2018). Social explanation: Structures, stories, and ontology. A reply to Díaz León, Saul, and Sterken. *Disputatio*, *10*(50), 245–73.

Haslanger, S. (2020). Going on, not in the same way. In A. Burgess, H. Cappelen, & D. Plunkett (Eds.)., *Conceptual engineering and conceptual ethics* (pp. 230–60). Oxford: Oxford University Press.

Hassan, J., Grogan, S., Clark-Carter, D., Richards, H., & Yates, V. M. (2009). The individual health burden of acne: Appearance-related distress in male and female adolescents and adults with back, chest and facial acne. *Journal of Health Psychology*, *14*(8), 1105–18.

Haueis, P. (2021). A generalized patchwork approach to scientific concepts. *British Journal for Philosophy of Science*. https://doi.org/10.1086/716179.

Hausman, D. M. (2011). Is an overdose of paracetamol bad for one's health? *British Journal for the Philosophy of Science*, *62*(3), 657–68.

Hausman, D. M. (2012). Health, naturalism, and functional efficiency. *Philosophy of Science*, *79*(4), 519–41.

Hausman, D. M. (2014). Health and functional efficiency. *Journal of Medicine and Philosophy*, *39*(6), 634–47.

Hausman, D. M. (2015). *Valuing health: Well-being, freedom, and suffering*. Oxford: Oxford University Press.

Hausman, D. M. (2017). Health and wellbeing. In M. Solomon, J. R. Simon, & H. Kinkaid (Eds.), *The Routledge companion to the philosophy of medicine* (pp. 27–35). London: Routledge.

Headey, B., Kelley, J., & Wearing, A. (1993). Dimensions of mental health: Life satisfaction, positive affect, anxiety and depression. *Social Indicators Research*, *29*, 63–82. https://doi.org/10.1007/BF01136197.

Heathwood, C. (2021). *Happiness and well-being*. Cambridge: Cambridge University Press.

Helgeson, V. S. (2003). Social support and quality of life. *Quality of Life Research 12* (Suppl1), 25–31. https://doi.org/10.1023/A:1023509117524.

Horgan, O., & MacLachlan, M. (2004). Psychosocial adjustment to lower-limb amputation: A review. *Disability and Rehabilitation*, *26*(14-15), 837–50. DOI: 10.1080/09638280410001708869.

Horowitz, D., & Wakefield, J. (2007). *The loss of sadness*. Oxford: Oxford University Press.

Hunt, B., & Whitman, S. (2015). Black/White health disparities in the United States and Chicago: 1990–2010. *Journal of Racial and Ethnic Health Disparities*, *2*(1), 93–100.

Hurst, L., Mahtani, K., Pluddemann, A., Lewis, S., Harvey, K., Briggs, A., Boyle, A., Bajwa, R., Haire, K., Entwistle, A., Handa, A., & Heneghan, C. (2019). *Defining value-based healthcare in the NHS: CEBM report*. -the-nhs.

Idler, E.L. (2003), Discussion: Gender differences in self-rated health, in mortality, and in the relationship between the two. *The Gerontologist*, *43*(3), 372–5. https://doi.org/10.1093/geront/43.3.372.

Ilie, G., Bradfield, J., Moodie, L., Lawen, T., Ilie, A., Lawen, Z., Blackman, C., Gainer, R., & Rutledge, R. (2019). The role of response-shift in studies assessing quality of life outcomes among cancer patients: A systematic review. *Frontiers in Oncology*, *9*, 783. https://doi.org/10.3389/fonc.2019.00783.

Im, E. O. (2007). Ethnic differences in cancer pain experience. *Nursing Research*, *56*(5), 296.

Institute for Health Metrics Evaluation. (2010). *Global burden of diseases, injuries, and risk factors profile: Germany*. Seattle, WA.

International Association for the Study of Pain. (2020, July 16). IASP announces revised definition of pain. -of-pain.

Isaksson, A., Ahlström, G., & Gunnarsson, L. (2005). Quality of life and impairment in patients with multiple sclerosis. *Journal of Neurology, Neurosurgery & Psychiatry*, *76*(1), 64–9. https://jnnp.bmj.com/content/76/1/64

Ito, N., Ishiguro, M., Uno, M., Kato, S., Shimizu, S., Obata, R., Tanaka, M., Tokunaga, K., Nagano, M., Sugihara, K., & Kazuma, K. (2012). Prospective longitudinal evaluation of quality of life in patients with permanent colostomy after curative resection for rectal cancer: A preliminary study. *Journal of Wound, Ostomy and Continence Nursing*, *39*(2), 172–7. DOI: 10.1097/WON.0b013e3182456177.

Jackson, F. (1998). *From metaphysics to ethics: A defence of conceptual analysis*. Oxford: Clarendon Press.

Jenkins, K. (2016). Amelioration and inclusion: Gender identity and the concept of woman. *Ethics*, *126*(2), 394–421.

Jenkins, K. (2022). How to be a pluralist about gender categories. In R. Halwani, J. M. Held, N. McKeever, & A. Soble (Eds.), *The philosophy of sex: Contemporary readings*. 8th edition (pp. 233–59). Lanham, MD: Rowman & Littlefield Publishers.

Jenkins, K. (2023). *Ontology and oppression: Race, gender, and social reality*. Oxford: Oxford University Press.

Jensen, P. S., Mrazek, D., Knapp, P. K., Steinberg, L., Pfeffer, C., Schowalter, J., & Shapiro, T. (1997). Evolution and revolution in child psychiatry: ADHD as a disorder of adaptation. *Journal of the American Academy of Child & Adolescent Psychiatry, 36*(12), 1672–81.

Jensen, M. P., Romano, J. M., Turner, J. A., Good, A. B., & Wald, L. H. (1999). Patient beliefs predict patient functioning: further support for a cognitive-behavioural model of chronic pain. *Pain, 81*(1–2), 95–104. https://doi.org/10.1016/S0304-3959(99)00005-6.

Joffe, B., & Zimmet, P. (1998). The thrifty genotype in type 2 diabetes. *Endocrine, 9*(2), 139–41.

Johnson, H. M. (2005). *Too late to die young*. New York: Picador Press.

Joseph, K. S., Boutin, A., Lisonkova, S., Muraca, G. M., Razaz, N., John, S.... & Schisterman, E. (2021). Maternal mortality in the United States: Recent trends, current status, and future considerations. *Obstetrics and Gynecology, 137*(5), 763.

Kamp, H. (1975). Two theories about adjectives. In E. L. Keenan (Ed.), *Formal semantics of natural language* (pp. 123–55). Cambridge: Cambridge University Press.

Kaplan, J. M. (2010). When socially determined categories make biological realities: understanding Black/White health disparities in the US. *The Monist, 93*(2), 281–97.

Kaplan, R. M., Schmidt, S. M., & Cronan, T. A. (2000). Quality of well-being in patients with fibromyalgia. *Journal of Rheumatology, 27*(3), 780–5. https://lup.lub.lu.se/search/publication/203e6f79-478d-4285-8d72-b29545382ba8.

Keefe, P. R. (2021). *Empire of pain: The secret history of the Sackler Dynasty*. New York: Penguin Random House Press.

Keil, G., & Stoecker, R. (2016). Disease as a vague and thick cluster concept. In G. Keil, L. Keuck, & R. Hauswald (Eds.), *Vagueness in psychiatry: International perspectives in philosophy and psychiatry*. Oxford: Oxford University Press.

Kennedy, C. (1999). *Projecting the adjective: The syntax and semantics of gradability and comparison*. Garland, NY. (1997 University of California Santa Cruz PhD thesis).

Kennedy, C. (2007). Vagueness and grammar: The semantics of relative and absolute gradable predicates. *Linguistics and Philosophy, 30*, 1–45.

Kennedy, P., & Rogers, B. A. (2000). Anxiety and depression after spinal cord injury: A longitudinal analysis. *Archives of Physical Medicine and Rehabilitation, 81*(7), 932–7. https://doi.org/10.1053/apmr.2000.5580.

Khader, S. (2011). *Adaptive preference and women's empowerment*. Oxford: Oxford University Press.

Kiani, A.N., & Petri, M. (2010). Quality-of-life measurements versus disease activity in systemic lupus erythematosus. *Current Rheumatology Reports, 12*, 250–8. https://doi.org/10.1007/s11926-010-0114-1.

Kim, J., Lee, C. & Ji, M. (2018). Investigating the domains of life satisfaction in middle-aged, late middle-aged, and older adults with a physical disability. *Journal of Developmental and Physical Disability, 30*, 639–52. https://doi.org/10.1007/s10882-018-9609-x.

Kim, H. J., Hong, S., & Kim, M. (2015). Living arrangement, social connectedness, and life satisfaction among Korean older adults with physical disabilities: The results from the national survey on persons with disabilities. *Journal of Developmental and Physical Disability 27*, 307–21. https://doi.org/10.1007/s10882-014-9418-9.

Kingma, E. (2007). What is it to be healthy? *Analysis*, *67*(2), 128–33.

Kingma, E. (2010). Paracetamol, poison, and polio: Why Boorse's account of function fails to distinguish health and disease. *British Journal for the Philosophy of Science*, *61*(2), 241–64.

Kingma, E. (2016). Situation specific disease and dispositional function: Table 1. *British Journal for the Philosophy of Science*, *67*(2), 391–404.

Kinney, W. B., & Coyle, C. P. (1992). Predicting life satisfaction among adults with physical disabilities. *Archives of Physical Medicine and Rehabilitation*, *73*(9), 863–9. https://doi.org/10.5555/uri:pii:000399939290160X.

Klein, C. (2015). *What the body commands: The imperative theory of pain*. Cambridge, MA: MIT Press.

Knabe, A., & Rätzel, S. (2011). Quantifying the psychological costs of unemployment: The role of permanent income. *Applied Economics*, *43*(21), 2751–63. DOI: 10.1080/00036840903373295.

Koelmel, E., Hughes, A. J., Alschuler, K. N., & Ehde, D. M. (2007). Resilience mediates the longitudinal relationships between social support and mental health outcomes in multiple sclerosis. *Archives of Physical Medicine and Rehabilitation*, *98*(6), 1139–48. https://doi.org/10.1016/j.apmr.2016.09.127.

Kohler, M., Clarenbach, C. F., Böni, L., Brack, T., Russi, E. W., & Bloch, K. E. (2005). Quality of life, physical disability, and respiratory impairment in Duchenne muscular dystrophy. *American Journal of Respiratory and Critical Care Medicine*, *178*(8), 1032–6. https://doi.org/10.1164/rccm.200503-322OC.

Koivumaa-Honkanen, H., Kaprio, J., Honkanen, R., Viinamäki, H., & Koskenvuo, M. (2004). Life satisfaction and depression in a 15-year follow-up of healthy adults. *Social Psychiatry and Psychiatric Epidemiology*, *39*, 994–9. https://doi.org/10.1007/s00127-004-0833-6.

Krahn, G. L., Walker, D. K., & Correa-De-Araujo, R. (2015). Persons with disabilities as an unrecognized health disparity population. *American Journal of Public Health*, *105*(S2), S198–206.

Kralik, D., van Loon, A., & Visentin, K. (2006). Resilience in the chronic illness experience. *Educational Action Research*, *14*(2), 187–201.

Krokavcova, M., van Dijk, J. P., Nagyova, I., Rosenberger, J., Gavelova, M., Middel, B., Gdovinova, Z., & Groothoff, J. W. (2008). Social support as a predictor of perceived health status in patients with multiple sclerosis. *Patient Education and Counseling*, *73*(1), 159–65. https://doi.org/10.1016/j.pec.2008.03.019.

Kukla, Q. (as Rebecca) (2014). Medicalization, "normal function," and the definition of health. In, J. D. Arras, E. Fenton, & R. Kukla (Eds.), *The Routledge companion to bioethics*. London: Routledge.

Kukla, Q. (as Rebecca) (2019). Infertility, epistemic risk, and disease definitions. *Synthese*, *196*(11), 4409–28.

Kukla, Q. (2022). Philosophy of population health: Philosophy for a new public health era by Sean Valles: Healthism and the weaponization of "health". *Studies in History and Philosophy of Science*, Part A 91, 316–9.

Kuriya, B., Gladman, D. D., Ibañez, D., & Urowitz, M.B. (2008). Quality of life over time in patients with systemic lupus erythematosus. *Arthritis & Rheumatism*, *59*(2), 181–5. https://doi.org/10.1002/art.23339.

Kwi-Ok, J., & Nan-Young, L. (2008). A study on the pain, depression, life satisfaction of the chronic low back pain patients. *Journal of Muscle and Joint Health*, *15*(1), 73–87. https://koreascience.kr/article/JAKO200820055491791.page.

Kwok, W., & Bhuvanakrishna, T. (2014). The relationship between ethnicity and the pain experience of cancer patients: A systematic review. *Indian Journal of Palliative Care*, *20*(3), 194.

Laborde, J. M., & Powers, M. J. (1980). Satisfaction with life for patients undergoing hemodialysis and patients suffering from osteoarthritis. *Research in Nursing and Health*, *3*(1), 19–24. https://doi.org/10.1002/nur.4770030105.

LaBrada, E. (2016). Categories we die for: Ameliorating gender in analytic feminist philosophy. *PMLA*, *131*(2), 449–59.

Layes, A., Asada, Y., & Kepart, G. (2012). Whiners and deniers—what does self-rated health measure? *Social Science & Medicine (1982)*, *75*(1), 1–9. https://doi.org/10.1016/j.socscimed.2011.10.030.

Lee, M. A., Walker, R. W., Hildreth, A. J., & Prentice, W. M. (2006). Individualized assessment of quality of life in idiopathic Parkinson's disease. *Movement Disorders: Official Journal of the Movement Disorder Society*, *21*(11), 1929–34.

Lenert, L. A., Treadwell, J. R., & Schwartz, C. E. (1999). Associations between health status and utilities implications for policy. *Medical Care*, *37*(5), 479–89.

Le Souëf, P. N., Goldblatt, J., & Lynch, N. R. (2000). Evolutionary adaptation of inflammatory immune responses in human beings. *The Lancet*, *356*(9225), 242–4.

Levy, N., Sturgess, J., & Mills, P. (2018). "Pain as the fifth vital sign" and dependence on the "numerical pain scale" is being abandoned in the US: Why? *British Journal of Anaesthesia*, *120*(3), 435–8. https://doi.org/10.1016/j.bja.2017.11.098.

Lewis, D. (1971). Counterparts of persons and their bodies. *Journal of Philosophy*, *68*(7), 203–11.

Lewis, D. (1983). Individuation by acquaintance and by stipulation. *Philosophical Review*, 3–32.

Lewis, D. (1986). *On the plurality of worlds*. Oxford: Blackwell.

Lewis, T. T., Everson-Rose, S. A., Colvin, A., Matthews, K., Bromberger, J. T., & Sutton-Tyrrell, K. (2009). Interactive effects of race and depressive symptoms on calcification in African American and white women. *Psychosomatic Medicine*, *71*(2), 163–70.

Li, L., & Moore, D. (1998). Acceptance of disability and its correlates. *The Journal of Social Psychology*, *138*(1), 13–25. DOI: 10.1080/00224549809600349.

Lilienfeld, S. O., & Marino, L. (1995). Mental disorder as a Roschian concept: A critique of Wakefield's "harmful dysfunction" analysis. *Journal of Abnormal Psychology*, *104*(3), 411–20.

Lindauer, M. (2020). Conceptual engineering as concept preservation. *Ratio*, *33*(3), 155–62.

Lindauer, M., Mayorga, M., Greene, J. D., Slovic, P., Västfjäll, D., & Singer, P. (2020). Comparing the effect of rational and emotional appeals on donation behavior. *Judgment and Decision Making*, *15*(3), 413–20.

Livingston, G. (2018, April 25). The changing profile of unmarried parents: A growing share are living with a partner. Pew Research Center. www.pewresearch.org/social-trends/2018/04/25/the-changing-profile-of-unmarried-parents.

Livneh, H. (2001). Psychosocial adaptation to chronic illness and disability: A conceptual framework. *Rehabilitation Counseling Bulletin*, *44*(3), 151–60. https://doi.org/10.1177/003435520104400305.

Livneh, H., & Antonak, R. F. (1997). *Psychosocial adaptation to chronic illness and disability*. Boston, MA: Aspen Publishers.

Livneh, H., & Martz, E. (2003). Psychosocial adaptation to spinal cord injury as a function of time since injury. *International Journal of Rehabilitation Research*, *26*(3), 191–200. https://journals.lww.com/intjrehabilres/Abstract/2003/09000/Psychosocial_adaptation_to_spinal_cord_injury_as_a.5.aspx.

Logue, H. (2022). Gender fictionalism. *Ergo: An Open Access Journal of Philosophy*, 8.

Lombardo, P., Jones, W., Wang, L., Shen, X., & Goldner, E. (2018). The fundamental association between mental health and life satisfaction: Results from successive waves of a Canadian national survey. *BMC Public Health*, 18. https://doi.org/10.1186/s12889-018-5235-x.

Lopez de Sa, D. (in progress). Terminological injustice.

Lorem, G., Cook, S., Leon, D.A., Emaus, N., & Schirmer, H. (2020). Self-reported health as a predictor of mortality: A cohort study of its relation to other health measurements and observation time. *Scientific Reports*, 10, 4886. https://doi.org/10.1038/s41598-020-61603-0.

Lucas, R. E. (2007). Adaptation and the set-point model of subjective well-being: Does happiness change after major life events? *Current Directions in Psychological Science*, 16(2), 75–9. https://doi.org/10.1111/j.1467-8721.2007.00479.x.

Luy, M., & Minagawa, Y. (2014). Gender gaps: Life expectancy and proportion of life in poor health. *Health Reports*, 25(12), 12–19.

MacKay, D. (2017). Calculating QALYs: Liberalism and the value of health states. *Economics and Philosophy*, 33(2), 259–85.

Madsen, S. G., Danneskiold-Samsøe, B., Stockmarr, A., & Bartels, E. M. (2016). Correlations between fatigue and disease duration, disease activity, and pain in patients with rheumatoid arthritis: A systematic review. *Scandinavian Journal of Rheumatology*, 45(4), 255–61. https://doi.org/10.3109/03009742.2015.1095943.

Mailhan, L., Azouvi, P., & Dazord, A. (2005). Life satisfaction and disability after severe traumatic brain injury. *Brain Injury*, 19(4), 227–38. DOI: 10.1080/02699050410001720149.

Mancini, A. D., Bonanno, G. A., & Clark, A. E. (2011). Stepping off the hedonic treadmill: Individual differences in response to major life events. *Journal of Individual Differences*, 32(3), 144–52. https://doi.org/10.1027/1614-0001/a000047.

Maus, T. (2010). Imaging the back pain patient. *Physical Medicine and Rehabilitation Clinics of North America*, 21(4), 725–66. https://doi.org/10.1016/j.pmr.2010.07.004.

McCabe, M. P., Stokes, M., & McDonald, E. (2009). Changes in quality of life and coping among people with multiple sclerosis over a 2 year period. *Psychology, Health & Medicine*, 14(1), 86–96.

Mcelhone, K., Abbott, J., & Teh, L.-S. (2006). A review of health related quality of life in systemic lupus erythematosus. *Lupus*, 15(10), 633–43. https://doi.org/10.1177/0961203306071710.

McKitrick, J. (2015). A dispositional account of gender. *Philosophical Studies*, 172(10), 2575–89.

McNamee, P., & Mendolia, S. (2014). The effect of chronic pain on life satisfaction: Evidence from Australian data. *Social Science & Medicine*, 121, 65–73. https://doi.org/10.1016/j.socscimed.2014.09.019.

Mendlowicz, M. V., & Stein, M. B. (2000). Quality of life in individuals with anxiety disorders. *American Journal of Psychiatry*, 157(5), 669–82. https://ajp.psychiatryonline.org/doi/full/10.1176/appi.ajp.157.5.669.

Merleau-Ponty, M. (2012). *The phenomenology of perception*. London: Routledge. [Original publication 1945].

Michalak, E. E., Yatham, L. N., & Lam, R. W. (2005). Quality of life in bipolar disorder: A review of the literature. *Health and Quality of Life Outcomes*, 3. https://doi.org/10.1186/1477-7525-3-72.

Middleton, J. W., Dayton, A., Walsh, J., Rutkowski, S. B., Leong, G., & Duong, S. (2012). Life expectancy after spinal cord injury: A 50-year study. *Spinal Cord*, 50(11), 803–11.

Millikan, R. G. (1989a). In defense of proper functions. *Philosophy of Science, 56*(6), 288–302.

Millikan, R. G. (1989b). Biosemantics. *Journal of Philosophy, 86*(6), 281–97.

Moller, D. (2007). Love and death. *Journal of Philosophy, 104*(6), 301–16.

Moore, R. A., Straube, S., Paine, J., Phillips, C. J., Derry, S., & McQuay, H. J. (2010). Fibromyalgia: Moderate and substantial pain intensity reduction predicts improvement in other outcomes and substantial quality of life gain. *Pain, 149*(2), 360–4. https://doi.org/10.1016/j.pain.2010.02.039.

Moseley, G. L. (2007). Reconceptualising pain according to modern pain science. *Physical Therapy Reviews, 12*(3), 169–78.

Motl, R. W., McAuley, E., Snook, E. M., & Gliottoni, R. C. (2009). Physical activity and quality of life in multiple sclerosis: Intermediary roles of disability, fatigue, mood, pain, self-efficacy and social support. *Psychology, Health & Medicine, 14*(1), 111–24. DOI: 10.1080/13548500802241902.

Muhkerjee, S. (2010). *Emperor of all maladies: A biography of cancer.* New York: Scibner.

Münster, K., Schmidt, L., & Helm, P. (1992). Length and variation in the menstrual cycle: A cross-sectional study from a Danish county. *BJOG: An International Journal of Obstetrics & Gynaecology, 99*(5), 422–42.

Murata, C., Kondo, T., Tamakoshi, K., Yatsuya, H., & Toyoshima, H. (2006). Determinants of self-rated health: Could health status explain the association between self-rated health and mortality? *Archives of Gerontology and Geriatrics, 43*(3), 369–80. https://doi.org/10.1016/j.archger.2006.01.002.

Neander, K. (1991a). Functions as selected effects: The conceptual analyst's defense. *Philosophy of Science, 58*(2), 168–84.

Neander, K. (1991b). The teleological notion of "function." *Australasian Journal of Philosophy, 69*(4), 454–68.

Neander, K. (1995). Misrepresenting and malfunctioning. *Philosophical Studies, 79*(2), 109–41.

NEJM Catalyst. (2017, January 1). What is value-based healthcare? https://catalyst.nejm.org/doi/full/10.1056/CAT.17.0558.

Nguyen, C. T. (2020). Games: Agency as art. USA: Oxford University Press.

Nguyen, C. T. (2021). The Seductions of Clarity. *Royal Institute of Philosophy* Supplements 89, 227–55.

Nolan, D., Restall, G., & West, C. (2005). Moral fictionalism versus the rest. *Australasian Journal of Philosophy, 83*(3), 307–30.

Nordenfelt, L. (2006). The concepts of health and illness revisited. *Medicine, Health Care and Philosophy, 10*(1), 5–10.

Nordenfelt, L. (2018). Functions and health: Towards a praxis oriented concept of health. *Biological Theory, 13*(1), 10–16.

Nussbaum, M. (1993). Non-relative virtues: An Aristotelian approach. In M. Nussbaum & A. Sen (Eds.), *The quality of life* (pp. 242–69). Oxford: Oxford University Press.

Nussbaum, M. (2001a). Symposium on Amartya Sen's philosophy: 5 adaptive preferences and women's options. *Economics and Philosophy, 17,* 67–88.

Nussbaum, M. (2001b). *Women and human development: The capabilities approach.* Cambridge: Cambridge University Press.

Nussbaum, M. (2006). *Frontiers of justice: Disability, nationality, species membership.* Cambridge, MA: Belknap Press.

O'Keefe, J. H., Patil, H. R., Lavie, C. J., Magalski, A., Vogel, R. A., & McCullough, P. A. (2012). Potential adverse cardiovascular effects from excessive endurance exercise. *Mayo Clinic Proceedings, 87*(6), 587–95. https://doi.org/10.1016/j.mayocp.2012.04.005.

O'Sullivan, S. (2021). *The sleeping beauties: And other stories of mystery illness*. London: Pantheon Press.

Olatunji, B. O., Cisler, J. M., & Tolin, D. F. (2007). Quality of life in the anxiety disorders: A meta-analytic review. *Clinical Psychology Review*, 27(5), 572–81. https://doi.org/10.1016/j.cpr.2007.01.015.

Oliver, M. (2013). The social model of disability: Thirty years on. *Disability & Society*, 28(7), 1024–6.

Osterweis, M., Kleinman, A., & Mechanic, D. (Eds.). (1987). *Pain and disability: Clinical, behavioral, and public policy perspectives*. Washington, DC: National Academies Press.

Østlie, K., Magnus, P., Skjeldal, O. H., Garfelt, B., & Tambs, K. (2011). Mental health and satisfaction with life among upper limb amputees: A Norwegian population-based survey comparing adult acquired major upper limb amputees with a control group. *Disability and Rehabilitation*, 33(17–18), 1594–1607. DOI: 10.3109/09638288.2010.540293.

Oswald, A. & Powdthavee, N. (2008). Does happiness adapt? A longitudinal study of disability with implications for economists and judges. *Journal of Public Economics*, 92(5–6), 1061–77. https://EconPapers.repec.org/RePEc:eee:pubeco:v:92:y:2008:i:5-6:p:1061-1077.

Pace, F., Molteni, P., Bollani, S., Sarzi-Puttini, P., Stockbrügger, R., Bianchi Porro, G., & Drossman, D. A. (2003). Inflammatory bowel disease versus irritable bowel syndrome: A hospital-based, case-control study of disease impact on quality of life. *Scandinavian Journal of Gastroenterology*, 38(10), 1031–8. https://doi.org/10.1080/00365520310004524.

Palmer, J. R., Cozier, Y. C., & Rosenberg, L. (2022). Research on health disparities: Strategies and findings from the Black Women's Health Study. *American Journal of Epidemiology*, kwac022. Advance online publication. https://doi.org/10.1093/aje/kwac022.

Pavot, W., & Diener, E. (2008). The satisfaction with life scale and the emerging construct of life satisfaction. *The Journal of Positive Psychology*, 3(2), 137–52.

Pierce, C. A., & Hanks, R. A. (2006). Life satisfaction after traumatic brain injury and the World Health Organization model of disability. *American Journal of Physical Medicine & Rehabilitation*, 85(11), 889–98. DOI: 10.1097/01.phm.0000242615.43129.ae.

Plunkett, D. (2015). Which concepts should we use? Metalinguistic negotiations and the methodology of philosophy. *Inquiry: An Interdisciplinary Journal of Philosophy*, 58(7–8), 828–74.

Plunkett, D., & Sundell, T. (2013a). Disagreement and the semantics of normative and evaluative terms. *Philosophers Imprint*, 13(23), 1–37.

Plutynski, A. (2018). *Explaining cancer*. Oxford: Oxford University Press.

Poon, J.-L., Zhou, Z.-Y., Doctor, J. N., Wu, J., Ullman, M. M., Ross, C., Riske, B., Parish, K. L., Lou, M., Koerper, M. A., Gwadry-Sridhar, F., Forsberg, A. D., Curtis, R. G., & Johnson, K. A. (2012). Quality of life in haemophilia A: Hemophilia utilization group study Va (HUGS-Va). *Haemophilia*, 18(5), 699–707. https://doi.org/10.1111/j.1365-2516.2012.02791.x.

Pouwer, F., & Hermanns, N. (2009). Insulin therapy and quality of life: A review. *Diabetes/ Metabolism Research and Reviews*, 25(S1), S4–10.

Preston-Roedder, R., & Preston-Roedder, E. (2017). Grief and recovery. In A. Gotlib (Ed.), *The moral psychology of sadness*. Lanham, MD: Rowman & Littlefield International.

Priest, G. (1987). *In contradiction*. Oxford: Oxford University Press.

Raffman, D. (1994). Vagueness without paradox. *Philosophical Review*, 103(1), 41–74.

Raja, S. N., Carr, D. B., Cohen, M., Finnerup, N. B., Flor, H., Gibson, S.... & Vader, K. (2020). The revised IASP definition of pain: Concepts, challenges, and compromises. *Pain*, 161(9), 1976.

Rapaport, M. H., Clary, C., Fayyad, R., & Endicott, J. (2005). Quality-of-life impairment in depressive and anxiety disorders. *American Journal of Psychiatry*, *162*(6), 1171–8. https://doi.org/10.1176/appi.ajp.162.6.1171.

Reiss, F. (2013). Socioeconomic inequalities and mental health problems in children and adolescents: A systematic review. *Social Science & Medicine*, *90*, 24–31. https://doi.org/10.1016/j.socscimed.2013.04.026.

Richards, H. M., Reid, M. E., & Watt, G. C. M. (2002). Socioeconomic variations in responses to chest pain: Qualitative study. *BMJ*, *324*(7349), 1308.

Richardson, K. (2022). Exclusion and erasure: Two types of ontological opression. *Ergo: An Open Access Journal of Philosophy*, *9*.

Riis, J., Loewenstein, G., Baron, J., Jepson, C., Fagerlin, A., & Ubel, P. A. (2005). Ignorance of hedonic adaptation to hemodialysis: A study using ecological momentary assessment. *Journal of Experimental Psychology: General*, *134*(1), 3–9. https://doi.org/10.1037/0096-3445.134.1.3.

Roberts, E., Wessely, S., Chalder, T., Chang, C. K., & Hotopf, M. (2016). Mortality of people with chronic fatigue syndrome: A retrospective cohort study in England and Wales from the South London and Maudsley NHS Foundation Trust Biomedical Research Centre (SLaM BRC) Clinical Record Interactive Search (CRIS) register. *The Lancet*, *387*(10028), 1638–43.

Rosenberger, P. H., Jokl, P., & Ickovics, J. (2006). Psychosocial factors and surgical outcomes: An evidence-based literature review. *JAAOS-Journal of the American Academy of Orthopaedic Surgeons*, *14*(7), 397–405.

Rosengren, L., Brogårdh, C., Jacobsson, L., & Lexell, J. (2016). Life satisfaction and associated factors in persons with mild to moderate Parkinson's disease. *NeuroRehabilitation*, *39*(2), 285–94. https://content.iospress.com/articles/neurorehabilitation/nre1359.

Rosengren, L., Forsberg, A., Brogårdh, C., & Lexell, J. (2021). Life satisfaction and adaptation in persons with Parkinson's disease: A qualitative study. *International Journal of Environmental Research and Public Health*, *18*(6), 3308.

Rosqvist, K., Hagell, P., Odin, P., Ekström, H., Iwarsson, S., & Nilsson, M. H. (2017). Factors associated with life satisfaction in Parkinson's disease. *Acta Neurologica Scandinavica*, *136*(1), 64–71. https://doi.org/10.1111/ane.12695.

Salaffi, F., Sarzi-Puttini, P., Girolimetti, R., Atzeni, F., Gasparini, S., & Grassi, W. (2009). Health-related quality of life in fibromyalgia patients: A comparison with rheumatoid arthritis patients and the general population using the SF-36 health survey. *Clinical and Experimental Rheumatology*, *27*(5 Suppl 56), S67–74.

Salmon, M., Blanchin, M., Rotonda, C., Guillemin, F., & Sébille, V. (2017). Identifying patterns of adaptation in breast cancer patients with cancer-related fatigue using response shift analyses at subgroup level. *Cancer Medicine*, *6*(11), 2562–75.

Sánchez-Rodríguez, E., Aragonès, E., Jensen, M. P., Tomé-Pires, C., Rambla, C., López-Cortacans, G., & Miró, J. (2020). The role of pain-related cognitions in the relationship between pain severity, depression, and pain interference in a sample of primary care patients with both chronic pain and depression. *Pain Medicine (Malden, Mass.)*, *21*(10), 2200–11. https://doi.org/10.1093/pm/pnz363.

Sarzi-Puttini, P., Fiorini, T., Panni, B., Turiel, M., Cazzola, M., & Atzeni, F. (2002). Correlation of the score for subjective pain with physical disability, clinical and radiographic scores in recent onset rheumatoid arthritis. *BMC Musculoskeletal Disorders*, *3*(18). https://doi.org/10.1186/1471-2474-3-18.

Sassoon, G. W. (2012). A typology of multidimensional adjectives. *Journal of Semantics*, *30*(3), 335–80.

Saul, J. (2006). Philosophical analysis and social kinds. *Proceedings of the Aristotelian Society, Supplementary Volumes, 80,* 89–143.

Scher, C., Meador, L., Van Cleave, J. H., & Reid, M. C. (2018). Moving beyond pain as the fifth vital sign and patient satisfaction scores to improve pain care in the 21st century. *Pain Management Nursing: Official Journal of the American Society of Pain Management Nurses, 19*(2), 125–9. https://doi.org/10.1016/j.pmn.2017.10.010.

Schnittker, J., & Bacak, V. (2014). The increasing predictive validity of self-rated health. *PLoS ONE, 9*(1), e84933. https://doi.org/10.1371/journal.pone.0084933.

Schoenfield, M. (2016). Moral vagueness is ontic vagueness. *Ethics, 126*(2), 257–82.

Schrag, A. (2006). Quality of life and depression in Parkinson's disease. *Journal of the Neurological Sciences, 248*(1–2), 151–7.

Schroeder, S. A. (2013). Rethinking health: Healthy or healthier than? *British Journal for the Philosophy of Science, 64*(1), 131–59.

Schwartz, P. H. (2007). Defining dysfunction: Natural selection, design, and drawing a line. *Philosophy of Science, 74*(3), 364–85.

Schwartz, C. E., & Sprangers, M. A. (Eds.). (2000). *Adaptation to changing health: Response shift in quality-of-life research.* Washington, DC: American Psychological Association.

Schwartz, C., & Frohner, R. (2005). Contribution of demographic, medical, and social support variables in predicting the mental health dimension of quality of life among people with multiple sclerosis. *Health & Social Work, 30*(3), 203–12. https://doi.org/10.1093/hsw/30.3.203.

Schwartz, C. E., Andresen, E. M., Nosek, M. A., Krahn, G. L., & RRTC Expert Panel on Health Status Measurement. (2007). Response shift theory: Important implications for measuring quality of life in people with disability. *Archives of Physical Medicine and Rehabilitation, 88*(4), 529–36.

Sen, A. (1985). *Commodities and capabilities.* Oxford: Oxford University Press.

Sen, A. (1990). Justice: Means vs. freedoms. *Philosophy and Public Affairs, 19*(2), 111–21.

Sen, A. (1993). Capability and well-being. In M. Nussbaum & A. Sen (Eds.), *The quality of life* (pp. 30–53). Oxford: Oxford University Press.

Seres, G., Kovács, Z., Kovács, Á., Kerékgyártó, O., Sárdi, K., Demeter, P.... & Túry, F. (2008). Different associations of health related quality of life with pain, psychological distress and coping strategies in patients with irritable bowel syndrome and inflammatory bowel disorder. *Journal of Clinical Psychology in Medical Settings, 15*(4), 287–95.

Shakespeare, T. (2006). The social model of disability. *The Disability Studies Reader, 2,* 197–204.

Sharma, L. (2021). Osteoarthritis of the knee. *New England Journal of Medicine, 384*(1), 51–9.

Silvers, A. (1994). "Defective" agents: Equality, difference and the tyranny of the normal: equality, normality and ability. *Journal of Social Philosophy, 25*(s1), 154–75.

Silvers, A. (2017). Disability and normality. In M. Solomon et al. (Eds.), *The Routledge companion to the philosophy of medicine* (pp. 36–47). London: Routledge.

Sinclair, V. G., & Blackburn, D. S. (2008). Adaptive coping with rheumatoid arthritis: The transforming nature of response shift. *Chronic Illness, 4*(3), 219–30.

Sironi, M., & Clerici, M. (2010). The hygiene hypothesis: An evolutionary perspective. *Microbes and Infection, 12*(6), 421–7.

Smith, D. M., Langa, K. M., Kabeto, M. U., & Ubel, P. A. (2005). Health, wealth, and happiness: Financial resources buffer subjective well-being after the onset of a disability. *Psychological Science, 16*(9), 663–6. https://doi.org/10.1111/j.1467-9280.2005.01592.x.

Smith, K. W., Avis, N. E. & Assmann, S. F. (1999). Distinguishing between quality of life and health status in quality of life research: A meta-analysis. *Quality of Life Research, 8,* 447–59. https://doi.org/10.1023/A:1008928518577.

Social Security Administration. (2017). *Annual statistical report on the social security disability insurance program, 2016.* Washington, DC: U.S. Department of Health and Human Services.

Social Security Administration. (2021). *Annual statistical report on the social security disability insurance program, 2020.* Washington, DC: U.S. Department of Health and Human Services.

Sprangers, M. A., & Schwartz, C. E. (1999). Integrating response shift into health-related quality of life research: A theoretical model. *Social Science & Medicine, 48*(11), 1507–15.

Spuling, S. M., Wolff, J. K., & Wurm, S. (2017). Response shift in self-rated health after serious health events in old age. *Social Science & Medicine, 192,* 85–93.

Stålnacke, B. M. (2011). Life satisfaction in patients with chronic pain: Relation to pain intensity, disability, and psychological factors. *Neuropsychiatric Disease and Treatment, 7,* 683–9. https://doi.org/10.2147/NDT.S25321.

Stanley, J. (2005). *Knowledge and practical interests.* Oxford: Clarendon Press.

Stegenga, J. (2018). *Medical nihilism.* Oxford: Oxford University Press.

Stein, R., Medeiros, C. M., Rosito, G. A., Zimerman, L. I., & Ribeiro, J. P. (2002). Intrinsic sinus and atrioventricular node electrophysiologic adaptations in endurance athletes. *Journal of the American College of Cardiology, 39*(6), 1033–8. https://doi.org/10.1016/s0735-1097(02)01722-9.

Stephan, Y., Demulier, V., & Terracciano, A. (2012). Personality, self-rated health, and subjective age in a life-span sample: The moderating role of chronological age. *Psychology and Aging, 27*(4), 875–80. https://doi.org/10.1037/a0028301.

Stephan, Y., Sutin, A. R., Luchetti, M., Hognon, L., Canada, B., & Terracciano, A. (2020). Personality and self-rated health across eight cohort studies. *Social Science & Medicine, 263,* 113245. https://doi.org/10.1016/j.socscimed.2020.113245.

Steptoe, A., Deaton, A., & Stone, A. A. (2015). Subjective wellbeing, health, and ageing. *The Lancet, 385*(9968), 640–8.

Stewart, D. E., & Yuen, T. (2011). A systematic review of resilience in the physically ill. *Psychosomatics, 52*(3), 199–209.

Stutzer, A., & Frey, B. S. (2006). Does marriage make people happy, or do happy people get married? *The Journal of Socio-Economics, 35*(2), 326–47. https://doi.org/10.1016/j.socec.2005.11.043.

Thomas, D. R. (2004). Psychosocial effects of acne. *Journal of Cutaneous Medicine and Surgery, 8*(4), 3–5.

Thomasson, A. L. (2017). Metaphysical disputes and metalinguistic negotiation. *Analytic Philosophy, 58*(1), 1–28.

Thomasson, A. L. (2019). The ontology of social groups. *Synthese, 196*(12), 4829–45.

Thompson, J. J. T. (1998). The statue and the clay. *Noûs, 32,* 148–73.

Thompson, M. (1995). The representation of life. In R. Hursthouse, G. Lawrence, & W. Quinn (Eds.), *Virtues and reasons: Philippa Foot and moral theory* (pp. 247–96). Oxford: Clarendon Press.

Thompson, M. (2004). Apprehending human form. In A. O'Hear (Ed.), *Modern moral philosophy* (pp. 47–74). Cambridge: Cambridge University Press.

Thompson, N. W., Mockford, B. J., & Cran, G. W. (2001). Absence of the palmaris longus muscle: A population study. *The Ulster Medical Journal, 70*(1), 22–4.

Turner, R., O'Sullivan, E., & Edelstein, J. (2012). Hip dysplasia and the performing arts: Is there a correlation? *Current Reviews in Musculoskeletal Medicine, 5*(1), 39–45.

Tyc, V. L. (1992). Psychosocial adaptation of children and adolescents with limb deficiencies: A review. *Clinical Psychology Review, 12*(3), 275–91. https://doi.org/10.1016/02727358(92)90138-X.

U.K. Prospective Diabetes Study Group. (1999). Quality of life in type 2 diabetic patients is affected by complications but not by intensive policies to improve blood glucose or blood pressure control (UKPDS 37). *Diabetes Care, 22*(7), 1125–36. https://doi.org/10.2337/diacare.22.7.1125.

Vagaska, E., Litavcova, A., Srotova, I., Vlckova, E., Kerkovsky, M., Jarkovsky, J., Bednarik, J., & Adamova, B. (2019). Do lumbar magnetic resonance imaging changes predict neuropathic pain in patients with chronic non-specific low back pain? *Medicine, 98*(17). DOI: 10.1097/MD.0000000000015377.

Valles, S. A. (2018). *Philosophy of population health: Philosophy for a new public health era.* London: Routledge.

van Ittersum, M. W., van Wilgen, C. P., Hilberdink, W. K. H. A., Groothoff, J. W., & van der Schans, C. P. (2009). Illness perceptions in patients with fibromyalgia. *Patient Education and Counseling, 74*(1), 53–60. https://doi.org/10.1016/j.pec.2008.07.041.

van Leeuwen, C. M., Post, M. W., Hoekstra, T., van der Woude, L. H., de Groot, S., Snoek, G. J., Mulder, D. G., & Lindeman, E. (2011). Trajectories in the course of life satisfaction after spinal cord injury: Identification and predictors. *Archives of Physical Medicine and Rehabilitation, 92*(2), 207–13. https://doi.org/10.1016/j.apmr.2010.10.011.

Venkatapuram, S. (2011). *Health justice: An argument from the capabilities approach.* Cambridge: Polity Press.

Venkatapuram, S. (2013). Health, vital goals, and central human capabilities. *Bioethics, 27*(5), 271–9.

Vergne-Salle, P., Pouplin, S., Trouvin, A. P., Bera-Louville, A., Soubrier, M., Richez, C., Javier, R. M., Perrot, S., & Bertin, P. (2020). The burden of pain in rheumatoid arthritis: Impact of disease activity and psychological factors. *European Journal of Pain (London, England), 24*(10), 1979–89. https://doi.org/10.1002/ejp.1651.

Ville, I., Ravaud, J.-F., & Tetrafigap Group (2001). Subjective well-being and severe motor impairments: The Tetrafigap survey on the long-term outcome of tetraplegic spinal cord injured persons. *Social Science & Medicine, 52*(3), 369–84. https://doi.org/10.1016/S02779536(00)00140-4.

Viola, S., & Moncrieff, J. (2016). Claims for sickness and disability benefits owing to mental disorders in the UK: Trends from 1995 to 2014. *BJPsych Open, 2*(1), 18–24. https://doi.org/10.1192/bjpo.bp.115.002246.

Vojta, C., Kinosian, B., Glick, H., Altshuler, L., & Bauer, M. S. (2001). Self-reported quality of life across mood states in bipolar disorder. *Comprehensive Psychiatry, 42*(3), 190–5. https://doi.org/10.1053/comp.2001.23143.

Wakefield, J. C. (1992). Disorder as harmful dysfunction: A conceptual critique of DSM-III-R's definition of mental disorder. *Psychological Review, 99*(2), 232–47.

Wakefield, J. C. (1995). Dysfunction as a value-free concept: A reply to Sadler and Agich. *Philosophy, Psychiatry, and Psychology, 2*(3), 233–46.

Wakefield, J. C. (2014). The biostatistical theory versus the harmful dysfunction analysis, part 1: Is part dysfunction a sufficient condition for medical disorder? *Journal of Medicine and Philosophy, 39*(6), 648–82.

Walid, M. S., Donahue, S. N., Darmohray, D. M., Hyer, L. A., Jr, & Robinson, J. S., Jr (2008). The fifth vital sign—what does it mean? *Pain Practice: The Official Journal of the World*

Institute of Pain, 8(6), 417–22. https://doi.org/10.1111/j.1533-2500.2008.00222.x. (Retraction published 2009, 9(3), 245.)

Walker, B. F., French, S. D., Grant, W., & Green, S. (2011). A Cochrane review of combined chiropractic interventions for low-back pain. *Spine*, 36(3), 230–42. DOI: 10.1097/BRS.0b013e318202ac73.

Watson, N., & Vehmas, S. (Eds.). (2019). *Routledge handbook of disability studies*. London: Routledge.

Wendell, S. (2001). Unhealthy disabled: Treating chronic illnesses as disabilities. *Hypatia*, 16(4), 17–33.

Whitcomb, I. (2022). When chronic pain becomes who you are. *Slate Magazine*. June 5, 2022. https://slate.com/technology/2022/06/chronic-pain-identity-spoonies-support-recovery.html.

White, K. P., Nielson, W. R., Harth, M., Ostbye, T., & Speechley, M. (2002). Chronic widespread musculoskeletal pain with or without fibromyalgia: Psychological distress in a representative community adult sample. *The Journal of Rheumatology*, 29(3), 588–94. www.jrheum.org/content/29/3/588.short.

Whitmann, I. (2022, June 5). When chronic pain becomes who you are. *Slate*. https://slate.com/technology/2022/06/chronic-pain-identity-spoonies-support-recovery.html?fbclid=IwAR1Aj-llhiSWPbcpP1XJZEHL5SwmFB2_WwlpEy441eKhhY2Mz2DtSjTN9rc.

Widmann, R. F., Laplaza, J. F., Bitan, F. D., Brooks, C. E., & Root, L. (2002). Quality of life in osteogenesis imperfecta. *International Orthopaedics*, 26, 3–6. https://doi.org/10.1007/s002640100292.

Wilken, U., & Breucker, G. (2000). *Mental health in the workplace, situation analysis: Germany*. Geneva: International Labor Office.

Williams, J. R. G. (2014). Decision-making under indeterminacy. *Philosophers' Imprint*, 14, 1–34.

Williams, J. R. G. (2016). Conflicting values. *Ratio*, 29(4), 412–33.

Wilson, V., & Jones, J. (2018, February 22). Working harder or finding it harder to work: Demographic trends in annual work hours show an increasingly fractured workforce. Economic Policy Institute. www.epi.org/publication/trends-in-work-hours-and-labor-market-disconnection.

Wolfe, F., Michaud, K., Li, T., & Katz, R. S. (2010). EQ-5D and SF-36 quality of life measures in systemic lupus erythematosus: Comparisons with rheumatoid arthritis, noninflammatory rheumatic disorders, and fibromyalgia. *The Journal of Rheumatology*, 37(2), 296–304. https://doi.org/10.3899/jrheum.090778.

Yelin, E., Meenan, R., Nevitt, M., & Epstein, W. (1980). Work disability in rheumatoid arthritis: Effects of disease, social, and work factors. *Annals of Internal Medicine*, 93(4), 551–6. https://doi.org/10.7326/0003-4819-93-4-551.

Zatarain, E., & Strand, V. (2006). Monitoring disease activity of rheumatoid arthritis in clinical practice: Contributions from clinical trials. *Nature Reviews Rheumatology*, 2, 611–18. https://doi.org/10.1038/ncprheum0246.

Zgierska, A., Miller, M., & Rabago, D. (2012). Patient satisfaction, prescription drug abuse, and potential unintended consequences. *Journal of the American Medical Association*, 307(13), 1377–8. DOI: 10.1001/jama.2012.419.

Index

For the benefit of digital users, indexed terms that span two pages (e.g., 52–53) may, on occasion, appear on only one of those pages.